AF539915

ENVIRONMENTAL CHEMISTRY

ENVIRONMENTAL CHEMISTRY

by
Dr Reza Marandi
Dr Hoda Pasdar

2012

SBS Publishers & Distributors Pvt. Ltd.
New Delhi

ISBN - 9788189741884

First Published in India in 2012

Published by:
SBS PUBLISHERS & DISTRIBUTORS PVT. LTD.
2/9, Ground Floor, Ansari Road, Darya Ganj,
New Delhi - 110002, INDIA
Tel: 23289119, 41563911
Email: mail@sbspublishers.com

Printed in India by Anvi Digital.

Preface

Environmental Chemistry is the use of chemistry to understand the interactions of macro-scale environmental systems. The subject area called environmental chemistry is the scientific study of the chemical and biochemical phenomena which occur in natural places. It can be defined as the study of the sources, reactions, transport, effects, and fates of chemical species in the air, soil, and water environments; and the effect of human activity on these. Environmental chemistry is an interdisciplinary science that includes atmospheric, aquatic and soil chemistry, as well as heavily relying on analytical chemistry and being related to environmental and other areas of science. Environmental chemistry involves first understanding how the uncontaminated environment works, which chemicals in what concentrations are present naturally, and with what effects. Without this it would be impossible to accurately study the effects humans have on the environment through the release of chemicals. Environmental chemists draw on a range of concepts from chemistry and various environmental sciences to assist in their study of what is happening to a chemical species in the environment. Important general concepts from chemistry include understanding chemical reactions and equations, solutions, units, sampling, and analytical techniques. So, environmental chemistry is the chemistry of the natural environment, and of pollutant chemicals in nature. Environmental chemistry encompasses many different topics. It may involve a study of Freon reactions in the stratosphere or an analysis of toxic Kepone deposits in ocean sediments. It also covers the chemistry and biochemistry of volatile and soluble organometallic compounds biosynthesized by anaerobic bacteria. Environmental chemistry is the study of the sources, reactions, transport, effects, and fates of chemical species in water, soil, and air environments.

This publication manages to address most of the prominent issues related to environmental chemistry in contemporary world. It provides readers with a complete overview of the said subject with a completely new perspective. In sum, this book serves the purpose of an up-to-date reference book on the said subject of environmental chemistry. This book will be helpful to students, scholars, doctoral candidates, policy analysts, NGO practitioners, etc. interested in the chemistry studies. This publication provides unique resources for understanding the very many areas of environmental chemistry, such as:

- Environmental Chemistry Cycles
- Electrochemistry
- Chemical Thermodynamics
- Analytical Environmental Chemistry
- Applied Environmental Chemistry
- Green Chemistry
- Environmental Methods for Chemical Analysis
- Radiological and Toxicological Chemistry
- Atmospheric Chemistry and Air Pollution
- Geochemistry, Soil Chemistry and Soil Pollution
- Aquatic Chemistry
- Water Pollution and Wastewater Treatment
- Environmental Nanotechnology
- Hazardous Wastes Disposal and Treatment
- Pharmacology, Hospital Wastes Control and Treatment
- Health, Safety and Environment
- Theoretical Chemistry
- Quantum Chemistry
- Computational Chemistry

This book is designed to provide a resource to teachers and students of environmental chemistry and to those interested in finding out what environmental chemistry is. This publication hopes to foster environmental chemistry education and to facilitate communication among those who teach environmental chemistry.

Dr. Reza Marandi
Dr. Hoda Pasdar

Contents

1

Introduction to Environmental Chemistry

1.1 Introduction

Environmental chemistry is the scientific study of the chemical and biochemical phenomena that occur in natural places. It should not be confused with green chemistry, which seeks to reduce potential pollution at its source. It can be defined as the study of the sources, reactions, transport, effects, and fates of chemical species in the air, soil, and water environments; and the effect of human activity on these. Environmental chemistry is an interdisciplinary science that includes atmospheric, aquatic and soil chemistry, as well as heavily relying on analytical chemistry and being related to environmental and other areas of science.

Environmental chemistry involves first understanding how the uncontaminated environment works, which chemicals in what concentrations are present naturally, and with what effects. Without this it would be impossible to accurately study the effects humans have on the environment through the release of chemicals.

Fig. 1.1: Forests contain many examples of what environmental chemistry encompasses.

Environmental chemists draw on a range of concepts from chemistry and various environmental sciences to assist in their study of what is happening to a chemical species in the environment. Important general concepts from chemistry include understanding chemical reactions and equations, solutions, units, sampling, and analytical techniques.

Contamination

A contaminant is a substance present in nature due to human activity, that would not otherwise be there. The term contaminant is often used interchangeably with *pollutant*, which is a substance that has a detrimental impact on the environment it is in. Whilst a contaminant is sometimes defined as a substance present in the environment as a result of human activity, but without harmful effects, it is sometimes the case that toxic or harmful effects from contamination only become apparent at a later date.

The "medium" (*e.g.* soil) or organism (*e.g.* fish) affected by the pollutant or contaminant is called a *receptor*, whilst a *sink* is a chemical medium or species that retains and interacts with the pollutant.

Environmental Indicators

Chemical measures of water quality include dissolved oxygen (DO), chemical oxygen demand (COD), biochemical oxygen demand (BOD), total dissolved solids (TDS), and pH.

Applications

Environmental chemistry is used by the Environment Agency (in England and Wales), the Environmental Protection Agency (in the United States) the Association of Public Analysts, and other environmental agencies and research bodies around the world to detect and identify the nature and source of pollutants. These can include:

- Heavy metal contamination of land by industry. These can then be transported into water flows and be taken up by living organisms.
- Nutrients such as nitrate and phosphate leaching from agricultural land into water courses, which can lead to algal blooms and eutrophication.

Methods

Quantitative chemical analysis is a key part of environmental chemistry, since it provides the data that frame most environmental studies.

Common analytical techniques used for quantitative determinations in environmental chemistry include classical wet chemistry, such as gravimetric, titrimetric and electrochemical methods. More sophisticated approaches are used in the determination of trace metals and organic compounds. Metals are commonly measured by atomic spectroscopy and mass spectrometry: Atomic Absorption Spectrophotometry (AA) and Inductively Coupled Plasma Atomic Emission (ICP-AES) or Inductively Coupled Plasma Mass Spectrometric (ICP-MS) techniques. Organic compounds are commonly measured also using mass spectrometric methods, such as Gas chromatography-mass spectrometry (GC/MS) and Liquid chromatography-mass spectrometry (LC/MS). Non-MS methods using GCs and LCs having universal or specific detectors are still staples in the arsenal of available analytical tools.

Other parameters often measured in environmental chemistry are radiochemicals. These are pollutants which emit radioactive materials, such as alpha and beta particles, posing danger to human health and the environment. Particle counters and Scintillation counters are most commonly used for these measurements. Bioassays and immunoassays are utilized for toxicity evaluations of chemical effects on various organisms.

1.2 Environmental Cycles

Definition

A natural process in which elements are continuously cycled in various forms between different compartments of the environment (*e.g.*, air, water, soil, organisms).

Examples include the carbon, nitrogen and phosphorus cycles (nutrient cycles) and the water cycle.

The *carbon cycle* includes the uptake of carbon dioxide by plants through, its ingestion by animals and its release to the atmosphere through respiration and decay of organic materials. Human activities like the burning of fossil fuels contribute to the release of carbon dioxide in the atmosphere.

The *nitrogen cycle* involves the uptake of nitrogen form the atmosphere by a process called fixation which is carried out by microbes or industrial processes. Decomposition of biological waste by microbes can return nitrogen to the atmosphere. Nitrogen is mainly used by humans as a fertilizer in farmlands, but its excessive usage can lead to serious problems (such as eutrophication).

The *phosphorus cycle* involves the uptake of phosphorus by organisms.

weathering processes can make it available to biological systems. After decomposition of biological waste, it can accumulate in large amounts in soils and sediments. Phosphorus is used by humans as a fertilizer in farmlands and in detergents. Overuse of phosphorus can lead to eutrophication.

The *water cycle* is the process by which water travels in a sequence from the air (condensation) to the earth (precipitation) and returns to the atmosphere (evaporation). It is also referred to as the hydrologic cycle.

Human use of water can transform the water cycle through irrigation or the construction of dams, for example.

1.3 Carbon Cycle

Diagram of the carbon cycle. The black numbers indicate how much carbon is stored in various reservoirs, in billions of tons ("GtC" stands for GigaTons of Carbon and figures are circa 2004). The purple numbers indicate how much carbon moves between reservoirs each year. The sediments, as defined in this diagram, do not include the ~70 million GtC of carbonate rock and kerogen.

The carbon cycle is the biogeochemical cycle by which carbon is exchanged between the biosphere, geosphere, hydrosphere, and atmosphere of the Earth.

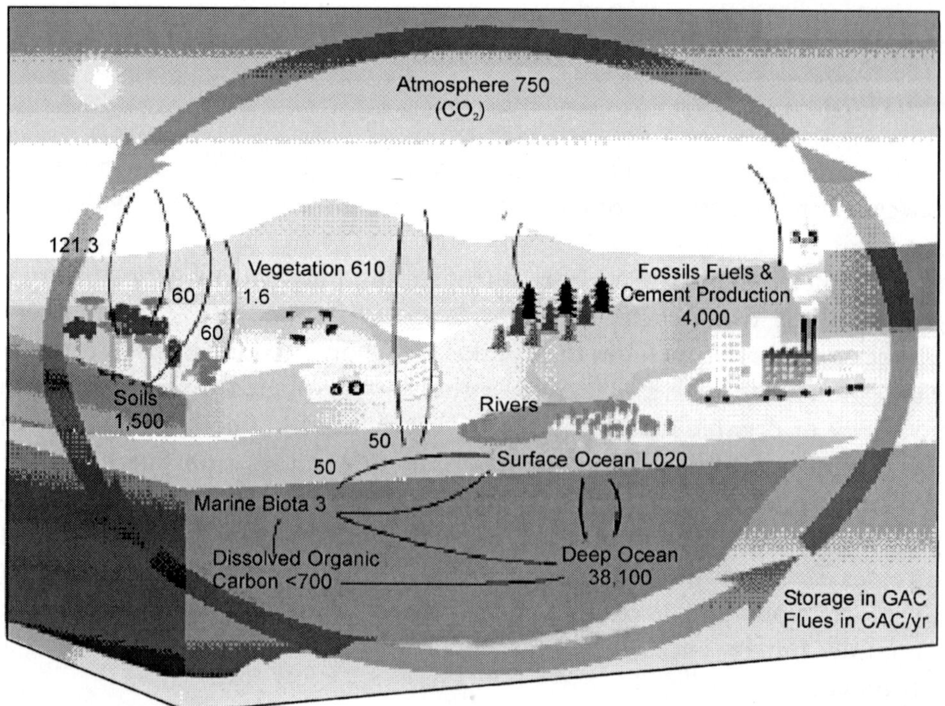

Fig. 1.2: Carbon cycle.

The cycle is usually thought of as four major reservoirs of carbon interconnected by pathways of exchange. These reservoirs are:

- The atmosphere.
- The terrestrial biosphere, which is usually defined to include fresh water systems and non-living organic material, such as soil carbon.
- The oceans, including dissolved inorganic carbon and living and non-living marine biota.
- The sediments including fossil fuels.

The annual movements of carbon, the carbon exchanges between reservoirs, occur because of various chemical, physical, geological, and biological processes. The ocean contains the largest active pool of carbon near the surface of the Earth, but the deep ocean part of this pool does not rapidly exchange with the atmosphere.

The *global carbon budget* is the balance of the exchanges (incomes and losses) of carbon between the carbon reservoirs or between one specific loop (*e.g.*, atmosphere? biosphere) of the carbon cycle. An examination of the carbon budget of a pool or reservoir can provide information about whether the pool or reservoir is functioning as a source or sink for carbon dioxide.

In the Atmosphere

Carbon exists in the Earth's atmosphere primarily as the gas carbon dioxide (CO_2). Although it is a very tiny percent of the atmosphere (approximately 0.04% on a molar basis, though rising), it plays an important role in supporting life. Other gases containing carbon in the atmosphere are methane and chlorofluorocarbons (the latter is entirely anthropogenic). The overall atmospheric concentration of these greenhouse gases has been increasing in recent decades, contributing to global warming.

Carbon is taken from the atmosphere in several ways:

- When the sun is shining, plants perform photosynthesis to convert carbon dioxide into carbohydrates, releasing oxygen in the process. This process is most prolific in relatively new forests where tree growth is still rapid. The effect is strongest in deciduous forests during spring leafing out. This is visible as an annual signal in the Keeling curve of measured CO_2 concentration. Northern hemisphere spring predominates, as there is far more land in temperate latitudes in that hemisphere than in the southern.
- Forests store 86 per cent of the planet's above-ground carbon and 73

per cent of the planet's soil carbon.

- At the surface of the oceans towards the poles, seawater becomes cooler and more carbonic acid is formed as CO_2 becomes more soluble. This is coupled to the ocean's thermohaline circulation which transports dense surface water into the ocean's interior (see the entry on the solubility pump).
- In upper ocean areas of high biological productivity, organisms convert reduced carbon to tissues, or carbonates to hard body parts such as shells and tests. These are, respectively, oxidized (soft-tissue pump) and redissolved (carbonate pump) at lower average levels of the ocean than those at which they formed, resulting in a downward flow of.
- The weathering of silicate rock. Carbonic acid reacts with weathered rock to produce bicarbonate ions. The bicarbonate ions produced are carried to the ocean, where they are used to make marine carbonates. Unlike dissolved CO_2 in equilibrium or tissues which decay, weathering does not move the carbon into a reservoir from which it can readily return to the atmosphere.

Carbon can be released back into the atmosphere in many different ways:

- Through the respiration performed by plants and animals. This is an exothermic reaction and it involves the breaking down of glucose (or other organic molecules) into carbon dioxide and water.
- Through the decay of animal and plant matter. Fungi and bacteria break down the carbon compounds in dead animals and plants and convert the carbon to carbon dioxide if oxygen is present, or methane if not.
- Through combustion of organic material which oxidizes the carbon it contains, producing carbon dioxide (and other things, like water vapor). Burning fossil fuels such as coal, petroleum products, and natural gas releases carbon that has been stored in the geosphere for millions of years. Burning agrofuels also releases carbon dioxide.
- Production of cement. Carbon dioxide is released when limestone (calcium carbonate) is heated to produce lime (calcium oxide), a component of cement.
- At the surface of the oceans where the water becomes warmer, dissolved carbon dioxide is released back into the atmosphere.
- Volcanic eruptions and metamorphism release gases into the atmosphere. Volcanic gases are primarily water vapor, carbon dioxide and sulfur dioxide. The carbon dioxide released is roughly equal to the amount removed by silicate weathering; so the two processes,

which are the chemical reverse of each other, sum to roughly zero, and do not affect the level of atmospheric carbon dioxide on time scales of less than about 100,000 yr.

- Forests and crops in the process of growing absorbs lots of carbon, while old and stable forest consumes as much CO_2 during the day as they produce during the night.

In the Biosphere

Around 1,900 gigatons of carbon are present in the biosphere. Carbon is an essential part of life on Earth. It plays an important role in the structure, biochemistry, and nutrition of all living cells.

- Autotrophs are organisms that produce their own organic compounds using carbon dioxide from the air or water in which they live. To do this they require an external source of energy. Almost all autotrophs use solar radiation to provide this, and their production process is called photosynthesis. A small number of autotrophs exploit chemical energy sources in a process called chemosynthesis. The most important autotrophs for the carbon cycle are trees in forests on land and phytoplankton in the Earth's oceans. Photosynthesis follows the reaction $6CO_2 + 6H_2O \rightarrow C_6H_{12}O_6 + 6O_2$
- Carbon is transferred within the biosphere as heterotrophs feed on other organisms or their parts (*e.g.*, fruits). This includes the uptake of dead organic material (detritus) by fungi and bacteria for fermentation or decay.
- Most carbon leaves the biosphere through respiration. When oxygen is present, aerobic respiration occurs, which releases carbon dioxide into the surrounding air or water, following the reaction $C_6H_{12}O_6 + 6O_2 \rightarrow 6CO_2 + 6H_2O$. Otherwise, anaerobic respiration occurs and releases methane into the surrounding environment, which eventually makes its way into the atmosphere or hydrosphere (*e.g.*, as marsh gas or flatulence).
- Burning of biomass (*e.g.* forest fires, wood used for heating, anything else organic) can also transfer substantial amounts of carbon to the atmosphere
- Carbon may also be circulated within the biosphere when dead organic matter (such as peat) becomes incorporated in the geosphere. Animal shells of calcium carbonate, in particular, may eventually become limestone through the process of sedimentation.
- Much remains to be learned about the cycling of carbon in the deep

ocean. For example, a recent discovery is that larvacean mucus houses (commonly known as "sinkers") are created in such large numbers that they can deliver as much carbon to the deep ocean as has been previously detected by sediment traps. Because of their size and composition, these houses are rarely collected in such traps, so most biogeochemical analyses have erroneously ignored them.

Carbon storage in the biosphere is influenced by a number of processes on different time-scales: while net primary productivity follows a diurnal and seasonal cycle, carbon can be stored up to several hundreds of years in trees and up to thousands of years in soils. Changes in those long term carbon pools (*e.g.* through de- or afforestation or through temperature-related changes in soil respiration) may thus affect global climate change.

In the Ocean

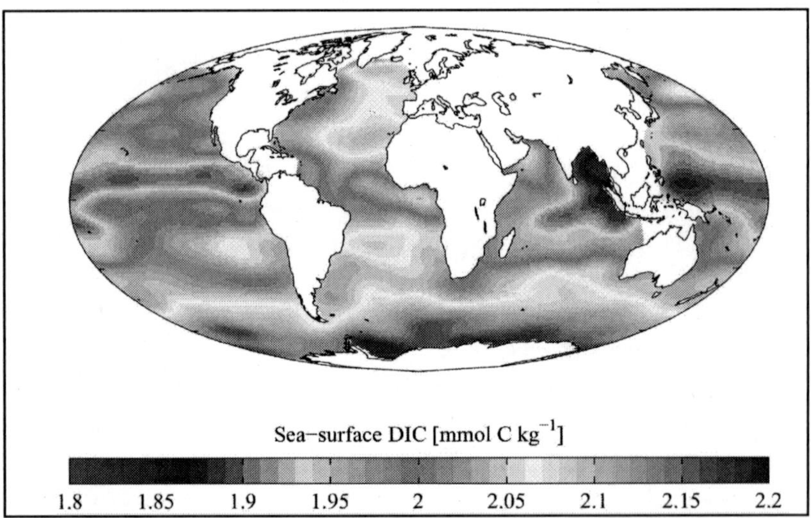

Fig. 1.3: "Present day" (1990s) sea surface dissolved inorganic carbon concentration (from the GLODAP climatology).

The oceans contain around 36,000 gigatonnes of carbon, mostly in the form of bicarbonate ion (over 90%, with most of the remainder being carbonate). Inorganic carbon, that is carbon compounds with no carbon-carbon or carbon-hydrogen bonds, is important in its reactions within water. This carbon exchange becomes important in controlling pH in the ocean and can also vary as a source or sink for carbon. Carbon is readily exchanged between the atmosphere and ocean. In regions of oceanic upwelling, carbon is released to the atmosphere. Conversely, regions of downwelling transfer carbon (CO_2)

from the atmosphere to the ocean. When CO_2 enters the ocean, it participates in a series of reactions which are locally in equilibrium:

Solution

$$CO_2 \text{ (atmospheric)} \rightarrow CO_2 \text{ (dissolved)}$$

Conversion to carbonic acid:

$$CO_2 \text{ (dissolved)} + H_2O \rightarrow H_2CO_3$$

First ionization:

$$H_2CO_3 \rightarrow H^+ + HCO_3^- \text{ (bicarbonate ion)}$$

Second ionization:

$$HCO_3^- \rightarrow H^+ + CO_3^- \text{ (carbonate ion)}$$

This set of reactions, each of which has its own equilibrium coefficient determines the form that inorganic carbon takes in the oceans. The coefficients, which have been determined empirically for ocean water, are themselves functions of temperature, pressure, and the presence of other ions (especially borate). In the ocean the equilibria strongly favor bicarbonate. Since this ion is three steps removed from atmospheric CO_2, the level of inorganic carbon storage in the ocean does not have a proportion of unity to the atmospheric partial pressure of CO_2. The factor for the ocean is about ten: that is, for a 10 per cent increase in atmospheric CO_2, oceanic storage (in equilibrium) increases by about 1 per cent, with the exact factor dependent on local conditions. This buffer factor is often called the "Revelle Factor", after Roger Revelle.

In the oceans, bicarbonate can combine with calcium to form limestone (calcium carbonate, $CaCO_3$, with silica), which precipitates to the ocean floor. Limestone is the largest reservoir of carbon in the carbon cycle. The calcium comes from the weathering of calcium-silicate rocks, which causes the silicon in the rocks to combine with oxygen to form sand or quartz (silicon dioxide), leaving calcium ions available to form limestone.

1.4 PBL Model

1. *Read and analyze the problem scenario.* Check your understanding of the scenario by discussing it within your group. A group effort will probably be more effective in deciding what the key factors are

in this situation. Because this is a real problem solving situation, your group will need to actively search for the information necessary to solve the problem.

2. *List what is known.* Start a list in which you write down everything you know about this situation. Begin with the information contained in the scenario. Add knowledge that group members bring. (You may want a column of things people think they know, but are not sure!)
3. *Develop a problem statement.* A problem statement should come from your analysis of what you know. In one or two sentences you should be able to describe what it is that your group is trying to solve, produce, respond to, or find out. The problem statement may have to be revised as new information is discovered and brought to bear on the situation.
4. *List what is needed.* Prepare a list of questions you think need to be answered to solve the problem. Record them under a second list titled: "What do we need to know?" Several types of questions may be appropriate. Some may address concepts or principles that need to be learned in order to address the situation. Other questions may be in the form of requests for more information. These questions will guide searches that may take place on-line, in the library, or in other out-of-class searches.
5. *List possible actions.* List recommendations, solutions, or hypotheses under the heading: "What should we do?" List actions to be taken, *e.g.*, question an expert, get on-line data, visit library.
6. *Analyze information.* Analyze information you have gathered. You may need to revise the problem statement. You may identify more problem statements. At this point, your group will likely formulate and test hypotheses to explain the problem. Some problems may not require hypotheses, instead a recommended solution or opinion (based on your research data) may be appropriate.
7. *Present findings.* Prepare a report in which you make recommendations, predictions, inferences, or other appropriate resolution of the problem based on your data and background. Be prepared to support your recommendation.

Note: The steps in this model may have to be visited several times. Steps two through five may be conducted concurrently as new information becomes available. As more information is gathered, the problem statement may be refined or altered.

1.5 Electrochemistry

Electrochemistry is a branch of chemistry that studies chemical reactions which take place in a solution at the interface of an electron conductor (a metal or a semiconductor) and an ionic conductor (the electrolyte), and which involve electron transfer between the electrode and the electrolyte or species in solution.

If a chemical reaction is driven by an external applied voltage, as in electrolysis, or if a voltage is created by a chemical reaction as in a battery, it is an *electrochemical* reaction. Chemical reactions where electrons are transferred between molecules are called oxidation/reduction (redox) reactions. In general, electrochemistry deals with situations where oxidation and reduction reactions are separated in space or time, connected by an external electric circuit to understand each process.

History

16th to 18th Century Developments

The 16th century marked the beginning of electrical understanding. During that century the English scientist William Gilbert spent 17 years experimenting with magnetism and, to a lesser extent, electricity. For his work on magnets, Gilbert became known as the "*Father of Magnetism.*" He discovered various methods for producing and strengthening magnets.

In 1663 the German physicist Otto von Guericke created the first electric generator, which produced static electricity by applying friction in the machine. The generator was made of a large sulfur ball cast inside a glass globe, mounted on a shaft. The ball was rotated by means of a crank and a static electric spark was produced when a pad was rubbed against the ball as it rotated. The globe could be removed and used as source for experiments with electricity.

By the mid 18th century the French chemist Charles François de Cisternay du Fay discovered two types of static electricity, and that like charges repel each other whilst unlike charges attract. Du Fay announced that electricity consisted of two fluids: "*vitreous*" (from the Latin for "*glass*"), or positive, electricity; and "*resinous,*" or negative, electricity. This was the *two-fluid theory* of electricity, which was to be opposed by Benjamin Franklin's *one-fluid theory* later in the century.

Charles-Augustin de Coulomb developed the law of electrostatic attraction in 1781 as an outgrowth of his attempt to investigate the law of electrical repulsions as stated by Joseph Priestley in England.

Italian physicist Alessandro Volta showing his *"battery"* to French emperor Napoleon Bonaparte in the early 19th century.

In the late 18th century the Italian physician and anatomist Luigi Galvani marked the birth of electrochemistry by establishing a bridge between chemical reactions and electricity on his essay *"De Viribus Electricitatis in Motu Musculari Commentarius"* (Latin for Commentary on the Effect of Electricity on Muscular Motion) in 1791 where he proposed a *"nerveo-electrical substance"* on biological life forms.

In his essay Galvani concluded that animal tissue contained a here-to-fore neglected innate, vital force, which he termed *"animal electricity,"* which activated nerves and muscles spanned by metal probes. He believed that this new force was a form of electricity in addition to the *"natural"* form produced by lightning or by the electric eel and torpedo ray as well as the *"artificial"* form produced by friction (*i.e.*, static electricity).

Galvani's scientific colleagues generally accepted his views, but Alessandro Volta rejected the idea of an *"animal electric fluid,"* replying that the frog's legs responded to differences in metal temper, composition, and bulk. Galvani refuted this by obtaining muscular action with two pieces of the same material.

19th Century

In 1800, William Nicholson and Johann Wilhelm Ritter succeeded in decomposing water into hydrogen and oxygen by electrolysis. Soon thereafter Ritter discovered the process of electroplating. He also observed that the amount of metal deposited and the amount of oxygen produced during an electrolytic process depended on the distance between the electrodes. By 1801 Ritter observed thermoelectric currents and anticipated the discovery of thermoelectricity by Thomas Johann Seebeck.

By the 1810s William Hyde Wollaston made improvements to the galvanic pile. Sir Humphry Davy's work with electrolysis led to the conclusion that the production of electricity in simple electrolytic cells resulted from chemical action and that chemical combination occurred between substances of opposite charge. This work led directly to the isolation of sodium and potassium from their compounds and of the alkaline earth metals from theirs in 1808.

Hans Christian Ørsted's discovery of the magnetic effect of electrical currents in 1820 was immediately recognized as an epoch-making advance, although he left further work on electromagnetism to others. André-Marie Ampère quickly repeated Ørsted's experiment, and formulated them mathematically.

In 1821, Estonian-German physicist Thomas Johann Seebeck demonstrated the electrical potential in the juncture points of two dissimilar metals when there is a heat difference between the joints.

In 1827 the German scientist Georg Ohm expressed his law in this famous book *"Die galvanische Kette, mathematisch bearbeitet"* (The Galvanic Circuit Investigated Mathematically) in which he gave his complete theory of electricity.

In 1832 Michael Faraday's experiments led him to state his two laws of electrochemistry. In 1836 John Daniell invented a primary cell in which hydrogen was eliminated in the generation of the electricity. Daniell had solved the problem of polarization. In his laboratory he had learned that alloying the amalgamated zinc of Sturgeon with mercury would produce a better voltage.

William Grove produced the first fuel cell in 1839. In 1846, Wilhelm Weber developed the electrodynamometer. In 1866, Georges Leclanché patented a new cell which eventually became the forerunner to the world's first widely used battery, the zinc carbon cell.

Svante August Arrhenius published his thesis in 1884 on *Recherches sur la conductibilité galvanique des électrolytes* (Investigations on the galvanic conductivity of electrolytes). From his results the author concluded that electrolytes, when dissolved in water, become to varying degrees split or dissociated into electrically opposite positive and negative ions.

In 1886 Paul Héroult and Charles M. Hall developed a successful method to obtain aluminium by using the principles described by Michael Faraday.

In 1894 Friedrich Ostwald concluded important studies of the electrical conductivity and electrolytic dissociation of organic acids.

Walther Hermann Nernst developed the theory of the electromotive force of the voltaic cell in 1888. In 1889, he showed how the characteristics of the current produced could be used to calculate the free energy change in the chemical reaction producing the current. He constructed an equation, known as Nernst Equation, which related the voltage of a cell to its properties.

In 1898 Fritz Haber showed that definite reduction products can result from electrolytic processes if the potential at the cathode is kept constant. In 1898 he explained the reduction of nitrobenzene in stages at the cathode and this became the model for other similar reduction processes.

The 20th Century and Recent Developments

In 1902, The Electrochemical Society (ECS) was founded.

In 1909, Robert Andrews Millikan began a series of experiments to determine the electric charge carried by a single electron.

In 1923, Johannes Nicolaus Brønsted and Thomas Martin Lowry published essentially the same theory about how acids and bases behave, using an electrochemical basis.

Arne Tiselius developed the first sophisticated electrophoretic apparatus in 1937 and some years later he was awarded the 1948 Nobel Prize for his work in protein electrophoresis.

A year later, in 1949, the International Society of Electrochemistry (ISE) was founded.

By the 1960s-1970s quantum electrochemistry was developed by Revaz Dogonadze and his pupils.

Principles

Redox Reactions

Electrochemical processes involve redox reactions where an electron is transferred to or from a molecule or ion changing its oxidation state. This reaction can occur through the application of an external voltage or through the release of chemical energy.

Oxidation and Reduction

The atoms, ions, or molecules involved in an electrochemical reaction are characterized by the number of electrons each has compared to its number of protons called its *oxidation state* and is denoted by a + or a -. Thus the superoxide ion, O_2^-, has an *oxidation state* of -1. An atom or ion that gives up an electron to another atom or ion has its oxidation state increase, and the recipient of the negatively charged electron has its oxidation state decrease. Oxidation and reduction always occur in a paired fashion such that one species is oxidized when another is reduced. This paired electron transfer is called a redox reaction.

For example when atomic sodium reacts with atomic chlorine, sodium donates one electron and attains an oxidation state of +1. Chlorine accepts the electron and its oxidation state is reduced to -1. The sign of the oxidation state (positive/negative) actually corresponds to the value of each ion's electronic charge. The attraction of the differently charged sodium and chlorine ions is the reason they then form an ionic bond.

The loss of electrons from an atom or molecule is called oxidation, and the gain of electrons is reduction. This can be easily remembered through the use of mnemonic devices. Two of the most popular are *"OIL RIG"* (Oxidation

Is Loss, Reduction Is Gain) and "*LEO*" the lion says "*GER*" (Lose Electrons: Oxidization, Gain Electrons: Reduction). For cases where electrons are shared (covalent bonds) between atoms with large differences in electronegativity, the electron is assigned to the atom with the largest electronegativity in determining the oxidation state.

The atom or molecule which loses electrons is known as the *reducing agent*, or *reductant*, and the substance which accepts the electrons is called the *oxidizing agent*, or *oxidant*. The oxidizing agent is always being reduced in a reaction; the reducing agent is always being oxidized. Oxygen is a common oxidizing agent, but not the only one. Despite the name, an oxidation reaction does not necessarily need to involve oxygen. In fact, a fire can be fed by an oxidant other than oxygen; fluorine fires are often unquenchable, as fluorine is an even stronger oxidant (it has a higher electronegativity) than oxygen.

For reactions involving oxygen, the gain of oxygen implies the oxidation of the atom or molecule to which the oxygen is added (and the oxygen is reduced). For example, in the oxidation of octane by oxygen to form carbon dioxide and water, both the carbon in the octane and the oxygen begin with an oxidation state of 0. In forming CO_2 the carbon loses four electrons to become C^{4+} and the oxygens each gain two electrons to be O^{2-}. In organic compounds, such as butane or ethanol, the loss of hydrogen implies oxidation of the molecule from which it is lost (and the hydrogen is reduced). This follows because the hydrogen donates its electron in covalent bonds with non-metals but it takes the electron along when it is lost. Conversely, loss of oxygen or gain of hydrogen implies reduction.

Balancing Redox Reactions

Electrochemical reactions in water are better understood by balancing redox reactions using the Ion-Electron Method where H^+, OH^- ion, H_2O and electrons (to compensate the oxidation changes) are added to cell's half reactions for oxidation and reduction.

Acid Medium

In acid medium H+ ions and water are added to half reactions to balance the overall reaction. For example, when manganese reacts with sodium bismuthate.

Reaction unbalanced: $Mn^{2+}\,(aq) + NaBiO_3(s) \rightarrow Bi^{3+}\,(aq) + MnO^{4-}\,(aq)$

Oxidation: $4H_2O(l) + Mn^{2+}(aq) \rightarrow MnO_4^-(aq) + 8H^+(aq) + 5e^-$

Reduction: $2e^- + 6H^+ (aq) + BiO_3^-(s) \rightarrow Bi^{3+} (aq) + 3H_2O(l)$

Finally the reaction is balanced by multiplying the number of electrons from the reduction half reaction to oxidation half reaction and vice versa and adding both half reactions, thus solving the equation.

$$8H_2O(l) + 2Mn^{2+} (aq) \rightarrow 2MnO_4^-(aq) + 16H^+(aq) + 10e^-$$

$$10e^- + 30H^+ (aq)\ 5BiO^{3-}(s) \rightarrow 5Bi^{3+} (aq) + 15H_2O(l)$$

Reaction Balanced

$$14H_+(aq) + 2Mn^{2+}(aq) + 5NaBiO_3(s) \rightarrow 7H_2O(l) + 2MnO_4^-(aq) + 5Bi^{3+}(aq) + 5Na^+(aq)$$

Basic Medium

In basic medium OH^- ions and water are added to half reactions to balance the overall reaction. For example on reaction between Potassium permanganate and Sodium sulfite.

Reaction Unbalanced: $KMnO_4 + Na_2SO_3 + H_2O \rightarrow MnO_2 + Na_2SO_4 + KOH$

Reduction: $3e^- + 2H_2O + MnO_4^- \rightarrow MnO_2 + 4OH^-$

Oxidation: $2OH^- + SO_3^{2-} \rightarrow SO_4^{2-} + H_2O + 2e^-$

The same procedure as followed on acid medium by multiplying electrons to opposite half reactions solve the equation thus balancing the overall reaction.

$$6e^- + 4H_2O + 2MnO_4^- \rightarrow 2MnO_2 + 8OH^-$$

$$6OH^- + SO_3^{2-} \rightarrow 3SO_4^{2-} + 3H_2O + 6e^-$$

Equation Balanced

$$2KMnO_4 + 3Na_2SO_3 + H_2O \rightarrow 2MnO_2 + 3Na_2SO_4 + 2KOH$$

Neutral Medium

The same procedure as used on acid medium is applied, for example on balancing using electron ion method to complete combustion of propane.

Reaction unbalanced: $C_3H_8 + O_2 \rightarrow CO_2 + H_2O$

Reduction: $4H^+ + O_2 + 4e^- \rightarrow 2H_2O$

Oxidation: $6H_2O + C_3H_8 \rightarrow 3CO_2 + 20e^- + 20H^+$

As in acid and basic medium, electrons which were used to compensate oxidation changes are multiplied to opposite half reactions, thus solving the equation.

$$20H^+ + 5O_2 + 20e^- \rightarrow 10H_2O$$

$$6H_2O + C_3H_8 \rightarrow 3CO_2 + 20e^- 20H^+$$

Equation Balanced

$$C_3H_8 + 5O_2 \rightarrow 3CO_2 + 4H_2O$$

Electrochemical Cells

An electrochemical cell is a device that produces an electric current from energy released by a spontaneous redox reaction. This kind of cell includes the Galvanic cell or Voltaic cell, named after Luigi Galvani and Alessandro Volta, both scientists who conducted several experiments on chemical reactions and electric current during the late 18th century.

Electrochemical cells have two conductive electrodes (the anode and the cathode). The anode is defined as the electrode where oxidation occurs and the cathode is the electrode where the reduction takes place. Electrodes can be made from any sufficiently conductive materials, such as metals, semiconductors, graphite, and even conductive polymers. In between these electrodes is the electrolyte, which contains ions that can freely move.

The Galvanic cell uses two different metal electrodes, each in an electrolyte where the positively charged ions are the oxidized form of the electrode metal. One electrode will undergo oxidation (the anode) and the other will undergo reduction (the cathode). The metal of the anode will oxidize, going from an oxidation state of 0 (in the solid form) to a positive oxidation state and become an ion. At the cathode, the metal ion in solution will accept one or more electrons from the cathode and the ion's oxidation state is reduced to 0. This forms a solid metal that electrodeposits on the cathode. The two electrodes must be electrically connected to each other, allowing for a flow of electrons that leave the metal of the anode and flow through this connection to the ions at the surface of the cathode. This flow of electrons is an electrical current that can be used to do work, such as turn a motor or power a light.

A Galvanic cell whose electrodes are zinc and copper submerged in zinc sulfate and copper sulfate, respectively, is known as a Daniell cell.

Half reactions for a Daniell cell are these:

Zinc electrode (anode): $Zn(s) \rightarrow Zn^{2+} (aq) + 2e^-$

Copper electode (cathode): $Cu^{2+} (aq) + 2e^- \rightarrow Cu(s)$

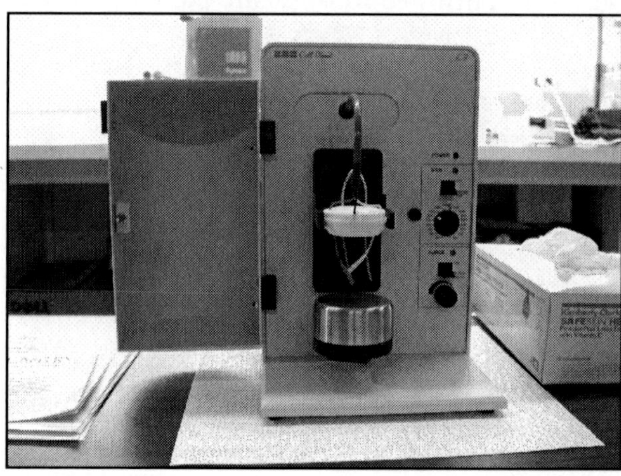

Fig. 1.4: Electrochemical Cells.

A modern cell stand for electrochemical research. The electrodes attach to high-quality metallic wires, and the stand is attached to a potentiostat/ galvanostat (not pictured). A shotglass-shaped container is aerated with a noble gas and sealed with the Teflon block.

In this example, the anode is zinc metal which oxidizes (loses electrons) to form zinc ions in solution, and copper ions accept electrons from the copper metal electrode and the ions deposit at the copper cathode as an electrodeposit. This cell forms a simple battery as it will spontaneously generate a flow of electrical current from the anode to the cathode through the external connection. This reaction can be driven in reverse by applying a voltage, resulting in the deposition of zinc metal at the anode and formation of copper ions at the cathode.

To provide a complete electric circuit, there must also be an ionic conduction path between the anode and cathode electrolytes in addition to the electron conduction path. The simplest ionic conduction path is to provide a liquid junction. To avoid mixing between the two electrolytes, the liquid junction can be provided through a porous plug that allows ion flow while reducing electrolyte mixing. To further minimize mixing of the electrolytes, a salt bridge can be used which consists of an electrolyte saturated gel in an inverted U-tube. As the negatively charged electrons flow in one direction around this circuit, the positively charged metal ions flow in the opposite direction in the electrolyte.

A voltmeter is capable of measuring the change of electrical potential between the anode and the cathode.

Electrochemical cell voltage is also referred to as electromotive force or emf.

A cell diagram can be used to trace the path of the electrons in the electrochemical cell. For example, here is a cell diagram of a Daniell cell:

$$Zn(s)\ |Zn^{2+}\ (1\ M)|\ |Cu^{2+}\ (1\ M)|\ Cu(s)$$

First, the reduced form of the metal to be oxidized at the anode (Zn) is written. This is separated from its oxidized form by a vertical line, which represents the limit between the phases (oxidation changes). The double vertical lines represent the saline bridge on the cell. Finally, the oxidized form of the metal to be reduced at the cathode, is written, separated from its reduced form by the vertical line. The electrolyte concentration is given as it is an important variable in determining the cell potential.

Standard Electrode Potential

To allow prediction of the cell potential, tabulations of standard electrode potential are available. Such tabulations are referenced to the standard hydrogen electrode (SHE). The standard hydrogen electrode undergoes the reaction:

$$2H^+(aq) + 2e^- \rightarrow H_2$$

Which is shown as reduction but, in fact, the SHE can act as either the anode or the cathode, depending on the relative oxidation/reduction potential of the other electrode/electrolyte combination. The term standard in SHE requires a supply of hydrogen gas bubbled through the electrolyte at a pressure of 1 atm and an acidic electrolyte with H+ activity equal to 1 (usually assumed to be [H+] = 1 mol/liter).

The SHE electrode can be connected to any other electrode by a salt bridge to form a cell. If the second electrode is also at standard conditions, then the measured cell potential is called the standard electrode potential for the electrode. The standard electrode potential for the SHE is zero, by definition. The polarity of the standard electrode potential provides information about the relative reduction potential of the electrode compared to the SHE. If the electrode has a positive potential with respect to the SHE, then that means it is a strongly reducing electrode which forces the SHE to be the anode (an example is Cu in aqueous $CuSO_4$ with a standard electrode potential of 0.337 V). Conversely, if the measured potential is negative, the electrode is more oxidizing than the SHE (such as Zn in $ZnSO_4$ where the standard electrode potential is -0.763 V).

Standard electrode potentials are usually tabulated as reduction potentials. However, the reactions are reversible and the role of a particular electrode in a cell depends on the relative oxidation/reduction potential of both electrodes. The oxidation potential for a particular electrode is just the negative of the reduction potential. A standard cell potential can be determined by looking up the standard electrode potentials for both electrodes (sometimes called half cell potentials). The one that is smaller will be the anode and will undergo oxidation. The cell potential is then calculated as the sum of the reduction potential for the cathode and the oxidation potential for the anode.

$$E^0_{cell} = E^0_{red}(cathode) - E^0_{red}(anode) = E^0_{red}(cathode) + E^0_{oxi}(anode)$$

For example, the standard electrode potential for a copper electrode is:

Cell diagram:

$$Pt(s)|H_2(1\,atm)|H^+(1\,M)||Cu^{2+}(1\,M)|Cu(s)$$

$$E^0_{cell} = E^0_{red}(cathode) - E^0_{red}(anode)$$

At standard temperature, pressure and concentration conditions, the cell's emf (measured by a multimeter) is 0.34 V. by definition, the electrode potential for the SHE is zero. Thus, the Cu is the cathode and the SHE is the anode giving

$$E_{cell} = E^0_{Cu^{2+}/Cu} - E^0_{H^+/H_2}$$

Or,

$$E^0_{Cu^{2+}/Cu} = 0.34V$$

Changes in the stoichiometric coefficients of a balanced cell equation will not change E^0_{red} value because the standard electrode potential is an intensive property.

Spontaneity of Redox Reaction

During operation of electrochemical cells, chemical energy is transformed into electrical energy and is expressed mathematically as the product of the cell's emf and the electrical charge transferred through the external circuit.

$$\text{Electrical energy} = E_{cell}C_{trans}$$

where E_{cell} is the cell potential measured in volts (V) and C_{trans} is the cell current integrated over time and measured in coulumbs (C). C_{trans} can also be determined by multiplying the total number of electrons transferred (measured in moles) times Faraday's constant, F = 96,485 C/mole.

The emf of the cell at zero current is the maximum possible emf. It is used to calculate the maximum possible electrical energy that could be obtained from a chemical reaction. This energy is referred to as electrical work and is expressed by the following equation:

$$W_{max} = W_{electrical} = -nFE_{cell}$$

where work is defined as positive into the system.

Since the free energy is the maximum amount of work that can be extracted from a system, one can write:

$$\Delta G = -nFE_{cell}$$

A positive cell potential gives a negative change in Gibbs free energy. This is consistent with the cell production of an electric current flowing from the cathode to the anode through the external circuit. If the current is driven in the opposite direction by imposing an external potential, then work is done on the cell to drive electrolysis.

A spontaneous electrochemical reaction (change in Gibbs free energy less than zero) can be used to generate an electric current, in electrochemical cells. This is the basis of all batteries and fuel cells. For example, gaseous oxygen (O_2) and hydrogen (H_2) can be combined in a fuel cell to form water and energy, typically a combination of heat and electrical energy.

Conversely, non-spontaneous electrochemical reactions can be driven forward by the application of a current at sufficient voltage. The electrolysis of water into gaseous oxygen and hydrogen is a typical example.

The relation between the equilibrium constant, K, and the Gibbs free energy for an electrochemical cell is expressed as follows:

$$\Delta G^0 = -RT \ln K = -nFE^0_{cell}$$

Rearranging to express the relation between standard potential and equilibrium constant yields

$$E^0_{cell} = \frac{RT}{nF}\ln K$$

Previous equation can use Briggsian logarithm as shown below:

$$E^0_{cell} = \frac{0.0592V}{n}\log K$$

Cell EMF Dependency on Changes in Concentration

Nernst Equation

The standard potential of an electrochemical cell require s standard conditions for all of the reactants. When reactant concentrations differ from standard conditions, the cell potential will deviate from the standard potential. In the 20th century German chemist Walther Hermann Nernst proposed a mathematical model to determine the effect of reactant concentration on electrochemical cell potential.

In the late 19th century Josiah Willard Gibbs had formulated a theory to predict whether a chemical reaction is spontaneous based on the free energy

$$\Delta G = \Delta G^0 + RT \ln Q.$$

Where:

ΔG = change in Gibbs free energy,
T = absolute temperature,
R = gas constant,
ln = natural logarithm, and
Q = reaction quotient.

Gibbs' key contribution was to formalize the understanding of the effect of reactant concentration on spontaneity.

Based on Gibbs' work, Nernst extended the theory to include the contribution from electric potential on charged species. As shown in the previous section, the change in Gibbs free energy for an electrochemical cell can be related to the cell potential. Thus, Gibbs' theory becomes

$$nF\Delta E = nF\Delta E^0 - RT \ln Q$$

Where:

n = number of electrons/mole product, F = Faraday constant (coulombs/mole), and ΔE = cell potential.

Finally, Nernst divided through by the amount of charge transferred to arrive at a new equation which now bears his name:

$$\Delta E = \Delta E^0 - \frac{RT}{nF} \ln Q$$

Assuming standard conditions (Temperature = 25°C) and R = $8.3145 \frac{J}{K_{mol}}$ the equation above can be expressed on Base—10 logarithm as shown below:

$$\Delta E = \Delta E^0 - \frac{0.0592\,V}{n} \log Q$$

Concentration Cells

A concentration cell is an electrochemical cell where the two electrodes are the same material, the electrolytes on the two half-cells involve the same ions, but the electrolyte concentration differs between the two half-cells.

For example an electrochemical cell, where two copper electrodes are submerged in two copper (II) sulfate solutions, whose concentrations are 0.05 M and 2.0 M, connected through a salt bridge. This type of cell will generate a potential that can be predicted by the Nernst equation. Both electrodes undergo the same chemistry (although the reaction proceeds in reverse at the cathode).

$$Cu^{2+}\,(aq) + 2e^- \rightarrow Cu(s)$$

Le Chatelier's principle indicates that the reaction is more favourable to reduction as the concentration of Cu^{2+} ions increases. Reduction will take place in the cell's compartment where concentration is higher and oxidation will occur on the more dilute side.

The following cell diagram describes the cell mentioned above:

$$Cu(s)\ |Cu^{2+}\ (0.05\ M)|\ |Cu^{2+}\ (2.0\ M)|\ Cu(s)$$

Where the half cell reactions for oxidation and reduction are:

Oxidation: Cu(s) → Cu2+ (0.05 M) + 2e–

Reduction: Cu^{2+} (2.0 M) + $2e^-$ → Cu(*s*)

Overallreaction: Cu^{2+} (2.0 M) → Cu^{2+} (0.05 M)

Where the cell's emf is calculated through Nernst equation as follows:

$$E = E^0 - \frac{0.0257\,V}{2} \ln \frac{\left[CuH^{2+}\right]_{diluted}}{\left[Cu^{2+}\right]_{Concentrated}}$$

E^0's value of this kind of cell is zero, as electrodes and ions are the same in both half-cells. After replacing values from the case mentioned, it is possible to calculate cell's potential:

$$E = 0 - \frac{0.0257V}{2} \ln \frac{0.05}{2.0} = 0.0474V.$$

However, this value is only approximate, as reaction quotient is defined in terms of ion activities which can be approximated with the concentrations as calculated here.

The Nernst equation plays an important role in understanding electrical effects in cells and organelles. Such effects include nerve synapses and cardiac beat as well as the resting potential of a somatic cell.

Battery

A battery is an electrochemical cell (sometimes several in series) used for chemical energy storage. Batteries are optimized to produce a constant electric current for as long as possible. Although the cells discussed previously are useful for theoretical purposes and some laboratory experiments, the large internal resistance of the salt bridge make them inappropriate battery technologies. Various alternative battery technologies have been commercialized as discussed next.

Dry Cell

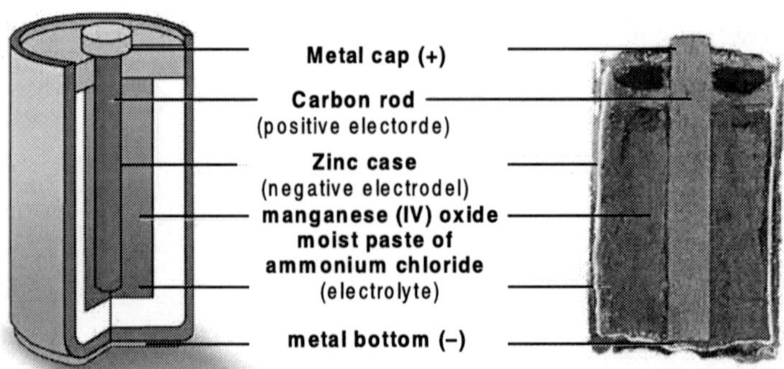

Fig. 1.5: Zinc carbon battery diagram.

Dry cells do not have a fluid electrolyte. Instead, they use a moist electrolyte paste. Leclanché's cell is a good example of this, where the anode is a zinc container surrounded by a thin layer of manganese dioxide and a moist electrolyte paste of ammonium chloride and zinc chloride mixed with starch. The cell's cathode is represented by a carbon bar inserted on the cell's electrolyte, usually placed in the middle.

Leclanché's simplified half reactions are shown below:

Anode: $Zn(s) \rightarrow Zn^{2+}(aq) + 2e^-$

Cathode : $2NH_4^+(aq) + 2MnO_2(s) + 2e^- \rightarrow Mn_2O_3(s) + 2NH_3(aq) + H_2O(l)$

Overall reaction: $Zn(s) + 2NH_4^+(aq) + 2MnO_2(s) \rightarrow Zn^{2+}(aq) + Mn_2O_3(s) + 2NH_3(aq) + H_2O(l)$

The voltage obtained from the *zinc-carbon battery* is around 1.5 V.

Mercury Battery

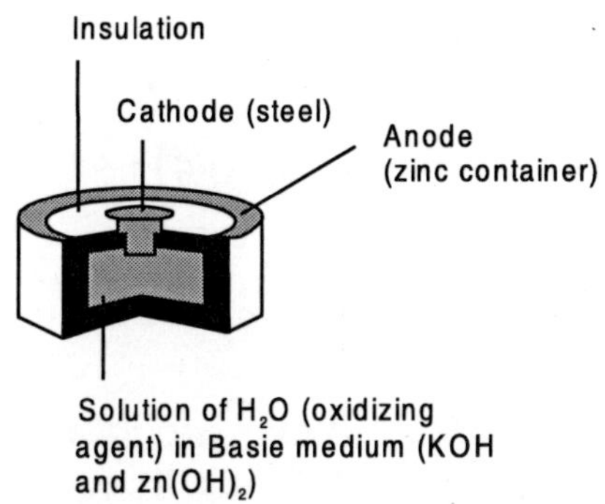

Fig 1.6: Cutaway view of a mercury battery.

The mercury battery has many applications in medicine and electronics. The battery consists of a steel—made container in the shape of a cylinder acting as the cathode, where an amalgamated anode of mercury and zinc is surrounded by a stronger alkaline electrolyte and a paste of zinc oxide and mercury(II) oxide.

Mercury battery half reactions are shown below:

Anode: $Zn(Hg) + 2OH^-(aq) \rightarrow ZnO(s) + H_2O(l) + 2e^-$

Cathode: $HgO(s) + H2O(l) + 2e^- \rightarrow Hg(l) + 2OH^-(aq)$

Overall reaction: $Zn(Hg) + HgO(s) \rightarrow ZnO(s) + Hg(l)$

There are no changes in the electrolyte's composition when the cell works. Such batteries provide 1.35 V of direct current.

Lead-acid Battery

Fig. 1.7: A sealed lead-acid battery.

The lead-acid battery used in automobiles, consists of a series of six identical cells assembled in series. Each cell has a lead anode and a cathode made from lead dioxide packed in a metal plaque. Cathode and anode are submerged in a solution of sulfuric acid acting as the electrolyte.

Lead-acid battery half cell reactions are shown below:

Anode: $Pb(s) + SO_4^{2-}(aq) \rightarrow PbSO_4(s) + 2e^-$

Cathode: $PbO_2(s) + 4H^+(aq) + SO_4^{2-}(aq) + 2e^- \rightarrow PbSO_4(s) + 2H_2(l)$

Overall reaction:

$$Pb(s) + PbO_2 + 4H^+(aq) + 2SO_4^{2-}(aq) \rightarrow 2PbSO_4(s) + 2H_2O(l)$$

At standard conditions, each cell may produce a potential of 2 V, hence overall voltage produced is 12 V. Differing from mercury and zinc-carbon batteries, lead-acid batteries are rechargeable. If an external voltage is supplied to the battery it will produce an electrolysis of the products in the overall reaction (discharge), thus recovering initial components which made the battery work.

Lithium Rechargeable Battery

Instead of an aqueous electrolyte or a moist electrolyte paste, a solid state battery operates using a solid electrolyte. Lithium polymer batteries are an example of this; a graphite bar acts as the anode, a bar of lithium cobaltate acts as the cathode, and a polymer, swollen with a lithium salt, allows the passage of ions and serves as the electrolyte. In this cell, the carbon in the anode can reversibly form a lithium-carbon alloy. Upon discharging, lithium ions spontaneously leave the lithium cobaltate cathode and travel through the polymer and into the carbon anode forming the alloy. This flow of positive lithium ions is the electrical current that the battery provides. By charging the cell, the lithium dealloys and travels back into the cathode. The advantage of this kind of battery is that Lithium possess the highest negative value of standard reduction potential. It is also a light metal and therefore less mass is required to generate 1 mole of electrons. Lithium ion battery technologies are widely used in portable electronic devices because they have high energy storage density and are rechargeable. These technologies show promise for future automotive applications, with new materials such as iron phosphates and lithium vanadates.

Flow Battery/Redox Flow Battery

Most batteries have all of the electrolyte and electrodes within a single housing. A flow battery is unusual in that the majority of the electrolyte, including dissolved reactive species, is stored in separate tanks. The electrolytes are pumped through a reactor, which houses the electrodes, when the battery is charged or discharged.

These types of batteries are typically used for large-scale energy storage (kWh—multi MWh). Of the several different types that have been developed, some are of current commercial interest, including the vanadium redox battery and zinc bromine battery.

Fuel Cells

Fossil fuels are used in power plants to supply electrical needs, however their conversion into electricity is an inefficient process. The most efficient electrical power plant may only convert about 40 per cent of the original chemical energy into electricity when burned or processed.

To enhance electrical production, scientists have developed fuel cells where combustion is replaced by electrochemical methods, similar to a battery but requiring continuous replenishment of the reactants consumed.

The most popular is the oxygen-hydrogen fuel cell, where two inert electrodes (porous electrodes of nickel and nickel oxide) are placed in an electrolytic solution such as hot caustic potash, in both compartments (anode and cathode) gaseous hydrogen and oxygen are bubbled into solution.

Oxygen-hydrogen fuel cell reactions are shown bellow:

Anode: $2H_2(g) \rightarrow 4H^+ + 4e^-$

Cathode: $O_2(g) + 4e^- + 4H^+ \rightarrow 2H_2O(l)$

Overall Reaction: $2H_2(g) + O_2(g) \rightarrow 2H_2O(l)$

The overall reaction is identical to hydrogen combustion. Oxidation and reduction take place in the anode and cathode separately. This is similar to the electrode used in the cell for measuring standard reduction potential which has a double function acting as electrical conductors providing a surface required to decomposition of the molecules into atoms before electron transferring, thus named electrocatalysts. Platinum, nickel, and rhodium are good electrocatalysts.

Corrosion

Corrosion is the term applied to metal rust caused by an electrochemical process. Most people are likely familiar with the corrosion of iron, in the form of reddish rust. Other examples include the black tarnish on silver, and red or green corrosion that may appear on copper and its alloys, such as brass. The cost of replacing metals lost to corrosion is in the multi-billions of dollars per year.

Iron Corrosion

For iron rust to occur the metal has to be in contact with oxygen and water, although chemical reactions for this process are relatively complex and not all of them are completely understood, it is believed the causes are the following:

1. Electron transferring (Reduction-Oxidation)
    1. One area on the surface of the metal acts as the anode, which is where the oxidation (corrosion) occurs. At the anode, the metal gives up electrons.

    $$Fe(s) \rightarrow Fe^{2+}\,(aq) + 2e^-$$

    2. Electrons are transferred from iron reducing oxygen in the atmosphere into water on the cathode, which is placed in another region of the metal.

    $$O_2(g) + 4H^+\,(aq)\ 4e^- \rightarrow 2H_2O(l)$$

    3. Global reaction for the process:

    $$2Fe(s)\ O_2(g)\ 4H^+\,(aq) \rightarrow 2Fe^{2+}\,(aq)\ 2H_2O(l)$$

    4. Standard emf for iron rusting:

    $$E^0 = E^0_{cathode} - E^0_{anode}$$

$$E^0 = 1.23V(-1.44V) = 1.67V.$$

Iron corrosion takes place on acid medium; H^+ ions come from reaction between carbon dioxide in the atmosphere and water, forming carbonic acid. Fe^{2+} ions oxides, following this equation:

$$4Fe^{2+}(aq) + O_2(g) + (4 + 2x)\,H_2O(l) \rightarrow 2Fe_2O_3.x\,H_2O + 8H^+(aq)$$

Iron(III) oxide hydrated is known as rust. Water associated with iron oxide it varies, thus chemical representation is presented as $Fe_2O_3.x\,H_2O$. The electric circuit works as passage of electrons and ions occurs, thus if an electrolyte is present it will facilitate oxidation, this explains why rusting is quicker on salt water.

Corrosion of Common Metals

Coinage metals, such as copper and silver, slowly corrode through use. A patina of green-blue copper carbonate forms on the surface of copper with exposure to the water and carbon dioxide in the air. Silver coins or cutlery that are exposed to high sulfur foods such as eggs or the low levels of sulfur species in the air develop a layer of black Silver sulfide.

Gold and platinum are extremely difficult to oxidize under normal circumstances, and require exposure to a powerful chemical oxidizing agent such as aqua regia.

Some common metals oxidize extremely rapidly in air. Titanium and aluminium oxidize instantaneouly in contact with the oxygen in the air. These metals form an extremely thin layer of oxidized metal on the surface. This thin layer of oxide protects the underlying layers of the metal from the air preventing the entire metal from oxidizing. These metals are used in applications where corrosion resistance is important. Iron, in contrast, has an oxide that forms in air and water, called rust, that does not stop the further oxidation of the iron. Thus iron left exposed to air and water will continue to rust until all of the iron is oxidized.

Prevention of Corrosion

Attempts to save a metal from becoming anodic are of two general types. Anodic regions dissolve and destroy the structural integrity of the metal.

While it is almost impossible to prevent anode/cathode formation, if a non-conducting material covers the metal, contact with the electrolyte is not possible and corrosion will not occur.

Metals are coated on its surface with paint or some other non-conducting coating. This prevents the electrolyte from reaching the metal surface if the coating is complete. Scratches exposing the metal will corrode with the region under the paint, adjacent to the scratch, to be anodic.

Other prevention is called *passivation* where a metal is coated with another metal such as a tin can. Tin is a metal that rapidly corrodes to form a monomolecular oxide coating that prevents further corrosion of the tin. The tin prevents the electrolyte from reaching the base metal, usually steel (iron). However, if the tin coating is scratched the iron becomes anodic and the can corrodes rapidly.

Sacrificial Anodes

A method commonly used to protect a structural metal is to attach a metal which is more anodic than the metal to be protected. This forces the structural metal to be cathodic, thus spared corrosion. It is called *"sacrificial"* because the anode dissolves and has to be replaced periodically.

Zinc bars are attached at various locations on steel ship hulls to render the ship hull cathodic. The zinc bars are replaced periodically. Other metals, such as magnesium, would work very well but zinc is the least expensive useful metal.

To protect pipelines, an ingot of buried or exposed magnesium (or zinc) is buried beside the pipeline and is connected electrically to the pipe above ground. The pipeline is forced to be a cathode and is protected from being oxidized and rusting. The magnesium anode is sacrificed. At intervals new ingots are buried to replace those lost.

Electrolysis

The spontaneous redox reactions of a conventional battery produce electricity through the different chemical potentials of the cathode and anode in the electrolyte. However, electrolysis requires an external source of electrical energy to induce a chemical reaction, and this process takes place in a compartment called electrolytic cell.

Electrolysis of Molten Sodium Chloride

When molten, the salt sodium chloride can be electrolyzed to yield metallic sodium and gaseous chlorine. Industrially this process takes place in a special cell named Down's cell. The cell is connected to an electrical power supply, allowing electrons to migrate from the power supply to the electrolytic cell.

Reactions that take place at Down's cell are the following:

Anode (oxidation): $2Cl^- \rightarrow Cl_2(g) + 2e^-$

Cathode (reduction): $2Na^+ (l) + 2e^- \rightarrow 2Na(l)$

Overall reaction: $2Na^+ + 2Cl^-(l) \rightarrow 2Na(l) + Cl_2(g)$

This process can yield large amounts of metallic sodium and gaseous chlorine, and is widely used on mineral dressing and metallurgy industries.

The EMF for this process is approximately -4 V indicating a (very) non-spontaneous process. In order for this reaction to occur the power supply should provide at least a potential of 4 V. However, larger voltages must be used for this reaction to occur at a high rate.

Electrolysis of Water

Water can be converted to its component elemental gasses, H_2 and O_2 through the application of an external voltage. Water doesn't decompose into hydrogen and oxygen spontaneously as the Gibbs free energy for the process at standard conditions is about 474.4 kJ. The decomposition of water into hydrogen and oxygen can be performed in an electrolytic cell. In it, a pair of inert electrodes usually made of platinum immersed in water act as anode and cathode in the electrolytic process. The electrolysis starts with the application of an external voltage between the electrodes. This process will not occur except at extremely high voltages without an electrolyte such as sodium chloride or sulfuric acid (most used 0.1 M).

Bubbles from the gases will be seen near both electrodes. The following half reactions describe the process mentioned above:

Anode (oxidation): $Na^+ (aq) + 1e^- \rightarrow O_2(g) + 4H^+ (aq) + 4e^-$

Cathode (reduction): $2H_2O (g) + 2e^- \rightarrow H_2(g) + 2OH^- (aq)$

Overall reaction: $2H2O(l) \rightarrow 2H_2(g) + O_2(g)$

Although strong acids may be used in the apparatus, the reaction will not net consume the acid. While this reaction will work at any conductive electrode at a sufficiently large potential, platinum catalyzes both hydrogen and oxygen formation, allowing for relatively mild voltages (~2V depending on the pH).

Electrolysis of Aqueous Solutions

Electrolysis in an aqueous is a similar process as mentioned in electrolysis of water. However, it is considered to be a complex process because the contents in solution have to be analyzed in half reactions, whether reduced or oxidized.

Electrolysis of a Solution of Sodium Chloride

The presence of water in a solution of sodium chloride must be examined in respect to its reduction and oxidation in both electrodes. Usually, water is electrolysed as mentioned in electrolysis of water yielding *gaseous oxygen in the anode* and gaseous hydrogen in the cathode. On the other hand, sodium chloride in water dissociates in Na^+ and Cl^- ions, anion will be attracted to the cathode, thus reducing the sodium ion. The cation will then be attracted to the anode oxidizing chloride ion.

The following half reactions describes the process mentioned:

1. *Cathode*: $Na^+ (aq) + 1e^- \rightarrow Na(s)$ $E^0_{red} = -2.71V$
2. *Anode*: $2Cl^-(aq) \rightarrow Cl_2(g) + 2e^-$ $E^0_{red} = +1.36V$
3. *Cathode*: $2H_2O(l) + 2e^+ \rightarrow H_2(g) + 2OH^- (aq)$ $E^0_{red} = -0.83V$
4. *Anode*: $2H_2O(l) \rightarrow O_2(g) + 4H^+ (aq) + 4e^-$

Reaction 1 is discarded as it has the most negative value on standard reduction potential thus making it less thermodynamically favorable in the process.

When comparing the reduction potentials in reactions 2 & 4, the reduction of chloride ion is favored. Thus, if the Cl^- ion is favored for reduction, then the water reaction is favored for oxidation producing gaseous oxygen, however experiments shown gaseous chlorine is produced and not oxygen.

Although the initial analysis is correct, there is another effect that can happen, known as the overvoltage effect. Additional voltage is sometimes required, beyond the voltage predicted by the #. This may be due to kinetic rather than thermodynamic considerations. In fact, it has been proven that the activation energy for the chloride ion is very low, hence favorable in kinetic terms. In other words, although the voltage applied is thermodynamically sufficient to drive electrolysis, the rate is so slow that to make the process proceed in a reasonable time frame, the voltage of the external source has to be increased (hence, overvoltage).

Finally, reaction 3 is favorable because it describes the proliferation of OH^- ions thus letting a probable reduction of H^+ ions less favorable an option.

The overall reaction for the process according to the analysis would be the following:

Anode (oxidation): $2Cl^-(aq) \rightarrow Cl_2(g)\ 2e^-$

Cathode (reduction): $2H_2O(l) + 2e^- \rightarrow H_2(g) + 2OH^-(aq)$

Overall reaction: $2H_2O + 2Cl^-\ (aq) \rightarrow H_2(g) + Cl_2(g) + 2OH^-(aq)$

As the overall reaction indicates, the concentration of chloride ions is reduced in comparison to OH^- ions (whose concentration increases). The reaction also shows the production of gaseous hydrogen, chlorine and aqueous sodium hydroxide.

Quantitative Electrolysis & Faraday's Laws

Quantitative aspects of electrolysis were originally developed by Michael Faraday in 1834. Faraday is also credited to have coined the terms *electrolyte*, electrolysis, among many others while he studied quantitative analysis of electrochemical reactions. Also he was an advocate of the law of conservation of energy.

First Law

Faraday concluded after several experiments on electrical current in non-spontaneous process, the mass of the products yielded on the electrodes was proportional to the value of current supplied to the cell, the length of time the current existed, and the molar mass of the substance analyzed.

In other words, the amount of a substance deposited on each electrode of an electrolytic cell is directly proportional to the quantity of electricity passed through the cell.

Below a simplified equation of Faraday's first law:

$$m = \frac{1}{96,485\left(C.mol^{-1}\right)} \cdot \frac{QM}{n}$$

Where,

m is the mass of the substance produced at the electrode (in grams),

Q is the total electric charge that passed through the solution (in coulombs),

n is the valence number of the substance as an ion in solution (electrons per ion),

M is the molar mass of the substance (in grams per mole).

Second Law

Faraday devised the laws of chemical electrodeposition of metals from solutions in 1857. He formulated the second law of electrolysis stating *"the amounts of bodies which are equivalent to each other in their ordinary chemical action have equal quantities of electricity naturally associated with them."* In other terms, the quantities of different elements deposited by a given amount of electricity are in the ratio of their chemical equivalent weights.

An important aspect of the second law of electrolysis is electroplating which together with the first law of electrolysis, has a significant number of applications in the industry, as when used to protect metals to avoid corrosion.

Applications

There are various extremely important electrochemical processes in both nature and industry, like the coating of objects with metals or metal oxides through electrodeposition and the detection of alcohol in drunken drivers through the redox reaction of ethanol. The generation of chemical energy through photosynthesis is inherently an electrochemical process, as is production of metals like aluminum and titanium from their ores. Certain diabetes blood sugar meters measure the amount of glucose in the blood through its redox potential.

The nervous impulses in neurons are based on electric current generated by the movement of sodium and potassium ions into and out of cells, and certain animals like eels can generate a powerful voltage from certain cells that can disable much larger animals.

Electrochemistry involves oxidation-reduction reactions that can be brought about by electricity or used to produce electricity.

Oxidation and reduction, which will be considered here as the loss and gain of electrons, occur in many chemical systems. The rusting of iron, the photosynthesis that takes place in the leaves of green plants, and the conversion of foods to energy in the body are all examples of chemical changes that involve the transfer of electrons from one chemical species to another. When such reactions can be made to cause electrons to flow through a wire or when a flow of electrons makes a redox reaction happen, the processes are referred to as *electrochemical changes*. The study of these changes is called *electrochemistry*.

The applications of electrochemistry are widespread. Batteries, which produce electrical energy by means of chemical reactions are in almost anything portable and electronic. In the laboratory, electrical measurements enable us to monitor chemical reactions of all sorts, even those in systems as tiny as a living cell. In industry, many important chemical—including liquid bleach (sodium hypochlorite) and lye (sodium hydroxide) are manufactured by electrochemical means. If it weren't for electrochemistry the important structural metals of aluminum and magnesium would only be laboratory curiosities and most people would see them only in museums.

Oxidation-Reduction Reactions

Oxidation-reduction reactions or redox reactions involve the transfer of electron density from one atom to another. Two example are the reaction of sodium with chlorine and the reaction of hydrogen with oxygen.

$$2\ Na^{\circ} + Cl_2 \rightarrow 2\ NaCl$$

$$2\ H_2 + O_2 \rightarrow 2\ H_2O$$

At first glance these reaction appear to be very different from each other. In NaCl ions have been formed, and that has certainly involved the transfer of electrons. The electron is transferred completely from a sodium to a chlorine atom as the Na^+ and Cl^- ions are created. But how about the reaction that produces water? Here we have the formation of a molecule held together by covalent bonds—bonds in which electrons are shared. How does this reaction involve a transfer of electrons? To answer this question, we have to look closely at the bonds in both the reactants and products.

Hydrogen and oxygen molecules are non-polar. This is because both atoms in an H_2 or O_2 molecule are the same, so the electronegativity difference between them is zero. In a non- polar molecule, the electron pair in the bond is shared equally, and neither atom carries a partial charge. Stated another way, each atom in an H_2 or O_2 molecule is electrically neutral. Now let's look at the product, water. The electronegativities of hydrogen and oxygen are quite different, oxygen being more electronegative than hydrogen. this means that the O-H bonds are quite polar, with the hydrogen carrying a substantial positive partial charge and the oxygen carrying a substantial negative partial charge.

Now we can look at what happens to the electrons around an atom during the reaction. A hydrogen atom begins with a zero positive charge in H_2 and finishes with a partial positive charge in H_2O. Similarly, oxygen begins with a zero partial charge in O_2 and finishes with a partial negative charge in H_2O. Thus, during the reaction there is a shift of electron density from a hydrogen atom to an oxygen atom, and it is in this sense that the reaction of H_2 and O_2 is similar to the reaction of Na with Cl_2.

Many chemical reactions involve (or at least appear to involve) a shift of electron density by one atom to another. Collectively, such reactions are called *oxidation-reduction reactions*, or simply redox reaction. The term *oxidation* refers to the loss of electrons by one reactant, and *reduction* refers to the gain of electrons by another. For example, the reaction between sodium and chlorine involves a loss of electrons by sodium (oxidation of sodium) and a gain of electrons of chlorine (reduction of chlorine).

$$Na^\circ \rightarrow Na^+ + e^- \text{ (oxidation)}$$

$$Cl_2 + 2e^- \rightarrow 2Cl^- \text{ (reduction)}$$

we say that the sodium is oxidized and the chlorine is reduced.

Oxidation and reduction always occur together. No substance is ever oxidized unless something else is reduced. Otherwise, electrons would appear as a product of the reaction, and this is never observed. During a redox reaction, then, some substance must accept the electrons that another substance loses. this electron-accepting substance is called the *oxidizing agent* because it helps something else to be oxidized. The substance that supplies the electrons is called the *reducing agent* because it helps something else be reduced. Sodium is a reducing agent, for example, when it supplies electrons to chlorine. In the process, sodium is oxidized. Chlorine is an oxidizing agent when it accepts electrons from the sodium, and when that happens, chlorine is reduced to chloride ion. One way to remember is:

The substance that is oxidized is the reducing agent.

Redox reactions are very common. Whenever you use a battery, a redox reaction occurs. The metabolism of foods, which supplies our bodies with energy, also occurs by a series of redox reactions that use oxygen to convert carbohydrates and fats to carbon dioxide and water. Ordinary household bleach works by oxidizing substances that stain fabrics, making them easier to remove from the fabric or rendering them colourless.

Liquid bleach contains hypochlorite OCl^-, as the oxidizing agent.

In 1791 Luigi Galvani discovered electrical activity in the nerves of the frogs that he was dissecting. He thought that electricity was of animal origin and could be found only in living tissues. A few years later, in 1800 Alessandro Volta discovered that electricity could be produced through inorganic means. In fact, by using small sheets of copper and zinc and cloth spacers soaked in an acid solution, he built a battery—the first apparatus capable of producing electricity. Naysayers were quick to predict that electricity would never serve a useful purpose. Obviously they were very wrong. Electricity has a central role in our lives and to this day Electrochemistry is a standard course of study.

While listening to lessons on Electrochemistry, many students may wonder why it was ever invented, if it was really ever necessary to invent it and if the world would be better off without it. With the small experiments that follow, we hope to make peace between these students and the study of Electrochemistry. These fun and simple experiments can teach the fundamental concepts of Electrochemistry without asking much of the student. As you will see, many of these demonstrations are easily adapted to various configurations and each can be done independently or as part of the full curriculum.

Porous Vase—An actual porous vase made for the purpose may be difficult to acquire. It is used to prevent the quick mixing of various solutions, while permitting the exchange of ions. For our purposes you can adapt a terracotta pot of the type used in gardening simply by plugging the hole in the bottom with molten wax and allowing it to cool. Another even more economical answer lies in constructing a barrier of paper. As shown in figure 4, roll the paper to form a cylinder and glue it in place on the bottom of the main container using a silicone adhesive such that liquids cannot pass between the two areas defined by the paper. A barrier of just one sheet would be too permeable, therefore use at least three layers of paper when building this device.

Distilled Water—Don't use de-ionized water in place of actual distilled water. Many times water sold for use in household irons is de-ionized rather than distilled water. Pharmacies sell actual distilled water. What is the difference? Many substances are soluble in water and a few of these substances separate into positive and negative ions in water. Generally these are made up of molecules that have ionic bonds, while non-ionic molecules remain intact, just in solution. For example, sugar dissolves easily in water and the sugar molecules remain intact as sugar. De-ionized water can have any number of dissolved substances that do not result in ions, yet are present nonetheless. Apart from this, a poor quality sample of de-ionized water can contain significant amounts of ions. On the other hand, distilled water is usually very pure—containing only actual water molecules.

Where to Obtain the Materials—You can buy the chemicals in a store that sells chemicals and laboratory equipment. Often, this type of store is located near universities. Some grocery or hardware store are able to get you the chemicals needed for these experiments. You can often purchase copper and zinc strips at a hobby retailer or hardware store. You can also ask a firm which works with sheets of these metals for some pieces, for example a gutter-pipe firm and a zinc galvanic treatments firm. For the first trials, you can also use a piece of copper and a zinc-galvanized nail or screw.

Cathode And Anode—In the course of these experiments, often I'll speak of cathodes and anodes. As you may notice, in some cases the cathode has been assigned a positive polarity and, in other cases, a negative one. This is due to the cathode being the electrode where the reaction of reduction occurs. In the case of a battery, this reduction occurs spontaneously and consumes electrons, so the cathode garners a positive charge. In the case of an electrolytic cell, the reduction is forced, producing electrons, so the cathode carries a negative charge. The anode will have in each case the polarity opposite the cathode's.

Solutions (Important)—The dilution of acids is dangerous. If water is added to a concentrated acid, it can explode violently causing severe injuries. Never pour water into concentrated acid. Always add the acid to the water. If you need to dilute an acid, get help from a Chemist. When making a solution of copper sulfate or zinc sulfate, add these chemicals to water rather than adding the water to the chemical.

Other Precautions—Do not allow any of the chemicals discussed in these experiments to get on your hands or skin. Do not put them in your mouth or eat them. Do not breathe any vapors from these chemicals. Do not keep them in bottles or containers that could be confused with a food or drink container. Do not leave these chemicals in such a place that it might be confused with a food or drink (such as on a kitchen table or counter or in the refrigerator). Store them in a separate, controlled place away from foods and out of the reach of children. Label each container clearly with the name of the contents and as a non-food item.

Conductors

Materials:

- — a battery
- — an electric light bulb and socket
- — three wires with alligator clips
- — a glass drinking cup or a beaker
- — distilled water
- — two electrodes of the same material (for example copper)

— table salt
— an ammeter

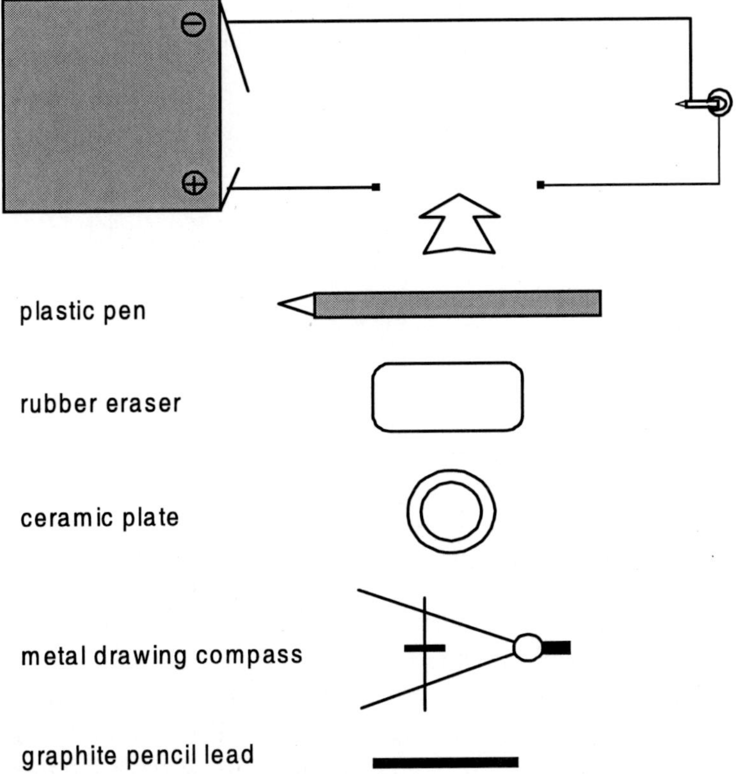

Fig. 1.8: Testing the conductivity of solid materials.

Using the various items listed above, construct the device shown in figure 1 (which is known as a continuity tester). As you can see, there is a break in the electrical path between the battery and the lamp. Using various household items (such as a plastic pen, an eraser, a ceramic plate, etc.) try to complete the circuit and see if the lamp lights.

You will notice that metallic items cause the lamp to light while most others do not. Try using the graphite (also known as the 'lead') from a pencil. You will notice that this material allows the bulb to light although perhaps not as brightly, depending on how long or short the graphite is. This is because the graphite, although a conductor presents a certain amount of resistance to the flow of electricity. Using this device, we can characterize solid materials as either conductors or non-conductors. Non conductors are also known as insulators. There are many solid materials that have intermediate properties between conductors and insulators, like graphite.

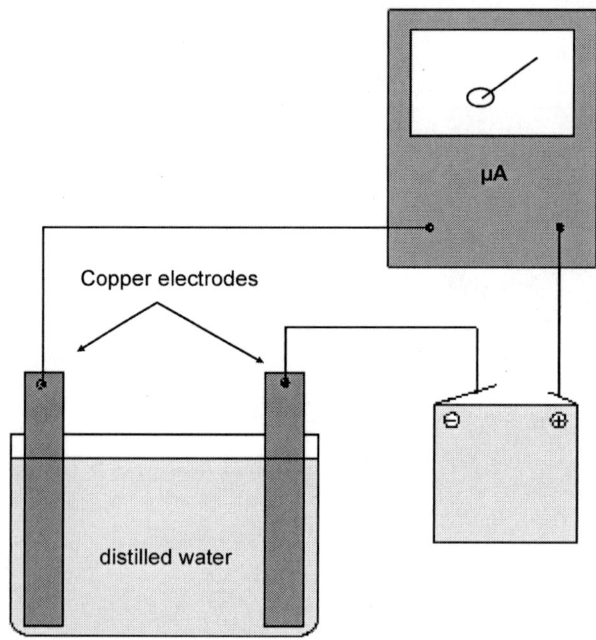

Fig. 1.9: Testing the conductive of water.

What about water? Let's modify the circuit we used earlier to now measure the conductivity of water (see figure 1.8). Instead of a light bulb to indicate the flow of electricity, now we will use an ammeter set to the microamp scale. This setup will be much more sensitive and will allow us to actually measure the electrical flow. Fill the container with *distilled water* and place the electrodes in the water as shown. Be careful not to touch the electrodes to each other. If there is any indication of current flow on the meter, it should be very small. This small current is due to the presence of H^+ and OH^-ions produced when a few of the water molecules dissociate and recompose spontaneously. The number of molecules doing this in any given moment is very small and this is why distilled water is a very poor conductor. Without disturbing the device, place a very small amount of salt in the water and with a plastic or glass utensil stir the water to mix the solution. As the salt crystals dissolve, the needle on the ammeter will move noticeably, indicating an increased conductivity of the water/salt solution. We used the same setup as before to help illustrate the similarity in the two devices, but you can also conduct the same experiment without the battery, but with an ohmmeter set to read resistance, as the ohmmeter has an internal battery and makes the setup less complicated.

What happened? Why did the salt make the water more conductive to electricity? Kitchen salt consists of sodium and chlorine molecules (NaCl).

The water not only separates the salt molecules one from another, but also separates the chlorine and sodium from each other. There is one change in the chlorine and sodium atoms as they separate. Elemental sodium weakly holds one excess electron and elemental chlorine has a strong attraction for a free electron, therefore as the two elements separate from each other the sodium gives up one electron to the chlorine. In this way, the sodium becomes electrically positive and the chlorine becomes negative. They have become *ions*. The molecules of salts, acids and bases become ions when dissolved in water. They dissociate into particles of opposite charges. It is the ions that render the water conductive to electricity. As a matter of fact, the positive ions migrate to the negative electrode and vice versa for the negative ions. In our experiment, the Na^+ and Cl^- ions made it possible for the distilled water to conduct electricity. This type of solution is called an *electrolyte*. Not only water, but other liquids as well can ionize various substances and result in electrolytic solutions.

If you add more salt to the solution you have been working with, you will see that the needle indicates increased current flow. The conductivity of a solution is proportional to the concentration of ions in the solution.

Let's try the same experiment again, but in place of salt we will use an acid—vinegar for example. Refill the original container with fresh distilled water and stir in a few drops of vinegar while watching the ammeter. Try another time with a base such as household ammonia. Compare the conductivity of tap water to that of distilled water. You will find that the tap water is a good electrical conductor. This is why it is wise to be very careful around electricity when your hands are wet. Many people have been killed by appliances like hair dryers when used around water.

You should note that in contrast to the conduction of electricity in a solid, such as a metal wire, current that passes through an electrolyte moves ions to the electrodes where they undergo chemical reactions. In a wire this does not occur. Additionally, reactions in solution tend to polarize the electrodes. This phenomenon is what causes a battery to wear out with use and the measurements that we make with our experiments to change over time. In commercial conductivity meters, the electrodes are made of a non-reactive metal and use an alternating current at low voltage.

How does electricity flow in a metal if there are no ions moving around? In metals, free electrons carry the current. In a metal wire, all the atoms share the electrons in their outermost electron shell and these electrons are free to move anywhere within the wire. If a voltage is applied across a metallic object, electrons enter one side, exit at the other and within the object there is an electrical current.

The passage of electrons in a conductor causes a rise in the oscillations of the atoms in the crystal lattice. This corresponds to a rise in the temperature of the object (the Joule Effect). This phenomenon is used to produce light in an incandescent lamp. The current of electrons flowing through a lamp filament causes a temperature rise of thousands of degrees. At this high temperature the filament emits intense light.

Why don't plastics, rubber and ceramics conduct electricity? This is due to the different molecular bond, called a *covalent* bond that binds these atoms to one another. In contrast to the *metallic bond* that we discussed in metals, a covalent bond does not allow the electrons to move freely within the body of the object. Electrons are exchanged only within the confines of each individual molecule and cannot move from molecule to molecule. Since there is no movement of electrons within the piece, and no movement of ions because the material is solid, there can be no flow of electrical charges through the material.

Despite what we have said about conductors and insulators, given a high enough voltage and a thin layer of the material in question, all materials will conduct electricity. Therefore, if we consider all materials to be conductors, it becomes necessary to know each material's *electrical conductivity*.

In this experiment we have seen that various solids will conduct electricity better than others will (metals vs. rubber, plastic, ceramics). We have also seen that some liquids are better conductors than others are. While a solid's conductivity is dependent upon the movement of electrons within the material's crystalline structure, a liquid's conductivity is dependent upon the movement of ions.

Batteries

The Lemon Battery

Materials:

- — a lemon
- — a strip of copper
- — a strip of zinc
- — a voltmeter
- — two cables with alligator clips
- — a thermometer or clock with an LCD display

Roll the lemon firmly with the palm of your hand on a tabletop or other hard surface in order to break up some of the small sacks of juice within the lemon. Insert the two metal strips deeply into the lemon, being careful that the strips not touch each other. Using the voltmeter, measure the voltage produced between the two strips. It should show to be about one volt.

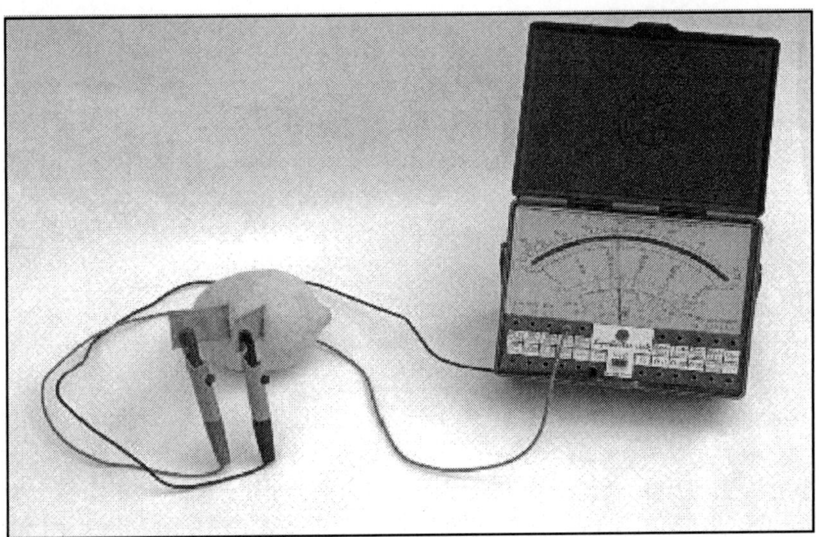

Fig. 1.10: The lemon batrries.

It would be nice to be able to illuminate a light bulb using your new lemon powered battery, but unfortunately it is not strong enough. If you were to try to light a bulb using this setup, the voltage across the strips would fall immediately to zero. Given this, if you want to demonstrate that the current produced by this battery is capable of powering something, try with a small device that uses an LCD display. A clock or a thermometer usually works well. An LCD display consumes an extremely small amount of current and your lemon battery is able to adequately drive this type of device. Remove any conventional battery that is in your clock or thermometer and power it with your lemon battery. You should see the device recommence functioning normally. If not, try swapping the polarity of the electricity from your lemon battery. This system allows you to demonstrate that the battery is producing energy even if you don't have a voltmeter.

How does this battery work? The Copper (Cu) atoms attract electrons more than do the Zinc (Zn) atoms. If you place a piece of copper and a piece of zinc in contact with each other, many electrons will pass from the zinc to the copper. As they concentrate on the copper, the electrons repel each other. When the force of repulsion between electrons and the force of attraction of electrons to the copper become equalized, the flow of electrons stops. Unfortunately there is no way to take advantage of this behavior to produce electricity because the flow of charges stops almost immediately. On the other hand, if you bathe the two strips in a conductive solution, and connect them externally with a wire, the reactions between the electrodes and the solution furnish the circuit with charges continually. In this way, the process that produces the electrical energy continues and becomes useful.

As a conductive solution, you can use any electrolyte, whether it be an acid, base or salt solution. The lemon battery works well because the lemon juice is acidic. Try the same setup with other types of solutions. As you may know, other fruits and vegetables also contain juices rich in ions and are therefore good electrical conductors. You are not then, limited to using lemons in this type of battery, but can make batteries out of every type of fruit or vegetable that you wish.

Like any battery, this type of battery has a limited life. The electrodes undergo chemical reactions that block the flow of electricity. The electromotive force diminishes and the battery stops working. Usually, what happens is the production of hydrogen at the copper electrode and the zinc electrode acquires deposits of oxides that act as a barrier between the metal and the electrolyte. This is referred to as the electrodes being polarized. To achieve a longer life and higher voltages and current flows, it is necessary to use electrolytes better suited for the purpose. Commercial batteries, apart from their normal electrolyte, contain chemicals with an affinity for hydrogen which combine with the hydrogen before it can polarize the electrodes.

The Daniell's Cell

Materials:

- a strip of copper
- a strip of zinc
- a large beaker, bowl or other suitable container
- a porous vase (as discussed in the Introduction)
- a plastic tube
- cotton
- 100g of Copper Sulfate ($CuSO_4$)
- 100g of Zinc Sulfate ($ZnSO_4$)
- 5g of Potassium Nitrate (KNO_3)
- 5g of Sodium Chloride (NaCl) if Potassium Nitrate is unavailable
- a liter of distilled water
- a voltmeter
- two cables with alligator clips

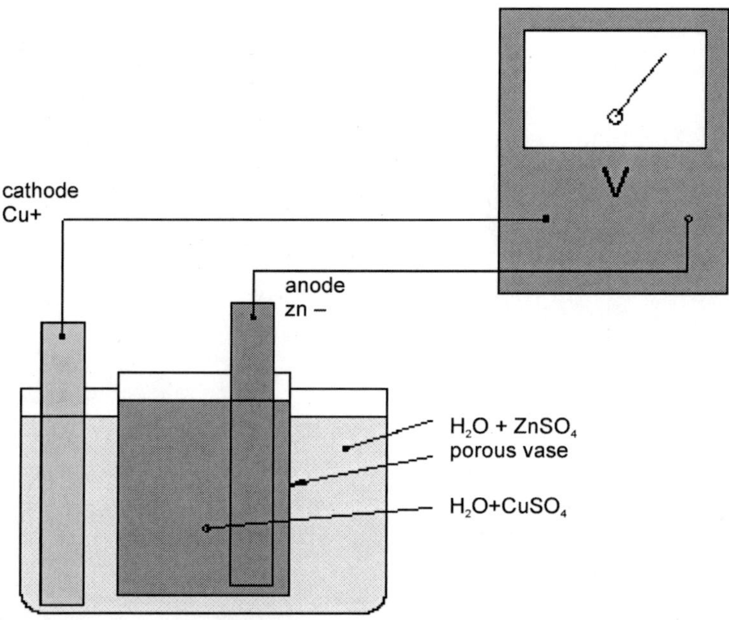

Fig. 1.11: Battery driven by electrolyte concentrations.

Prepare a concentrated solution of copper sulfate in distilled water and another solution of similar concentration of zinc sulfate in distilled water. For both of these solutions, use about 10-30 grams of dry chemical per 100°C of distilled water. Construct a setup as shown in figures 4 and 5. Pour the $CuSO_4$ solution in with the copper electrode and the $ZnSO_4$ solution in with the zinc electrode. When you measure the voltage across these electrodes, you should find it to be about 1.1 volts. Compared to the Lemon Battery, the Daniell's Cell puts out a higher power and lasts much longer. Even so, you would need electrodes with much greater surface area and more concentrated electrolyte to be able to power a very small light bulb with this device. Try instead a LED.

How does the Daniell's Cell work? As we have said, the reactions at the electrodes furnish charges that allow the battery to produce electrical current for extended periods. In the Daniell's Cell, the copper strip attracts electrons from the zinc strip. These electrons pass through the wires of our external circuit. As the copper electrode receives electrons, free positive ions in the solution arrive to equalize the charges. Positive copper ions (Cu^{++}) are attracted to the charged copper electrode where they receive two electrons and become neutral and deposit on the electrode in metallic form. The positive zinc ions (Zn^{++}) move to the porous vase. For each copper atom that is deposited on the copper electrode, a zinc atom goes into solution, giving up two electrons to the zinc electrode.

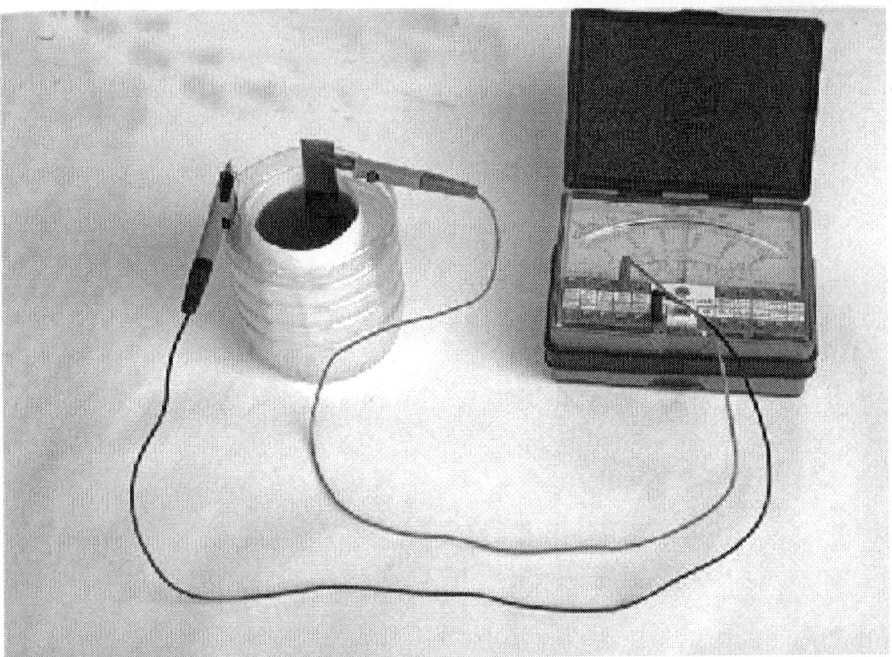

Fig. 1.12: Deniell's Cell, porous vase version.

The reactions at the electrodes can be represented by this formula:

$$Zn \rightarrow Zn^{++} \rightarrow + 2e^{-}$$
$$Cu^{++} + 2e^{-} \rightarrow Cu$$

These reactions result in the dissolution of zinc atoms in their ionic form, which corresponds to the deposition of copper ions in their metallic form:

$$Zn + Cu^{++} \rightarrow Zn^{++} + Cu$$

The electrons made available by the zinc atoms pass through the lamp filament, produce light through the joule effect and eventually reach the copper electrode. These electrons account for the current that is produced by the battery and is used by the lamp. If we didn't have the porous vase, the Cu^{++} ions would go directly to the zinc electrode and pick up free electrons, thereby bypassing the external circuit and stopping the current flow through the wires and lamp. The battery would no longer work. Because the copper electrode attracts electrons from the external circuit, it is considered the positive pole of the battery.

In a battery, there is always a flow of electrons in the external circuit (the electrical circuit or device) and a corresponding flow in the internal circuit (the electrolytic circuit). Like any battery, the Daniell Cell does not last forever, but only as long as there are Cu^{++} ions available and the zinc electrode is not consumed. In reality, the production of current diminishes as the concentration of the electrolyte bathing the zinc electrode increases and that bathing the copper electrode decreases. In fact, the positive ions produced by the zinc electrode need SO_4 ions to balance the charges. The exact opposite occurs in the copper solution, which becomes scarce of positive ions.

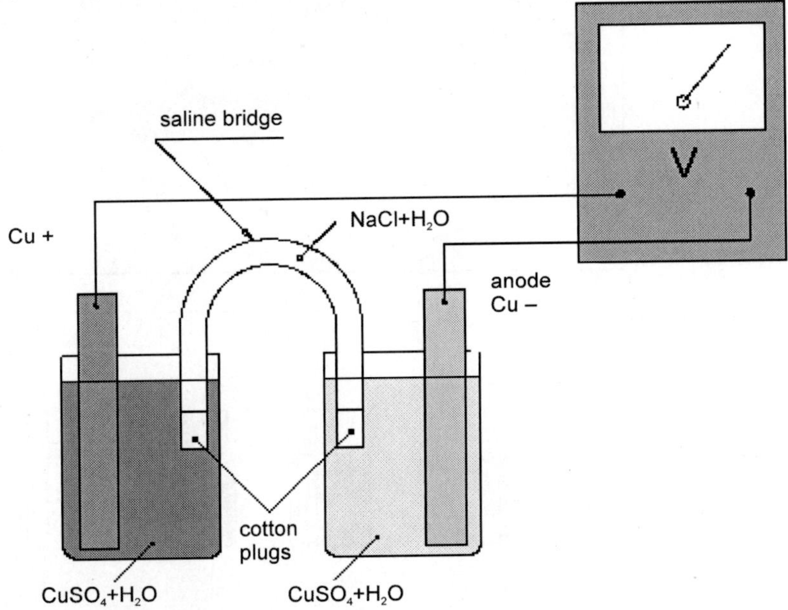

Fig. 1.14: Daniell's Cell saline bridge version.

Since the electromotive force of a battery is dependant not only upon the nature it's components, but also upon the concentration of it's electrolytes, the gradient of concentrations that results from the production of electricity causes the battery to generate lower and lower voltages and currents until finally it is considered dead. At the end, Zn^{++} ions finally reach the copper electrode, surrounding it and blocking any further movement of Cu^{++} ions by polarizing the electrode.

You can also build a Daniell Cell without a porous vase by using what is known as a Saline Bridge (fig. 6 and 7). This device is made by using a "U" shaped tube filled with saline solution. You can make one with the plastic tube called out in the materials list earlier. Fill the tube with a Potassium Nitrate (KNO_3) or Salt (NaCl) dissolved in distilled water (about 10 grams of chemical per 100 cc of water). Plug the ends with the cotton so that the saline solution will remain in the bridge and not mix with the electrolytes.

The bridge serves the same purpose as the porous vase, acting as a barrier between the two different electrolytes while allowing the flow of charges.

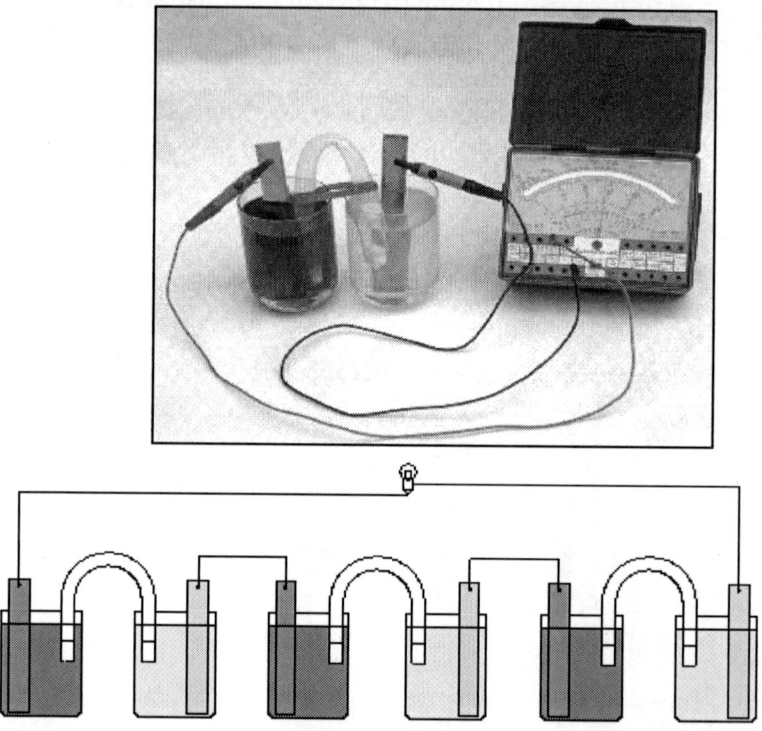

Fig. 1.15: Battries of three Daniell's cells.

If you want higher voltages, you can connect multiple Daniell Cells in series as shown in figure 1.15. Note that between one cell and another, the connection is a metallic wire rather than a saline bridge.

The Concentration Cell

Materials:

— two strips of copper
— a large beaker, bowl or other suitable container
— a porous vase (as discussed in the Introduction)
— a plastic tube (as used in previous experiment)
— cotton
— 50g of Copper Sulfate ($CuSO_4$)
— 1/2 liter of distilled water
— a voltmeter
— two cables with alligator clips

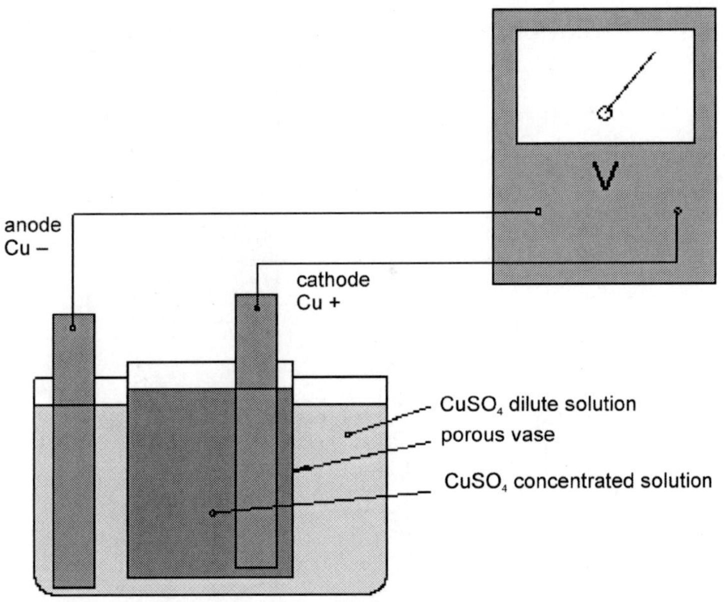

Fig. 1.16: Battery driven by electrolyte contratrations, porous vase version.

The setup of this experiment is very similar to the Daniell Cell's setup (Figs. 1.15 and 1.16). This battery takes advantage of the tendency of two solutions having different concentrations to reach the same concentration level. In contrast to the Daniell's Cell, we use only one material for the electrodes, in this case copper. As electrolyte, use copper sulfate in two concentrations, one using 30 grams of chemical in 100°C of water, the other using two grams of chemical in 100 cc of water. As you place the voltmeter in the circuit, it should indicate the voltage generated by the cell.

What is happening in this battery? The electrodes are undergoing a similar reduction reaction, but in two different concentrations. Due to this difference, there is a difference in potential (voltage) between the electrodes. At the anode, copper atoms dissolve into solution and give up electrons, while at the cathode, copper ions deposit on the electrode and acquire electrons.

$$Cu \rightarrow Cu^{++} + 2e^-$$

$$Cu^{++} + 2e^- \rightarrow Cu$$

You can build a smaller, simpler version of this battery by substituting a "U" shaped tube in place of the porous vase. In the center of the length of tubing, tightly press a quantity of cotton in order to form a barrier between the different concentrations of solutions and prevent premature mixing. Add the solutions and the electrodes. In this model the battery will not last very long, just until the solutions equalize across the cotton barrier.

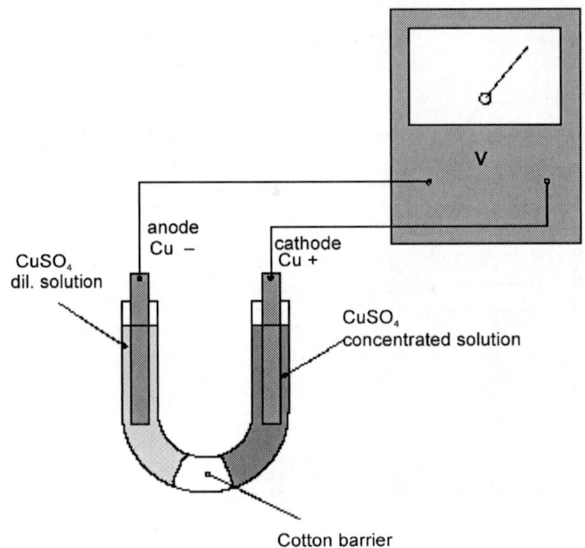

Fig. 1.17: Battery driven by electrolyte concentrations.

The Volta's Pile

Materials:

- — six strips or discs of copper
- — six strips or discs of zinc
- — filter paper
- — one of the following electrolytes:
- — lemon juice
- — vinegar
- — a solution of sodium chloride (salt water)
- — a solution of copper sulfate
- — a voltmeter
- — a thermometer or clock with an LCD display
- — two cables with alligator clips

In his famous experiment in the year 1800, Alessandro Volta used a solution of sulfuric acid as the electrolyte. In high concentrations, this acid can be very dangerous. If it gets on the skin or in the eyes, it can cause very serious burns and can permanently blind. Since we can build a functioning version with other types of electrolytes, in this experiment we will use a more benign substance, like those specified in the materials list—a solution of copper sulfate, for example. Despite this small modification in Volta's Pile, our intended result—the production of electricity by chemical means, can be again demonstrated. If you decide to try the experiment using sulfuric acid, use a low solution of it and an adult must be present to avoid any dangers.

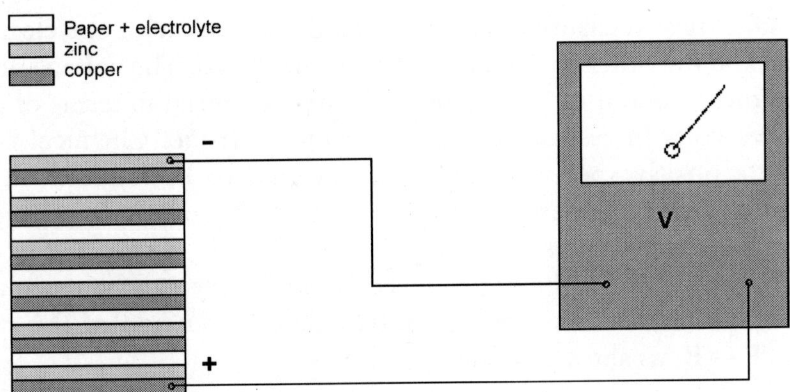

Fig. 1.18: Volta's pile.

Place each zinc disc on a copper disc. You should have six sets of Cu-Zn disc pairs. As shown in fig. 11, build a stack or "pile" of these pairs of discs, with a disc of electrolyte soaked filter paper separating each pair from its neighbors. Be careful that the solution does not dribble down the side of the stack as this can cause a short circuit between the elements of the pile. Note that the sequence of the elements is as follows: Cu, Zn, electrolyte, Cu, Zn, electrolyte, etc. When the device is all arranged as specified, measure the voltage between the bottom element of Cu and the top element of Zn. You should *see* 6.6 volts or 1.1 volt per pair of elements. The voltage generated is also dependent upon the electrolyte used and it's concentration. As you may have done with the lemon battery, try to power a small LCD electronic device such as a clock, thermometer or even a calculator.

This invention in 1800 spurred intense research in the field of electricity. About a century later, this research will have culminated in electric lights, the telephone and radio receivers in millions of homes. Today, electricity is an important part in every moment of our lives.

Measuring Potentials of Reduction

Another method of explaining the function of a battery lies in the oxidation the transfer of electrons from one chemical element to another. The tendency of an element to acquire or give up electrons is measured as electrical potential compared to that of a special hydrogen electrode, which is considered by convention to have zero electrical potential at 25°C. At this electrode, the following reaction occurs:

$$2H^+ + 2e^- <==> H_2$$

To make these measurements, use is made of a cell with one electrode of hydrogen and the other of the material to be measured. The voltage produced indicates the potential of reduction of the new material in terms of positive or negative volts in respect to the hydrogen electrode. Chemical elements that have a positive potential of reduction tend to be reduced, that is to acquire electrons, while elements that have a negative potential tend to oxidize, that is to give up electrons.

At this point, the voltage produced by a battery can be calculated by finding the difference between the potentials of reduction of the two half cells: $E = E_1—E_2$ as shown below.

s	*reaction*	*potential of reduction (V)*
E_1	$Cu^{++} + 2e^- = Cu$	+ 0.342
E_2	$Zn^{++} + 2e^- = Zn$	- 0.762

In this case, the voltage generated would be:

$E = + 0,342—(-0.762)$
$E = + 1.104\ V.$

You can make a battery from many different materials. It is possible to calculate in advance the voltage to expect from various materials by using the standard potential of reduction for the materials in question. This information can be found in any chemistry textbook.

Let's get started with the experiment.

Materials:

— a strip of copper
— various materials to try as electrodes
— two beakers or appropriate containers
— plastic tubing
— cotton

— 50 g. of copper sulfate ($CuSO_4$)
— 50 g. of a sulfate of the same element as the electrode you want to try
— 5 g. potassium nitrate (KNO_3)
— 5g. sodium chloride (NaCl) as an alternative to the potassium nitrate
— 1/2 liter of distilled water
— a voltmeter (it is best to use a digital model because the high input impedance will not affect the voltage as much as a low impedance analog model)
— a cable with alligator clips
— sandpaper

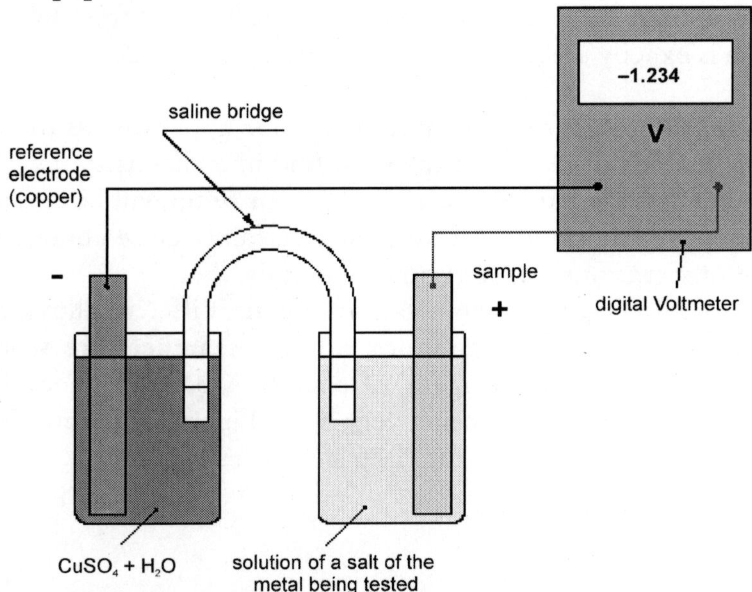

Fig. 1.19: Measuring potentials of reduction of respect to copper.

Construct a setup such as that shown in fig. 1.19, very similar to the Daniell's Cell with the saline bridge. Fill the container having the copper conductor with copper sulfate solution in the concentration of 1M. Fill the other container with the appropriate solution also in the concentration of 1M. Connect the negative side of the voltmeter to the copper electrode and the positive side to the electrode of test material and measure the voltage.

In order to get valid test measurements, it is necessary to have clean electrodes. Use the sandpaper to clean off any oxides or any other coatings or contamination so that the surfaces of the electrodes are clean and bright. Some metals oxidize very easily, such as aluminum, titanium, and magnesium. In the case that you are using a metal that oxidizes quickly, clean the electrode with sandpaper just before placing it in the electrolyte and then, during the experiment, wait a few moments for the chemical reaction to remove the slight amount of oxide that formed while you were cleaning it. You will see the voltage slowly rise to a maximum value. Note this reading as it represents the potential of reduction of the material under test with respect to copper. If the voltage starts immediately to drop, use the highest observed reading as the maximum value.

The voltages that you record are relative to copper. In order to obtain the value relative to hydrogen, that is the standard value, you would factor out the standard value of copper (Cu/Cu^{++}). For example, if you measure the voltage between zinc and copper with this setup, you should see -1.1 volts output. By adding -1.1 and +.34 (the standard value for copper) you get -.76 volts which is exactly the potential of reduction for zinc (Zn/Zn^{++}) relative to hydrogen.

Compare the values you measure with your apparatus to the tables of standard potentials of reduction that you find in a chemistry text. You will probably find that the values you get using your setup will be close, but not exact matches to the standard values, and certainly close enough to make interesting observations about various materials.

Try many different materials—various metals, different alloys, and even conductive plastics, rubber, ceramics. Some researchers are working on batteries that have electrodes made of plastic. A battery made of plastics would have the advantage of being very light. Lightweight batteries would be necessary to realize lightweight electric vehicles.

Galvanic Deposition

Materials:

- two copper strips
- one zinc strip
- one iron strip
- one stainless steel strip
- a container
- 50 g. copper sulfate ($CuSO_4$)
- 1/2 liter of distilled water
- a battery
- an ammeter
- two cables with alligator clips
- an analytic balance
- sandpaper

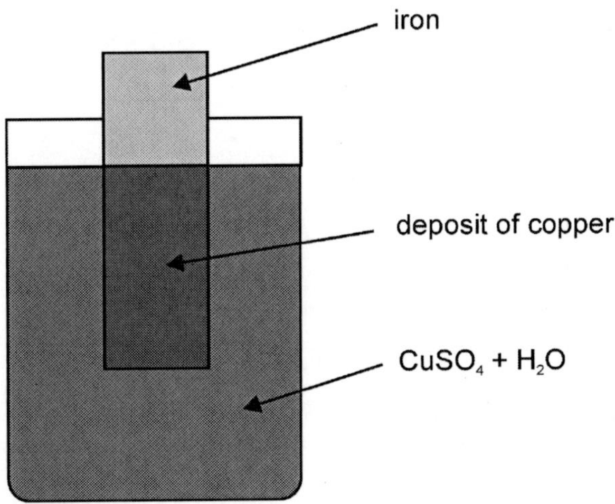

Fig. 1.20: Eletrolytic cell.

Oxireduction reactions involve the exchange of electrons between one chemical and another, always moving toward lower energy levels. In an electrolytic cell, the chemical processes involved can be reversed simply by supplying electrical energy to the cell instead of drawing it from the cell. Now we will use this principle to deposit metallic copper on an electrode.

Experiment 1: $CuSO_4$ in Solution and Copper Electrodes

Construct a setup such as that shown in figure 1.20. Use a solution of copper sulfate as the electrolyte (30g. of $CuSO_4$ per 100 cc of water) and strips of copper as the electrodes. In this cell, the sulfate ions remain in solution, while copper ions are deposited on the cathode and dissolved from the anode. As the copper is constantly being replenished into solution from the anode, eventually it will be consumed entirely and redeposited atom by atom on the cathode. Any impurities present in the anode will settle out of solution at the bottom of the container. The copper deposited at the cathode is very pure. In fact, this method is used in industrial environments to purify metals, the resultant being called "electrolytically pure". You can easily verify and monitor the progress of this transfer of metal from anode to cathode by weighing the electrodes before starting and at various times during the experiment. By changing the voltage across the cell, you can control the current flow. Make a chart of the rate of deposition on the cathode as compared to the current flow over fixed periods of time.

Experiment 2: $CuSO_4$ Solution and Strips of Zinc and Iron

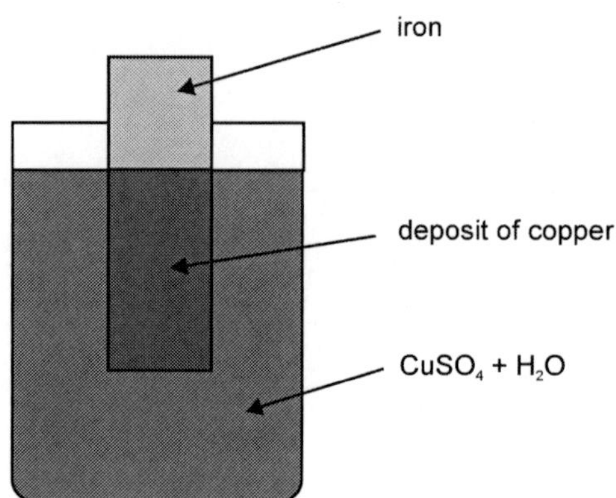

Fig. 1.21: Spontaneous deposition of copper on iron.

Perhaps you are wondering why we haven't been using a cathode of zinc which would make the deposition of copper easily visible. By virtue of it's bold color, copper would be easily visible on zinc which is a dull gray color. This way, we wouldn't need to use the scale to verify the deposition of copper. The problem is that the copper would begin depositing on the zinc before the current is even turned on. Copper ions from the copper sulfate would spontaneously deposit on the zinc strip. As you recall from earlier experiments, the copper atoms have a much stronger attraction for electrons than the zinc atoms. As the zinc strip is submerged in the solution containing copper ions, some zinc atoms are dissolved in solution as Zn^{++} ions and the same number of Cu^{++} ions are deposited on the zinc where they pick up the electrons lost by the zinc atoms and become electrically neutral. The zinc strip gets covered in a visible layer of copper. Normally zinc gets a deposit that appears dark in color, is of a powdery texture and has a dendritic, crystalline form. If you use an iron strip instead of zinc, you get a much more compact, metallic deposit of copper (fig. 1.21). Experiment freely with this phenomenon.

Experiment 3: $CuSO_4$ Solution and Cathode of Stainless Steel

Stainless steel is well adapted for use as a demonstrator of the deposition of copper on an electrode. It has a gray color that makes copper deposits easy to see and it will not spontaneously pick up copper deposits from copper sulfate solution. It is necessary to run a current through the solution in order to drive the copper deposits onto the stainless steel electrode. This is due to the fact that the stainless steel has a potential of reduction greater than that of copper. As shown in figure 13, use your strip of stainless steel as the cathode and apply electric current to the circuit. You will see that the portion of the cathode submerged in the electrolyte acquires a light coating of copper, as shown by the change in color.

In the course of these experiments, you can verify Faraday's law, which states that the quantity of electrons furnished is equal to the ionic charge transferred. The weight of the materials transformed is directly proportional to the amount of current supplied and the molecular weight of the ions and inversely proportional to the number of charges the material possesses in the ionic state. By way of example: a certain quantity of current will deposit a certain number of ions with a charge of 1, and half as many ions that have a charge of 2.

Electrolytic deposition is widely used in industry as a surface treatment on metals. These treatments can protect the underlying structure from oxidation, or can serve a decorative function. Jewelry makers deposit very thin layers of gold on base metals to achieve a product which is far less costly, yet has all the appearance of solid gold. In electronics, gold is widely used to coat electrical contacts and is valued for its resistance to oxidation and it's long-term stability. Galvanic techniques can also be used for other surface treatments of metals for decorative and protective purposes. Aluminum is often "anodized" which means that it is used as the anode in a process that oxidizes the surface, protecting it from corrosion and achieving various colorations all the way to black.

Conclusion

In the course of these experiments, you have become familiar with various concepts such as the conductivity of solutions by way of dissolved ions. you have seen redox reactions at work, and that they can be reversed by running current through the circuit. You have learned to measure potentials of reduction of various materials and to predict the voltage generated by a battery when you know it's material components. You can build a working battery from various materials, including lemons, tomatoes, potatoes, almost any fruit or vegetable. Up until now, batteries may have seemed mysterious, but now we are among the few that have a working knowledge of how and why they work. You can amaze your friends by building a battery out of almost anything, but more important than this, now the study and mastery of this subject will be easier and more meaningful.

1.6 Chemical Thermodynamics

In thermodynamics, *chemical thermodynamics* is the mathematical study of the interrelation of heat and work with chemical reactions or with a physical change of state within the confines of the laws of thermodynamics. Chemical thermodynamics can be generally thought of as the application of mathematical methods to the study of chemical questions and is concerned with the *spontaneity* of processes.

The structure of chemical thermodynamics is based on the first two laws of thermodynamics. Starting from the first and second laws of thermodynamics, four equations called the "fundamental equations of Gibbs" can be derived. From these four, a multitude of equations, relating the thermodynamic properties of the thermodynamic system can be derived using relatively simple mathematics. This outlines the mathematical framework of chemical thermodynamics.

History

In 1865, the German physicist Rudolf Clausius, in his *Mechanical Theory of Heat*, suggested that the principles of thermochemistry, *e.g.* such as the heat evolved in combustion reactions, could be applied to the principles of thermodynamics. Building on the work of Clausius, between the years 1873-76 the American mathematical physicist Willard Gibbs published a series of three papers, the most famous one being the paper *On the Equilibrium of Heterogeneous Substances*. In these papers, Gibbs showed how the first two laws of thermodynamics could be measured graphically and mathematically to determine both the thermodynamic equilibrium of chemical reactions as well as their tendencies to occur or proceed. Gibbs' collection of papers provided the first unified body of thermodynamic theorems from the principles developed by others, such as Clausius and Sadi Carnot.

During the early 20th century, two major publications successfully applied the principles developed by Gibbs to chemical processes, and thus established the foundation of the science of chemical thermodynamics. The first was the 1923 textbook *Thermodynamics and the Free Energy of Chemical Substances* by Gilbert N. Lewis and Merle Randall. This book was responsible for supplanting the chemical affinity for the term free energy in the English-speaking world. The second was the 1933 book *Modern Thermodynamics by the methods of Willard Gibbs* written by E. A. Guggenheim. In this manner, Lewis, Randall, and Guggenheim are considered as the founders of modern chemical thermodynamics because of the major contribution of these two books in unifying the application of thermodynamics to chemistry.

Overview

The primary objective of chemical thermodynamics is the establishment of a criterion for the determination of the feasibility or spontaneity of a given transformation. In this manner, chemical thermodynamics is typically used to predict the energy exchanges that occur in the following processes:

1. Chemical reactions
2. Phase changes
3. The formation of solutions

The following state functions are of primary concern in chemical thermodynamics:

- Internal energy (U)
- Enthalpy (H).
- Entropy (S)
- Gibbs free energy (G)

Most identities in chemical thermodynamics arise from application of the first and second laws of thermodynamics, particularly the law of conservation of energy, to these state functions.

Chemical Energy

It is the potential of a chemical substance to undergo a transformation through a chemical reaction or transform other chemical substances. Breaking or making of chemical bonds, involves energy, that may be either absorbed or evolved from a chemical system.

Energy that can be released (or absorbed) because of a reaction between a set of chemical substances is equal to the difference between the energy content of the products and the reactants. This change in energy is called the change in internal energy of a chemical reaction. It can be calculated using the formula

$$\Delta U^\circ = S(\Delta U_f^\circ{}_{products}) — S(\Delta U_f^\circ{}_{reactants})$$

Where $\Delta U_f^\circ{}_{reactants}$ is the internal energy of formation of the reactant molecules that can be calculated from the bond energies of the various chemical bonds of the molecules under consideration and $\Delta U_{f\ products}^{\circ}$ is the internal energy of formation of the product molecules. The internal energy change of a process is equal to the heat change if it is measured under conditions of constant volume, as in a closed rigid container such as a bomb calorimeter. However, under conditions of constant pressure, as in reactions in vessels open to the atmosphere, the heat change measured is not always equal to the internal energy change, because pressure-volume work also releases or absorbs energy. (The heat change at constant pressure is called the enthalpy change, in this case the enthalpy of formation).

Another useful term is the heat of combustion, it is the energy released due to a combustion reaction and often applied in the study of fuels. Food is similar to hydrocarbon fuel and carbohydrate fuels, and when it is oxidized, its caloric content is similar (though not assessed in the same way as a hydrocarbon fuel).

In *chemical thermodynamics* the term used for the chemical potential energy is chemical potential and for chemical transformation an equation most often used is Gibbs-Duhem equation

$$\sum_i N_i d\mu_1 = -SdT + Vdp$$

Chemical Reactions

In most cases of interest in chemical thermodynamics there are internal degrees of freedom and processes, such as chemical reactions and phase transitions, which always create entropy unless they are at equilibrium, or are maintained at a "running equilibrium" through "quasi-static" changes by being coupled to constraining devices, such as pistons or electrodes, to deliver and receive external work. Even for homogeneous "bulk" materials, the free energy functions depend on the composition, as do all the extensive thermodynamic potentials, including the internal energy. If the quantities $\{N_i\}$, the number of chemical species, are omitted from the formulae, it is impossible to describe compositional changes.

Gibbs Function

For a "bulk" (unstructured) system they are the last remaining extensive variables. For an unstructured, homogeneous "bulk" system, there are still various *extensive* compositional variables $\{N_i\}$ that G depends on, which specify the composition, the amounts of each chemical substance, expressed as the numbers of molecules present or (dividing by Avogadro's number), the numbers of moles

$$G = (T, P, \{Ni\})$$

For the case where only *PV* work is possible

$$dG = -SdT + VdP + \sum_i \mu_i dN_i$$

in which ì$_i$ is the chemical potential for the *i*-th component in the system

$$\mu_i = \left(\frac{\partial G}{\partial N_i}\right)_{T,P,N_{j\neq i},etc.}$$

The expression for d*G* is especially useful at constant *T* and *P*, conditions which are easy to achieve experimentally and which approximates the condition in living creatures

$$(dG)T, P = \sum_i \mu_i dN_i$$

Chemical Affinity

While this formulation is mathematically defensible, it is not particularly transparent since one does not simply add or remove molecules from a system. There is always a *process* involved in changing the composition; *e.g.*, a chemical reaction (or many), or movement of molecules from one phase (liquid) to another (gas or solid). We should find a notation which does not seem to imply that the amounts of the components (N_i} can be changed independently. All real processes obey conservation of mass, and in addition, conservation of the numbers of atoms of each kind. Whatever molecules are transferred to or from should be considered part of the "system".

Consequently we introduce an explicit variable to represent the degree of advancement of a process, a progress variable ξ for the *extent of reaction* (Prigogine & Defay, p. 18; Prigogine, pp. 4-7; Guggenheim, p. 37.62), and to the use of the partial derivative $\partial G / \partial \xi$ (in place of the widely used "ΔG", since the quantity at issue is not a finite change). The result is an understandable expression for the dependence of d*G* on chemical reactions (or other processes). If there is just one reaction

$$(dG)_{T,P} = \sum_\imath \mu_i dN_i$$

If we introduce the *stoichiometric coefficient* for the *i-th* component in the reaction

$$V_i = \partial N_i / \partial \xi$$

which tells how many molecules of *i* are produced or consumed, we obtain an algebraic expression for the partial derivative

$$\left(\frac{\partial G}{\partial \xi}\right)_{T,P} = \sum_i \mu_i v_i = -A$$

where, (De Donder; Progoine & Defay, p. 69; Guggenheim, pp. 37,240), we introduce a concise and historical name for this quantity, the "affinity", symbolized by A, as introduced by Théophile de Donder in 1923. The minus sign comes from the fact the affinity was defined to represent the rule that spontaneous changes will ensue only when the change in the Gibbs free energy of the process is negative, meaning that the chemical species have a positive affinity for each other. The differential for G takes on a simple form which displays its dependence on compositional change

$$(dG)_{T,P} = -A\, d\xi\ .$$

If there are a number of chemical reactions going on simultaneously, as is usually the case

$$(dG)_{T,P} = \sum_k A_k d\xi_k .$$

a set of reaction coordinates $\{\xi_j\}$, avoiding the notion that the amounts of the components $(N_i\}$ can be changed independently. The expressions above are equal to zero at thermodynamic equilibrium, while in the general case for real systems, they are negative, due to the fact that all chemical reactions proceeding at a finite rate produce entropy. This can be made even more explicit by introducing the reaction *rates* $d\xi_j/dt$. For each and every *physically independent process* (Prigogine & Defay, p. 38; Prigogine, p. 24)

$$A\xi \leq 0.$$

This is a remarkable result since the chemical potentials are intensive system variables, depending only on the local molecular milieu. They cannot "know" whether the temperature and pressure (or any other system variables) are going to be held constant over time. It is a purely local criterion and must hold regardless of any such constraints. Of course, it could have been obtained by taking partial derivatives of any of the other fundamental state functions, but nonetheless is a general criterion for (-T times) the entropy production from that spontaneous process; or at least any part of it that is not captured as external work.

We now relax the requirement of a homogeneous "bulk" system by letting the chemical potentials and the affinity apply to any locality in which a chemical reaction (or any other process) is occurring. By accounting for the entropy production due to irreversible processes, the inequality for dG is now replace by an equality

$$\mathrm{d}G = -SdT + V\,dP - \sum_k A_k d\xi_k + W'$$

or

$$dG_{T,P} = -\sum_k A_k d\xi_k + W'$$

Any decrease in the Gibbs function of a system is the upper limit for any isothermal, isobaric work that can be captured in the surroundings, or it may simply be dissipated, appearing as T times a corresponding increase in the entropy of the system and/or its surrounding. Or it may go partly toward doing external work and partly toward creating entropy. The important point is that the *extent of reaction* for a chemical reaction may be coupled to the displacement of some external mechanical or electrical quantity in such a way that one can advance only if the other one also does. The coupling may occasionally be *rigid*, but it is often flexible and variable.

Solutions

In solution chemistry and biochemistry, the Gibbs free energy decrease ($\partial G / G\xi$, in molar units, denoted cryptically by ΔG) is commonly used as a surrogate for (-T times) the entropy produced by spontaneous chemical reactions in situations where there is no work being done; or at least no "useful" work; *i.e.*, other than perhaps some $\pm PdV$. The assertion that all *spontaneous reactions have a negative* ΔG is merely a restatement of the fundamental thermodynamic relation, giving it the physical dimensions of energy and somewhat obscuring its significance in terms of entropy. When there is no useful work being done, it would be less misleading to use the

Legendre transforms of the entropy appropriate for constant T, or for constant T and P, the Massieu functions $-F/T$ and $-G/T$ respectively.

Non Equilibrium

Generally the systems treated with the conventional chemical thermodynamics are either at equilibrium or near equilibeium. Ilya Prigogine developed the thermodynamic treatment of open systems that are far from equilibrium. In doing so he has discovered phenomena and structures of completely new and completely unexpected types. His generalized, nonlinear and irreversible thermodynamics has found surprising applications in a wide variety of fields.

The non equilibrium thermodynamics has been applied for explaining how ordered structures *e.g.* the biological systems, can develop from disorder. Even if Onsager's relations are utilized, the classical principles of equilibrium in thermodynamics still show that linear systems close to equilibrium always develop into states of disorder which are stable to perturbations and cannot explain the occurrence of ordered structures.

Prigogine called these systems dissipative systems, because they are formed and maintained by the dissipative processes which take place because of the exchange of energy between the system and its environment and because they disappear if that exchange ceases. They may be said to live in symbiosis with their environment.

The method which Prigogine used to study the stability of the dissipative structures to perturbations is of very great general interest. It makes it possible to study the most varied problems, such as city traffic problems, the stability of insect communities, the development of ordered biological structures and the growth of cancer cells to mention but a few examples.

System Constraints

In this regard, it is crucial to understand the role of walls and other *constraints*, and the distinction between *independent* processes and *coupling*. Contrary to the clear implications of many reference sources, the previous analysis is not restricted to homogenous, isotropic bulk systems which can deliver only PdV work to the outside world, but applies even to the most structured systems. There are complex systems with many chemical "reactions" going on at the same time, some of which are really only parts of the same, overall process. An *independent* process is one that *could* proceed even if all others were unaccountably stopped in their tracks.

Understanding this is perhaps a "thought experiment" in chemical kinetics, but actual examples exist.

A gas reaction which results in an increase in the number of molecules will lead to an increase in volume at constant external pressure. If it occurs inside a cylinder closed with a piston, the equilibrated reaction can proceed only by doing work against an external force on the piston. The extent variable for the reaction can increase only if the piston moves, and conversely, if the piston is pushed inward, the reaction is driven backwards.

Similarly, a redox reaction might occur in an electrochemical cell with the passage of current in wires connecting the electrodes. The half-cell reactions at the electrodes are constrained if no current is allowed to flow. The current might be dissipated as joule heating, or it might in turn run an electrical device like a motor doing mechanical work. An automobile lead-acid battery can be recharged, driving the chemical reaction backwards. In this case as well, the reaction is not an independent process. Some, perhaps most, of the Gibbs free energy of reaction may be delivered as external work.

The hydrolysis of ATP to ADP and phosphate can drive the force times distance work delivered by living muscles, and synthesis of ATP is in turn driven by a redox chain in mitochondria and chloroplasts, which involves the transport of ions across the membranes of these cellular organelles. The coupling of processes here, and in the previous examples, is often not complete. Gas can leak slowly past a piston, just as it can slowly leak out of a rubber balloon. Some reaction may occur in a battery even if no external current is flowing. There is usually a coupling coefficient, which may depend on relative rates, which determines what percentage of the driving free energy is turned into external work, or captured as "chemical work", a misnomer for the free energy of another chemical process.

Quote

In the preface section to popular book *Basic Chemical Thermodynamics* by physical chemist Brian Smith, originally published in 1973, and now in the 5th edition, we find the following overview of the subject as it is perceived in college:

The first time I heard about chemical thermodynamics was when a second-year undergraduate brought me the news early in my freshman year. He told me a spine-chilling story of endless lectures with almost three-hundred numbered equations, all of which, it appeared, had to be committed to memory and reproduced in exactly the same form in subsequent examinations.

Not only did these equations contain all the normal algebraic symbols but in addition they were liberally sprinkled with stars, daggers, and circles so as to stretch even the most powerful of minds.

The scientific discipline that intersects the areas of chemistry and physics is commonly known as physical chemistry, and it is in that area that a thorough study of thermodynamics takes place. Physics concerns itself heavily with the mechanics of events in nature. Certainly changes in energy—however measured, whether it be heat, light, work, etc.—are clearly physical events that also have a chemical nature to them. Thermodynamics is the study of energy changes accompanying physical and chemical changes. The term itself clearly suggests what is happening—"thermo", from temperature, meaning energy, and "dynamics", which means the change over time. Thermodynamics can be roughly encapsulated with these five (5) topics:

1. Heat and Work
2. Energy
3. Enthalpy
4. Entropy
5. Free Energy

Heat and Work

Heat and work are both forms of energy. They are also related forms, in that one can be transformed into the other. Heat energy (such as steam engines) can be used to do work (such as pushing a train down the track). Work can be transformed into heat, such as might be experienced by rubbing your hands together to warm them up.

Work and heat can both be described using the same unit of measure. Sometimes the calorie is the unit of measure, and refers to the amount of heat required to raise one (1) gram of water one (1) degree Celsius. Heat energy is measured in kilocalories, or 1000 calories. Typically, we use the SI units of Joules (J) and kilojoules (kJ). One calorie of heat is equivalent to 4.187 J. You will also encounter the term specific heat, the heat required to raise one (1) gram of a material one (1) degree Celsius. Specific heat, given by the symbol "C", is generally defined as:

$$C = \frac{q}{M\Delta T}$$

Where:

C = specific heat in calories/gram-degrees Celsius

q = heat added in calories,
M = mass in grams
delta T = rise in temperature of the material in degrees Celsius.

The value of C for water is 1.00 calories/gram-degrees Celsius.

The values for specific heat that are reported in the literature are usually listed at a specific pressure and/or volume, and you need to pay attention to these settings when using values from textbooks in problems or computer models.

Example Problem

If a 2.34 g substance at 22 degrees Celsius with a specific heat of 3.88 cal/gÂ°C is heated with 124 cal of energy, what is the new temperature of the substance?

Answer:

$$\Delta T = \frac{q}{MC}$$

$$\text{DT} = \frac{(124)}{(2.34)\times(3.88)} = 13.7°\text{C}$$

$$\text{new T} = 22 + 13.7 = 35.7°\text{C}$$

Two other common heat variables are the heat of fusion and the heat of vaporization. Heat of fusion is the heat required to melt a substance at is normal melting temperature, while the heat of vaporization is the heat required to evaporate the substance at its normal boiling point.

Chemical work is primarily related to that of expansion. In physics, work is defined as:

$$w = \text{distance X (opposing force)}$$

Where:

w = work, in joules (N*m) (or calories, but we are using primarily SI units) distance is in meters opposing force is in Newton's (kg*m/s^2)

In chemical reactions, work is generally defined as:

$$w = \text{distance} \times \text{(area X pressure)}$$

The value of distance times area is actually the volume. If we imagine a reaction taking place in a container of some volume, we measure work by pressure times the change in volume.

$$w = dV \times P$$

Where:

dV is the change in volume, in liters
If dV = 0, then no work is done.

Example Problem: Calculate the work that must be done at standard temperature and pressure (STP is 0 degrees C and 1 atm) to make room for the products of the octane combustion:

$$2C_8H_{18} + 25O_2 \rightarrow 16CO_2 + 18H_2O$$

Answer

Knowing the 25 moles of gas are replaced by 34 moles of gas in this reaction, we can calculate a net increase of 9 moles of gas. Knowing the molar volume of an ideal gas at STP (22.4 L/mol), the change in volume and the work of expansion can be calculated.

$$dV = 9 \text{ moles} \times 22.4 \text{ L/mol} = 202 \text{ L}$$

Theexternal pressure is 1.0 atm (standard pressure), so the work required is:

$$w = dV \times P = 202\ L \times 1.00 \text{ atm} = 202 \text{ 1-atm}$$

Using the conversion factor of 1 L-atm = 101 J, the amount of work in joules is:

$$w = 202\ L\text{-atm} \times 101\ j/L\text{-atm} = 2000\ j, \text{ or } 2\ kj \text{ of energy}$$

Energy

You might remember the first law of thermodynamics: energy cannot be created or destroyed. Energy can only change form. Chemically, that usually means energy is converted to work, energy in the form of heat moves from one place to another, or energy is stored up in the constituent chemicals. You have seen how to calculate work. Heat is defined as that energy that is transferred as a result of a temperature difference between a system and its surroundings. Mathematically, we can look at the change in energy of a system as being a function of both heat and work:

$$dE = q - w$$

Where:

dE is the change in internal energy of a system, in joules
q is the heat flowing into the system in joules
w is the work being done by the system in joules

If q is positive, we say that the reaction is endothermic, that is, heat flows into the reaction from the outside surroundings. If q is negative, then the reaction is exothermic, that is, heat is given off to the external surroundings.

You might also remember the terms kinetic energy and potential energy. Kinetic energy is the energy of motion—the amount of energy in an object that is moving. Potential energy is stationary, stored energy. If you think of a ball sitting on the edge of a table, it has potential energy in the energy possible if it falls off the table. Potential energy can be transformed into kinetic energy if and when the ball actually rolls off the table and is in motion. The total energy of the system is defined as the sum of kinetic and potential energies.

In descriptions of the energy of a system, you will also see the phrase "state properties". A state property is a quantity whose value is independent of the past history of the substance. Typical state properties are altitude, pressure, volume, temperature, and internal energy.

Enthalpy

Enthalpy is an interesting concept: it is defined by its change rather than a single entity. A state property, the word enthalpy comes from the Greek "heat inside". If you have a chemical system that undergoes some kind of change but has a fixed volume, the heat output is equal to the change in internal energy ($q = dE$). We will define the enthalpy change, dH, of a system as being equal to its heat output at constant pressure:

$$dH = q \text{ at constant pressure}$$

Where:

$$dH = \text{change in enthalpy}$$

We define enthalpy itself as:

$$H = E + PV$$

Where:

$$H = \text{enthalpy}$$

$$E = \text{energy of the system}$$
$$PV = \text{pressure in atm times volume in liters}$$

You will not need to be able to calculate the enthalpy directly; in chemistry, we are only interested in the change in enthalpy, or dH.

$$dH = H_{final} - H_{initial} \text{ or } dH = H \text{ (products)} - H \text{ (reactants)}$$

Tables of enthalpies are generally given as dH values.

Example Problem: Calculate the dH value of the reaction:

$$HCl + NH_3 \rightarrow NH_4Cl$$

(dH values for HCl is –92.30; NH_3 is –80.29; NH_4Cl is –314.4)

Answer:

$$\Delta H = \Delta H_{products} - \Delta H_{reactants}$$

$$\Delta H_{products} = -314.4$$

$$\Delta H_{reactants} = -92.30 + (-80.29) = -172.59$$

$$\Delta H = -314.4 - 172.59 = 141.8$$

We can also represent enthalpy change with the equation:

$$\Delta H = dE + P\,dV$$

Where:

dV is the change in volume, in liters
P is the constant pressure

If you recall, work is defined as $P \times dV$, so enthalpy changes are simply a reflection of the amount of energy change (energy going in or out, endothermic or exothermic), and the amount of work being done by the reaction. For example, if $dE = -100\ kJ$ in a certain combustion reaction, but 10 kJ of work needs to be done to make room for the products, the change in enthalphy is:

$$dH = -100\ kJ + 10\ kJ = -90\ kJ$$

This is an exothermic reaction (which is expected with combustion), and 90 *kJ* of energy is released to the environment. Basically, you get warmer. Notice the convention used here—a negative value represents energy coming out of the system.

You can also determine dH for a reaction based on bond dissociation energies. Breaking bonds requires energy while forming bonds releases energy. In a given equation, you must determine what kinds of bonds are broken and what kind of bonds are formed. Use this information to calculate the amount of energy used to break bonds and the amount used to form bonds. If you subtract the amount to break bonds from the amount to form bonds, you will have the dH for the reaction.

Example Problem: Calculate dH for the reaction:

$$N_2 + H_2 \rightarrow 2NH_3$$

(The bond dissociation energy for N-N is 163; H-H is 436; N-H is 391)

Answer:

$$DH = DH_{products} - DH_{reactants}$$

To use the bond dissociation energies, we must determine how many bonds are in the products and the reactants. In N_2 there is 1 N-N bond and in $3H_2$ there are 3 H-H bonds.

$$DH_{products} = 6(391) = 2346$$

$$DH_{reactants} = 163 + 3(436) = 1471$$

$$DH = 2346 - 1471 = 875$$

Entropy

Entropy is a measure of the disorder of a system. Take your room as an example. Left to itself, your room will increase in entropy (*i.e.*, get messier) if no work (cleaning up) is done to contain the disorder. Work must be done to keep the entropy of the system low. Entropy comes from the second law of thermodynamics, which states that all systems tend to reach a state of equilibrium. The significance of entropy is that when a spontaneous change occurs in a system, it will always be found that if the total entropy change for everything involved is calculated, a positive value will be obtained. Simply, all spontaneous changes in an isolated chemical system occur with an increase in entropy. Entropy, like temperature, pressure, and enthalpy, is also a state property and is represented in the literature by the symbol "S". Like enthalpy, you can calculate the change of S (dS or delta S).

$$dS = S_{final} - S_{initial} \text{ or } dS = S(\text{products})—S(\text{reactants})$$

Where:

dS (or delta S) is change in entropy
S_{final} and $S_{initial}$ are the final and initial entropies, respectively

The following table shows the relationship between the state of a substance and its entropy:

State of substance	*Relative Entropy (S)*
gas	highest S
aqueous	high S
liquid	medium S
solid	lowest S

Free Energy

The free energy of a system, represented by the letter "G", is defined as the energy of a system that is free to do work at constant temperature and pressure. Mathematically, it is defined as:

$$G = H - TS$$

Where:

G is the free energy (sometimes called the Gibbs free energy, after its discoverer)
H is the enthalpy
T is the temperature
S is the entropy of the system.

You can also calculate the change in G the same way as you calculate the change in enthalpy or entropy:

$$dG = G(\text{products}) - G(\text{reactants})$$

Where:

dG (or delta G) is change in free energy

A pop-up calculator is available to calculate the enthalpy and Gibb's free energy changes in reactions.

Given a constant temperature and pressure, the direction of any spontaneous change is toward a lower free energy. The graphic below shows that during a reaction, the amount of free energy decreases until the reaction is at equilibrium. If the reaction goes towards completion, the free energy minimum occurs very close to the pure products part of the curve. In other words, the curve moves depending on the conditions of the reaction.

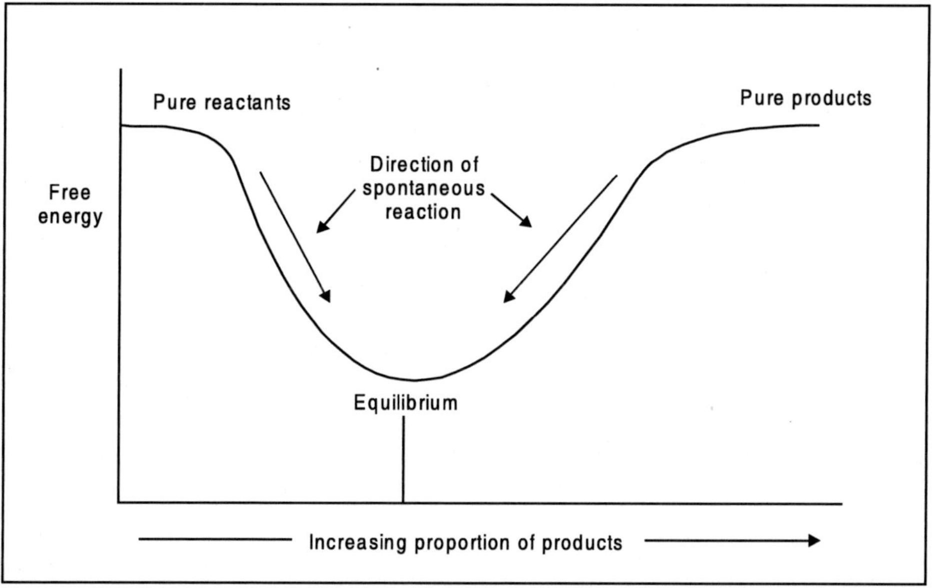

A table relating all of the state properties summarized above—enthalpy change, entropy change, and change in free energy—is shown below. A spontaneous reaction is one that occurs without any outside intervention. Processes that are spontaneous in one direction are non-spontaneous in the reverse direction.

Enthalpy Change	*Entropy Change*	*Spontaneous Reaction?*
Exothermic ($dH < 0$)	Increase ($dS > 0$)	Yes, $dG < 0$
Exothermic ($dH < 0$)	Decrease ($dS < 0$)	Only at low temps, if $\lvert T\,dS \rvert < \lvert dH \rvert$
Endothermic ($dH > 0$)	Increase ($dS > 0$)	Only at high temps, if $T\,dS > dH$
Endothermic ($dH > 0$)	Decrease ($dS < 0$)	No, $dG > 0$

Enthalpy Practice Problem: Given the following bond dissociation energies (H-C is 413 kJ/mol; H-H is 436 kJ/mol; C=C is 614 kJ/mol; C-C is 348 kJ/mol), determine dH for the reaction:

$$H_2C = CH_2(g) + H_2(g) \rightarrow H_3C - CH_3(g)$$

Enthalpy Solution

Entropy Practice Problem: Given the following entropy values (Al_2O_3(s) is 51.00; Al(s) is 28.32; H_2O(g) is 188.7; H_2(g) is 130.6), determine dS for the reaction:

$$Al_2O_3(s) + 3H_2(g) \rightarrow 2Al(s) + 3H_2O(g)$$

2
Analytical Environmental Chemistry

Research

Research is conducted in all of the traditional fields of chemistry: physical, theoretical, organic and inorganic—synthetic and mechanistic, analytical, biochemical, materials, and environmental. Multidisciplinary research is carried out both within the Department, and in cooperation with many of the Technion's other scientific and technological departments, and with other research institutions. Special emphasis is given to collaborative projects with leading research groups worldwide.

In 1996, 60 outside research grants provided close to two million dollars for funding research projects in the Department.

Current research activities of the academic staff extend over the following areas:

- synthesis and reaction mechanisms in organic and inorganic chemistry and organ metallic chemistry,
- heterogeneous and homogeneous catalysis,
- organ silicon chemistry,
- coordination chemistry,
- natural products,
- stereochemistry,
- photochemistry,
- bio-organic chemistry,
- enzymatic mechanisms and their use in synthesis,
- catalytic antibodies,
- heterocyclic compounds,
- modern electro analytical methods,
- ion exchange,
- X-ray crystallography of proteins, macromolecules and small molecules,

- analytical environmental chemistry,
- solid state chemistry and spectroscopy,
- phase-transition theory,
- chemical kinetics and molecular dynamics,
- quantum chemistry,
- computational chemistry and molecular modeling,
- resonance states,
- surface chemistry and spectroscopy,
- molecular beams,
- surface scattering,
- molecular spectroscopy,
- liquid state NMR,
- solid state NMR of proteins and polymers,
- electron paramagnetic resonance,
- non-linear optics,
- laser photo physics and photobiology,
- quantum optics,
- molecular electrooptics,
- batteries and solar cells,
- radiochemistry, and
- photo induced dynamics in strong laser fields.

For a more detailed picture of the research activities the reader is referred to the brief biographical sketch of each of our faculty members, including a concise outline of their research interests.

Facilities and Research Laboratories

The Department of Chemistry is well equipped with a wide variety of modern research instrumentation designed to meet the rapidly changing needs of teaching and research. Most of this instrumentation is available for hands-on use by students under the guidance of a professional staff. This enables the students to acquire rapidly an in-depth understanding of modern techniques and the high degree of expertise required to conduct independent research.

- Analytical Facilities,
- Environmental Research Laboratory,
- Surface Science—Chemistry and Physics of Surface,
- Lasers, Spectroscopy and Molecular Beams,

- Synthetic Organic, Inorganic and Organometallic Chemistry,
- Biology Related Laboratories,
- Computer Facilities,
- Technical Facilities, and
- Library.

Analytical Facilities

The analytical facilities at the Department of Chemistry currently include the following instruments:

- NMR spectrometers: *Bruker AV500 and AV300 (AM-400 and AC-200 (two units) still in use),*
- Fourier transform infrared spectrometer, *Nicolet, Impact 400,*
- Gas Chromatograph—Mass Spectrometer (GCMS), *Finnigan Magnum,*
- X-ray diffractometer, *Phillips*, equipped for low temperature measurements,
- Differential Scanning Calorimeter (DSC), and
- Atomic Absorption Spectrophotometer (AAS), flame and furnace.

The *National Center for Mass Spectrometry*, which is situated in the Department of Chemistry, operates the following instruments:

- An ion trap GCMS (*Finnigan ITS 40*),
- A triple quadrupole mass spectrometer (*Finnigan TSQ-70B*) with various ionization modes, and
- A double focusing high resolution mass spectrometer (*Varian-MAT 711*).

These capabilities are complemented by two modern research laboratories, the first for the crystallography of proteins (operating a *Rigaku R-Axis IIC X-ray diffractometer*), and the second comprising a fully equipped triple-resonance solid state NMR spectrometer (*Chemagnetics Infinity 300-MHz*).

Environmental Research Laboratory

Due to the increasing importance of environmental related issues a modern research laboratory has been established recently. This laboratory consists of: Laser Plasma Spectroscopy (LPS), Resonance Enhanced Multiphoton Ionization (REMPI), Reflection Time of Flight Mass-Spectrometer (RTOF-MS),

Portable Gas Chromatograph, Time Resolved Laser Induced Fluorimeter (TR-LIF) and other novel electroanalytical systems.

Surface Science—Chemistry and Physics of Surfaces

In recent years the Department of Chemistry has acquired modern equipment for advanced research in the fields of solid-state and of surface chemistry and physics.

The laboratories of surface chemistry and physics consist of:

- A scanning probe microscope for the studies of physical and chemical properties of surfaces with nanometer resolution,
- FTIR spectrometer for IR studies of solid surfaces and adsorbates,
- AES, LEED, and TPD facilities for the investigation of adsorption and other physico-chemical properties of surfaces, and
- An apparatus for hyper thermal molecular beam—surface scattering and thin film growth hemispherical electron energy analyzer for surface characterization by Auger and Electron Energy Loss (EEL) spectroscopies.

In addition, we have at our disposal research facilities at an adjacent building, the Solid State Institute (SSI), consisting of: facilities for ion implantation and ion beam analysis, equipment for surface studies by means of electron spectroscopy and secondary ion mass spectrometry, equipment for evaluation of semiconductor devices and crystal growth. Several members of the academic staff of the Chemistry Department have established research laboratories at the SSI in the areas of optical spectroscopy and surface science.

Lasers, Spectroscopy and Molecular Beams

Various research groups in the Department study chemistry by lasers, utilizing them in a broad range of applications. Several powerful pulsed Nd : YAG lasers are available (up to 220 MW/cm^2), as well as N_2 lasers (up to 2.5 MW/cm^2) and an Excimer laser. Some are equipped with harmonic generators and dye units for tuning the output wavelength. The available spectroscopic systems include classical and imaging spectrometers equipped with photomultipliers, Charged Coupled Devices (CCD's), Intensified Photodiode Array (IPDA) and Intensified CCD (ICCD) as detectors. These systems are utilized for molecular systems in solution and in supersonic molecular beams. Also available is a modulated beam mass spectrometer for studies of decay processes in isolated gas-phase molecules. A magneto-optical spectrometer which combines cryogenic systems, CW lasers, optical accessories and magnetic fields is utilized for studies of solid-state materials and devices.

An industrial Near Infra Red (NIR) spectroscopic system is also available as well as a portable (PC plugged in) UV-Visible fiber optical spectrometer for environmental applications.

Synthetic Organic, Inorganic and Organometallic Chemistry

Several advanced research laboratories focus on synthetic organic, inorganic, organometallic and catalysis chemistry. Gloveboxes, schlenk lines, high vacuum lines, greaseless vacuum lines, cryogenic spectrometry equipment, and cryoscopic molecular weight equipment for air-sensitive materials comprise some of the available equipment in these laboratories.

Biology-Related Laboratories

Several research laboratories in the Department focus on biology-oriented projects. Specialized equipment present in these laboratories includes fully computerized FPLC and HPLC protein purification systems, lyophilizer, spectrophotometers, electrophoresis apparatus, X-ray diffractometer for large molecules, triple-resonance solid-state NMR spectrometer, incubators, shakers, cold rooms and other equipment required for cell growth, purification and analysis of proteins.

Located in close proximity, the Departments of Biology and of Food Engineering & Biotechnology offer supplementary facilities to the Chemistry Department for work on biology-oriented projects. This equipment includes: a variety of low—and high-speed centrifuges, high-speed ultracentrifuges, fully computerized FPLC and HPLC protein purification systems, fully computerized fermenters (2, 10 and 50 liters), incubators, ultra low temperature freezers and more.

The cutting-edge National Center for Protein and Peptides Microanalysis situated at the Department of Biology, offers the following services: sequencing, internal and external radioactive sequencing, mass spectrometry of proteins and peptides, synthetic peptides, libraries of proteins and peptides, 2D-gel electrophoresis, database searches and analysis. Several faculty members of the Chemistry Department are associated with this center.

Computer Facilities

The Department of Chemistry is one of the most active in the Technion in scientific computing and as such is a major consumer of the services of the Technion's central computer facilities, which include parallel computers (*Silicon Graphics Power Challenge* and a 8-processor *Dec Alpha server*) as well as several other powerful Unix stations. Large-scale computation may be performed in the recently established National Supercomputer Center, which offers access to two different supercomputers (Cray J932 and IBM SP_2).

The Departmental Computer Center is equipped with 4-processor DEC *Alpha Server 4100 5/466*, two *Dec Alpha 600 5/266* workstations, and three *DEC Ultrix* workstations. Nine Silicon Graphics work-stations and dozens of X-terminals and personal computers are also available. Every office and laboratory in the department is connected via Ethernet to the Technion Local Area Network (LAN) and to the national and international networks and databases.

The *Minerva Center for Computational Quantum Chemistry*, recently established in the Department of Chemistry, includes advanced computers, state-of-the-art computational software packages and other resources to support research in this fast-developing field of chemistry.

Technical Facilities

The technical services are organized into well-equipped mechanical, electronic and glass-blowing units, as well as a support group for computer-related activities. These units are staffed by highly skilled engineers and technicians, who are responsible for all of the technical work carried out within the department. This includes the design and construction of new instrumentation, as well as maintenance of all scientific equipment.

Library

The Chemistry and Biology Library is located in the Chemistry Department Building. The library owns the complete collection of Chemical Abstracts, Beilstein and Gmelin as well as numerous encyclopedias, collective series and books. The library subscribes to more than 220 periodicals and owns a large collection of books dealing with a broad spectrum of topics in chemistry, biology, environmental science and related fields. A variety of on-line search services are available.

2.1 Analytical Chemistry

Analytical chemistry is the study of the chemical composition of natural and artificial materials. Unlike other major sub disciplines of chemistry such as inorganic chemistry and organic chemistry, analytical chemistry is not restricted to any particular type of chemical compound or reaction. Properties studied in analytical chemistry include geometric features such as molecular morphologies and distributions of species, as well as features such as composition and species identity. The contributions made by analytical chemists have played critical roles in the sciences ranging from the development of concepts and theories (pure science) to a variety of practical applications, such as biomedical applications, environmental monitoring, quality control of industrial manufacturing and forensic science (applied science).

Overview

Analytical chemistry is a sub discipline of chemistry that has the broad mission of understanding the chemical composition of all matter and developing the tools to elucidate such compositions. This differs from other sub disciplines of chemistry in that it is not intended to understand the physical basis for the observed chemistry as with physical chemistry and it is not intended to control or direct chemistry as is often the case in organic chemistry and it is not necessarily intended to provide engineering tactics as are often used in materials science. Analytical chemistry generally does not attempt to use chemistry or understand its basis; however, these are common outgrowths of analytical chemistry research. Analytical chemistry has significant overlap with other branches of chemistry, especially those that are focused on a certain broad class of chemicals, such as organic chemistry, inorganic chemistry or biochemistry, as opposed to a particular way of understanding chemistry, such as theoretical chemistry. For example the field of bioanalytical chemistry is a growing area of analytical chemistry that addresses all analytical questions in biochemistry, (the chemistry of life). Analytical chemistry and experimental physical chemistry, however, have a unique relationship in that they are very unrelated in their mission but often share the most in common in the tools used in experiments.

Analytical chemistry is particularly concerned with the questions of "what chemicals are present, what are their characteristics and in what quantities are they present?" These questions are often involved in questions that are more dynamic such as what chemical reaction an enzyme catalyzes or how fast it does it, or even more dynamic such as what is the transition state of the reaction. Although analytical chemistry addresses these types of questions it stops after they are answered. The next logical steps of understanding what it means, how it fits into a larger system, how can this result be generalized into theory or how it can be used are not analytical chemistry.

Since analytical chemistry is based on firm experimental evidence and limits itself to some fairly simple questions to the general public it is most closely associated with hard numbers such as how much lead is in drinking water.

Modern Analytical Chemistry

Modern analytical chemistry is dominated by instrumental analysis. There are so many different types of instruments today that it can seem like a confusing array of acronyms rather than a unified field of study. Many analytical chemists focus on a single type of instrument. Academics tend to either focus on new applications and discoveries or on new methods of analysis. The discovery of a chemical present in blood that increases the risk of cancer would be a discovery that an analytical chemist might be involved in. An effort to develop a new method might involve the use of a tunable laser to increase the specificity and sensitivity of a spectrometric method. Many methods, once developed, are kept purposely static so that data can be compared over long periods of time. This is particularly true in industrial quality assurance (QA), forensic and environmental applications. Analytical chemistry plays an increasingly important role in the pharmaceutical industry where, aside from QA, it is used in discovery of new drug candidates and in clinical applications where understanding the interactions between the drug and the patient are critical.

History

Much of early chemistry (1661~1900AD) was analytical chemistry since the questions of what elements and chemicals were present in the world around us and what are their fundamental natures is very much in the realm of analytical chemistry. There was also significant early progress in synthesis and theory which of course are not analytical chemistry. During this period significant analytical contributions to chemistry include the development of systematic elemental analysis by Justus von Liebig and systematized organic analysis based on the specific reactions of functional groups. The first instrumental analysis was flame emissive spectrometry developed by Robert Bunsen and Gustav Kirchhoff who discovered rubidium (Rb) and caesium (Cs) in 1860.

Most of the major developments in analytical chemistry take place after 1900. During this period instrumental analysis becomes progressively dominant in the field. In particular many of the basic spectroscopic and spectrometric techniques were discovered in the early 20th century and refined in the late 20th century. The separation sciences follow a similar time line of development and also become increasingly transformed into high performance instruments. In the 1970s many of these techniques began to be used together to achieve a complete characterization of samples. Starting in approximately the 1970s into the present day analytical chemistry has progressively become more inclusive of biological questions (bioanalytical chemistry), whereas it had previously been largely focused on inorganic or small organic molecules. The late 20th century also saw an expansion of the application of analytical chemistry from somewhat academic chemical questions to forensic, environmental, industrial and medical questions, such as in histology.

Types

Traditionally, analytical chemistry has been split into two main types, qualitative and quantitative:

Qualitative

- Qualitative inorganic analysis seeks to establish the presence of a given element or inorganic compound in a sample.
- Qualitative organic analysis seeks to establish the presence of a given functional group or organic compound in a sample.
- Quantitative analysis seeks to establish the amount of a given element or compound in a sample.

Approaches

Most modern analytical chemistry is categorized by two different approaches such as analytical targets or analytical methods. Analytical Chemistry (journal) reviews two different approaches alternatively in the issue 12 of each year.

By Analytical Targets

- Bioanalytical chemistry
- Material analysis
- Chemical analysis
- Environmental analysis
- Forensics

By Analytical Methods

- Spectroscopy

- Mass Spectrometry
- Spectrophotometry and Colorimetry
- Chromatography and Electrophoresis
- Crystallography
- Microscopy
- Electrochemistry

Traditional Analytical Techniques

Although modern analytical chemistry is dominated by sophisticated instrumentation, the roots of analytical chemistry and some of the principles used in modern instruments are from traditional techniques many of which are still used today. These techniques also tend to form the backbone of most undergraduate analytical chemistry educational labs. Examples include:

Titration

Titration involves the addition of a reactant to a solution being analyzed until some equivalence point is reached. Often the amount of material in the solution being analyzed may be determined. Most familiar to those who have taken college chemistry is the acid-base titration involving a color changing indicator. There are many other types of titrations, for example potentiometric titrations. These titrations may use different types of indicators to reach some equivalence point.

Gravimetry

Gravimetric analysis involves determining the amount of material present by weighing the sample before and/or after some transformation. A common example used in undergraduate education is the determination of the amount of water in a hydrate by heating the sample to remove the water such that the difference in weight is due to the water lost.

Inorganic Qualitative Analysis

Inorganic qualitative analysis generally refers to a systematic scheme to confirm the presence of certain, usually aqueous, ions or elements by performing a series of reactions that eliminate ranges of possibilities and then confirms suspected ions with a confirming test. Sometimes small carbon containing ions are included in such schemes. With modern instrumentation these tests are rarely used but can be useful for educational purposes and in field work or other situations where access to state-of-the-art instruments are not available or expedient.

Instrumental Analysis

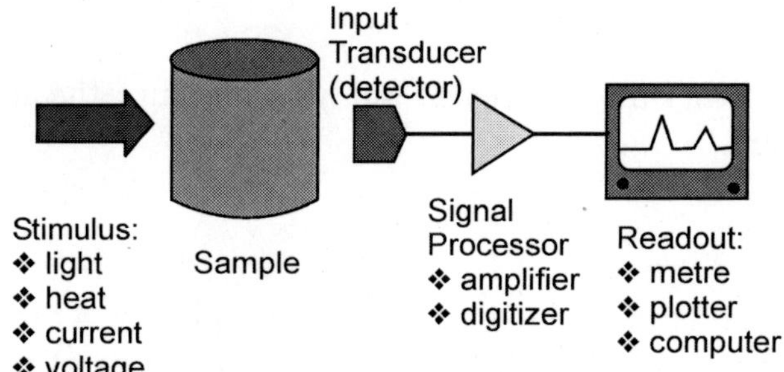

Fig. 2.1 Block diagram of an analytical instrument showing the stimulus and measurement of response.

Spectroscopy

Spectroscopy measures the interaction of the molecules with electromagnetic radiation. Spectroscopy consists of many different applications such as atomic absorption spectroscopy, atomic emission spectroscopy, ultraviolet-visible spectroscopy, x-ray fluorescence spectroscopy, infrared spectroscopy, Raman spectroscopy, nuclear magnetic resonance spectroscopy, photoemission spectroscopy, Mössbauer spectroscopy and so on.

Mass Spectrometry

Mass spectrometry measures mass-to-charge ratio of molecules using electric and magnetic fields. There are several ionization methods: electron impact, chemical ionization, electrospray, fast atom bombardment, matrix assisted laser desorption ionization, and others. Also, mass spectrometry is categorized by approaches of mass analyzers: magnetic-sector, quadrupole mass analyzer, quadrupole ion trap, Time-of-flight, Fourier transform ion cyclotron resonance, and so on.

Crystallography

Crystallography is a technique that characterizes the chemical structure of materials at the atomic level by analyzing the diffraction patterns of usually x-rays that have been deflected by atoms in the material. From the raw data the relative placement of atoms in space may be determined.

Electrochemical Analysis

Electrochemistry measures the interaction of the material with an electric field.

Thermal Analysis

Calorimetry and thermogravimetric analysis measure the interaction of a material and heat.

Separation

Separation processes are used to decrease the complexity of material mixtures. Chromatography and electrophoresis are representative of this field.

Hybrid Techniques

Combinations of the above techniques produce "hybrid" or "hyphenated" techniques. Several examples are in popular use today and new hybrid techniques are under development. For example, Gas chromatography-mass spectrometry, LC-MS, GC-IR, LC-NMR, CE-MS, and so on.

Hyphenated separation techniques refers to a combination of two (or more) techniques to detect and separate chemicals from solutions. Most often the other technique is some form of chromatography. Hyphenated techniques are widely used in chemistry and biochemistry. A slash is sometimes used instead of hyphen, especially if the name of one of the methods contains a hyphen itself.

Examples of hyphenated techniques:

1. LC-MS (or HPLC-MS)
2. HPLC/ESI-MS
3. LC-DAD
4. CE-MS
5. CE-UV
6. GC-MS

Microscopy

The visualization of single molecules, single cells, biological tissues and nano—micro materials is very important and attractive approach in analytical science. Also, hybridization with other traditional analytical tools is revolutionizing analytical science. Microscopy can be categorized into three different fields: optical microscopy, electron microscopy, and scanning probe microscopy. Recently, this field is rapidly progressing because of the rapid development of computer and camera industries.

Lab-on-a-Chip

Devices that integrate (multiple) laboratory functions on a single chip of only millimeters to a few square centimeters in size and that are capable of handling extremely small fluid volumes down to less than pico liters.

Methods and Data Analysis

Standard Curve

A standard method for analysis of concentration involves the creation of a calibration curve. This allows for determination of the amount of a chemical in a material by comparing the results of unknown sample to those of a series known standards. If the concentration of element or compound in a sample is too high for the detection range of the technique, it can simply be diluted in a pure solvent. If the amount in the sample is below an instrument's range of measurement, the method of addition can be used. In this method a known quantity of the element or compound under study is added, and the difference between the concentration added, and the concentration observed is the amount actually in the sample.

Internal Standards

Sometimes an internal standard is added at a known concentration directly to an analytical sample to aid in quantitation. The amount of analyte present is then determined relative to the internal standard as a calibrant.

Trends

Analytical chemistry research is largely driven by performance (sensitivity, selectivity, robustness, linear range, accuracy, precision, and speed), and cost (purchase, operation, training, time, and space). Among the main branches of contemporary analytical atomic spectrometry, the most widespread and universal are optical and mass spectrometry (*see* Prospects in Analytical Atomic Spectrometry). In the direct elemental analysis of solid samples, the new leaders are laser-induced breakdown and laser ablation mass spectrometry, and the related techniques with transfer of the laser ablation products into inductively coupled plasma. Advances in design of diode lasers and optical parametric oscillators promote developments in fluorescence and ionization spectrometry and also in absorption techniques where uses of optical cavities for increased effective absorption pathlength are expected to expand. Steady progress and growth in applications of plasma—and laser-based methods are noticeable. An interest towards the absolute (standardless) analysis has revived, particularly in the emission spectrometry.

A lot of effort is put in shrinking the analysis techniques to chip size. Although there are few examples of such systems competitive with traditional analysis techniques, potential advantages include size/portability, speed, and cost. (micro Total Analysis System (μ TAS) or Lab-on-a-chip). Microscale chemistry reduces the amounts of chemicals used.

Much effort is also put into analyzing biological systems. Examples of rapidly expanding fields in this area are:

- *Genomics*—DNA sequencing and its related research. Genetic fingerprinting and DNA microarray are very popular tools and research fields.
- *Proteomics*—the analysis of protein concentrations and modifications, especially in response to various stressors, at various developmental stages, or in various parts of the body.
- *Metabolomics*—similar to proteomics, but dealing with metabolites.
- *Transcriptomics*—mRNA and its associated field.
- *Lipidomics*—lipids and its associated field.
- *Peptidomics*—peptides and its associated field.
- *Metalomics*—similar to proteomics and metabolomics, but dealing with metal concentrations and especially with their binding to proteins and other molecules.

Analytical chemistry has played critical roles in the understanding of basic science to a variety of practical applications, such as biomedical applications, environmental monitoring, quality control of industrial manufacturing, forensic science and so on.

The recent developments of computer automation and information technologies have innervated analytical chemistry to initiate a number of new biological fields. For example, automated DNA sequencing machines were the basis to complete human genome projects leading to the birth of genomics. Protein identification and peptide sequencing by mass spectrometry opened a new field of proteomics. Furthermore, a number of ~omics based on analytical chemistry have become important areas in modern biology.

Also, analytical chemistry has been an indispensable area in the development of nanotechnology. Surface characterization instruments, electron microscopes and scanning probe microscopes enables scientists to visualize atomic structures with chemical characterizations.

Analytical chemistry is pursuing the development of practical applications and commercial instruments rather than elucidating scientific fundamentals. This may be an arguable difference from overlapping science areas such as physical chemistry and biophysics, although there isn't any distinct boundaries among disciplines in contemporary science and technology. However, this aspect may attract many engineers' interest; thus, it is not difficult to see papers from engineering departments in analytical chemistry journals.

Among active contemporary analytical chemistry research fields, micro total analysis system is considered as a great promise of revolutionary technology. In this approach, integrated and miniaturized analytical systems are being developed to control and analyze single cells and single molecules. This cutting-edge technology has a promising potential of leading a new revolution in science as integrated circuits did in computer developments.

2.2 Electroanalytical Method

Electroanalytical methods use both active and passive to study an analyte through the observation of potential (Volts) and/or current (Amps) of an electrochemical cell.

Variations

Potentiometry

Potentiometry measures the potential of a solution between two electrodes passively, affecting the solution very little in the process. The potential is then related to the concentration of one or more analytes. The cell structure used is often referred to as an electrode even though it contains two electrodes an *indicator electrode* and *reference electrode* (distinct form the reference electrode used in the three electrode system). Potentiometry is usually conducted in an ion selective way with a different electrode for each ion. The most common potentiometric electrode is the glass pH meter.

Coulometry

Coulometry uses applied current or potential to completely convert an analyte from one oxidation state to another. The in these experiments the total current passed is measured directly or indirectly to determine the number of electrons passed. Knowing the number electrons passed can indicate the concentration of the analyte or, when the concentration is known, the number of electrons involved with a redox couple. The bulk electrolysis also known as controlled potential coulometry or some hybrid of the two names is perhaps the most common form of coulometry.

Voltammetry

Voltammetry applies a constant and/or varying potential at an electrodes surface and measures the resulting current with a three electrode system. This method can reveal the reduction potential of an analyte and electrochemical reactivity among other things. This method in practical terms is nondestructive since only a very very small amount of the analyte is consumed at the two-dimensional surface of the working and auxiliary electrode. In practice the analyte solutions is usually disposed of since it is difficult to separate the anaylte from the bulk electrolyte and the experiment requires a small amount of analyte. A normal experiment involving 1-10 ml solution with an analyte concentration between 1-10 mM.

Polarometry

Polarometry is a subclass of voltammetry that employees a dropping mercury electrode as the working electrode and often uses the resulting mercury pool as the auxiliary electrode. Concern over the toxicity of mercury combined with the development of affordable, inert, easily cleaned, high quality electrodes made of materials such as noble metals and glass carbon has caused a great reduction of mercury electrodes.

Amperometry

Amperometry a subclass of voltammetry in which the electrode is held at constant potentials for various lengths of time. Mostly a historic distinction that still results in some confusion for example differential pulse voltammetry is also referred to as and *differential pulse amperometry*. This experiment can be seen as the combination of linear sweep voltammetry and chronoamperometry thus the confusion in what it should be called. One thing that distinguishes amperometry form other forms of voltammetry is that its common to sum the currents over a given time period rather than considering them at individual potentials. This summing can result in larger data sets and reduced error. Amperometric titration is a technique that would be considered amperometry since it measures the current, but would not be considered voltammetry since the entire solution is transformed during the experiment.

Constraints

Time Scale

The various techniques are each suited to investigating chemical phenomenon over different time scale window.

Concentration

The various elctrochemical methods can operate under different anaylte concentrations. Notably potentiometry may measure the widest range of analyte concentrations of any analytic method with fluorometry one of the next major competitor

2.3 Electrophoresis

Electrophoresis is the most known electrokinetic phenomenon. It was discovered by Reuss in 1809. He observed that clay particles dispersed in water migrate under influence of an applied electric field. There are detailed descriptions of Electrophoresis in many books on Colloid and Interface Science. There is an IUPAC Technical Report prepared by a group of most known world experts on the electrokinetic phenomena. Generally, electrophoresis is the motion of dispersed particles relative to a fluid under the influence of an electric field that is space uniform. Alternatively, similar motion in a space non-uniform electric field is called dielectrophoresis.

Electrophoresis occurs because particles dispersed in a fluid almost always carry an electric surface charge. An electric field exerts electrostatic Coulomb force on the particles through these charges. Recent molecular dynamics simulations, though, suggest that surface charge is not always necessary for electrophoresis and that even neutral particles can show electrophoresis due to the specific molecular structure of water at the interface.

The electrostatic Coulomb force exerted on a surface charge is reduced by an opposing force which is electrostatic as well. According to double layer theory, all surface charges in fluids are screened by a diffuse layer. This diffuse layer has the same absolute charge value, but with opposite sign from the surface charge. The electric field induces force on the diffuse layer, as well as on the surface charge. The total value of this force equals to the first mentioned force, but it is oppositely directed. However, only part of this force is applied to the particle. It is actually applied to the ions in the diffuse layer. These ions are at some distance from the particle surface. They transfer part of this electrostatic force to the particle surface through viscous stress. This part of the force that is applied to the particle body is called *electrophoretic retardation force.*

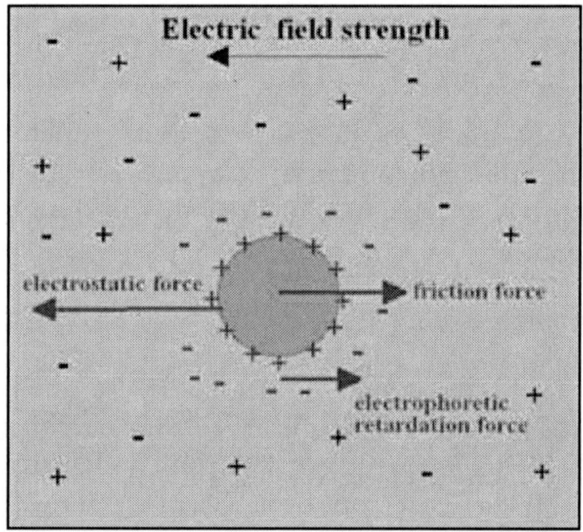

There is one more electric force, which is associated with deviation of the double layer from spherical symmetry and surface conductivity due to the excees ions in the diffuse layer. This force is called the *electrophoretic relaxation force.*

All these forces are balanced with hydrodynamic friction, which affects all bodies moving in viscous fluids with low Reynolds number. The speed of this motion v is proportional to the electric field strength E if the field is not too strong. Using this assumption makes possible the introduction of electrophoretic mobility μ_e as coefficient of proportionality between particle speed and electric field strength:

$$\mu e = \frac{v}{E}.$$

Multiple theories were developed during 20th century for calculating this parameter. Ref. 2 provides an overview.

Theory

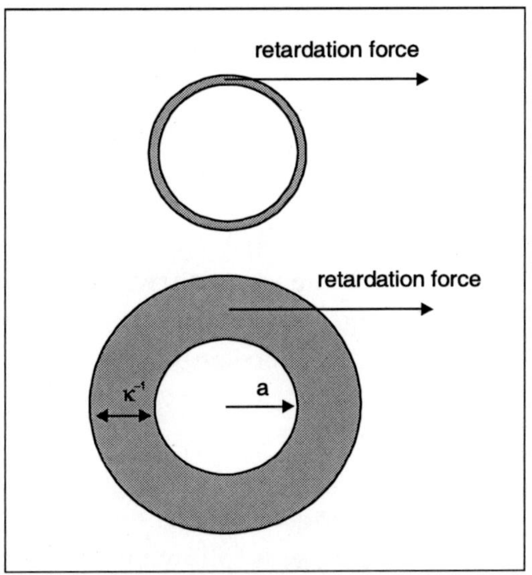

The most known and widely used theory of electrophoresis was developed by Smoluchowski in 1903.

$$\mu e = \frac{\varepsilon \varepsilon_0 \xi}{\eta}$$

where ε is the dielectric constant of the dispersion medium, ε_0 is the permittivity of free space (C^2 N m^{-2}), ξ is dynamic viscosity of the dispersion medium (Pa s), and æ is zeta potential (*i.e.*, the electrokinetic potential of the slipping plane in the double layer).

Smoluchowski theory is very powerful because it works for dispersed particles of any shape and any concentration, when it is valid. Unfortunately it has limitations of its validity. It follows, for instance, from the fact that it does not include Debye length κ^1. However, Debye length must be important for electrophoresis, as follows immediately from the Figure on the right. Increasing thickness of the DL leads to removing point of retardation force further from the particle surface. The thicker DL, the smaller retardation force must be.

Detail theoretical analysis proved that Smoluchowski theory is valid only for sufficiently thin DL, when Debye length is much smaller than particle radius *a*:

$$\kappa a >> 1$$

This model of "thin Double Layer" offers tremendous simplifications not only for electrophoresis theory but for many other electrokinetic theories. This model is valid for most aqueous systems because the Debye length is only a few nanometers there. It breaks only for nano-colloids in solution with ionic strength close to water

Smoluchowski theory also neglects contribution of surface conductivity. This is expressed in modern theory as condition of small Dukhin number

$$Du << 1$$

Creation of electrophoretic theory with wider range of validity was a purpose of many studies during 20th century.

One of the most known considers an opposite asymptotic case when Debye length is larger than particle radius:

$$\kappa a < 1$$

It is "thick Double Layer" model. Corresponding electrophoretic theory was created by Huckel in 1924. It yields following equation for electrophoretic mobility:

$$\mu_e = \frac{2\varepsilon\varepsilon_0\varsigma}{3\eta},$$

This model can be useful for some nano-colloids and non-polar fluids, where Debye length is much larger.

There are several analytical theories that incorporate surface conductivity and eliminate restriction of the small Dukhin number. Early pioneering work in that direction dates back to Overbeek and Booth

Modern, rigorous theories that are valid for any Zeta potential and often any κa, stem mostly from the Ukrainian (Dukhin, Shilov and others) and Australian (O'Brien, White, Hunter and others) Schools.

Historically the first one was Dukhin-Semenikhin theory. Similar theory was created 10 years later by O'Brien and Hunter. Assuming *thin Double Layer*, these theories would yield results that are very close to the numerical solution provided by O'Brien and White

2.4 Gel Electrophoresis

Gel electrophoresis apparatus—An agarose gel is placed in this buffer-filled box and electrical current is applied via the power supply to the rear. The negative terminal is at the far end (black wire), so DNA migrates toward the camera.

Gel electrophoresis is a technique used for the separation of deoxyribonucleic acid, ribonucleic acid, or protein molecules using an electric current applied to a gel matrix. It is usually performed for analytical purposes, but may be used as a preparative technique prior to use of other methods such as mass spectrometry, RFLP, PCR, cloning, DNA sequencing, or Southern blotting for further characterization.

Separation

The term "gel" in this instance refers to the matrix used to contain, then separate the target molecules. In most cases the gel is a crosslinked polymer whose composition and porosity is chosen based on the specific weight and composition of the target to be analyzed. When separating proteins or small nucleic acids (DNA, RNA, or oligonucleotides) the gel is usually composed of different concentrations of acrylamide and a cross-linker, producing different sized mesh networks of polyacrylamide. When separating larger nucleic acids (greater than a few hundred bases), the preferred matrix is purified agarose. In both cases, the gel forms a solid, yet porous matrix. Acrylamide, in contrast to polyacrylamide, is a neurotoxin and must be handled using Good Laboratory Practices to avoid poisoning.

"Electrophoresis" refers to the electromotive force (EMF) that is used to move the molecules through the gel matrix. By placing the molecules in wells in the gel and applying an electric current, the molecules will move through the matrix at different rates, usually determined by mass, toward the positive anode if negatively charged or toward the negative cathode if positively charged

Visualization

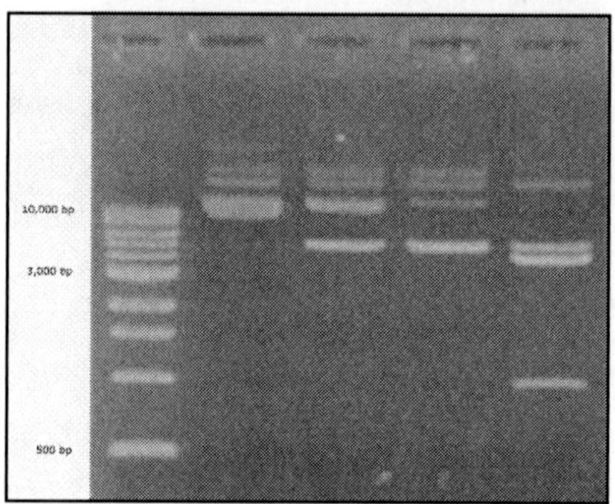

I *Agarose gel prepared for DNA analysis*—The first lane contains a *DNA ladder* for sizing, and the other four lanes show variously-sized DNA fragments that are present in some but not all of the samples.

After the electrophoresis is complete, the molecules in the gel can be stained to make them visible. Ethidium bromide, silver, or coomassie blue dye may be used for this process. Other methods may also be used to visualize the separation of the mixture's components on the gel. If the analyte molecules fluoresce under ultraviolet light, a photograph can be taken of the gel under ultraviolet lighting conditions. If the molecules to be separated contain radioactivity added for visibility, an autoradiogram can be recorded of the gel.

If several mixtures have initially been injected next to each other, they will run parallel in individual lanes. Depending on the number of different molecules, each lane shows separation of the components from the original mixture as one or more distinct bands, one band per component. Incomplete separation of the components can lead to overlapping bands, or to indistinguishable smears representing multiple unresolved components.

Bands in different lanes that end up at the same distance from the top contain molecules that passed through the gel with the same speed, which usually means they are approximately the same size. There are molecular weight size markers available that contain a mixture of molecules of known sizes. If such a marker was run on one lane in the gel parallel to the unknown samples, the bands observed can be compared to those of the unknown in order to determine their size. The distance a band travels is approximately inversely proportional to the logarithm of the size of the molecule.

Applications

Gel electrophoresis is used in forensics, molecular biology, genetics, microbiology and biochemistry. The results can be analyzed quantitatively by visualizing the gel with UV light and a gel imaging device. The image is recorded with a computer operated camera, and the intensity of the band or spot of interest is measured and compared against standard or markers loaded on the same gel. The measurement and analysis are mostly done with specialized software.

Depending on the type of analysis being performed, other techniques are often implemented in conjunction with the results of gel electrophoresis, providing a wide range of field-specific applications.

Nucleic Acids

In the case of nucleic acids, the direction of migration, from negative to positive electrodes, is due to the naturally-occurring negative charge carried by their sugar-phosphate backbone.

Double-stranded DNA fragments naturally behave as long rods, so their migration through the gel is relative to their radius of gyration, or, for non-cyclic fragments, size. Single-stranded DNA or RNA tend to fold up into molecules with complex shapes and migrate through the gel in a complicated manner based on their tertiary structure. Therefore, agents that disrupt the hydrogen bonds, such as sodium hydroxide or formamide, are used to denature the nucleic acids and cause them to behave as long rods again.

Gel electrophoresis of large DNA or RNA is usually done by agarose gel electrophoresis. See the "Chain termination method" page for an example of a polyacrylamide DNA sequencing gel.

Proteins

Proteins, unlike nucleic acids, can have varying charges and complex shapes, therefore they may not migrate into the gel at similar rates, or at all, when placing a negative to positive EMF on the sample. Proteins therefore, are usually denatured in the presence of a detergent such as sodium dodecyl sulfate/sodium dodecyl phosphate (SDS/SDP) that coats the proteins with a negative charge. Generally, the amount of SDS bound is relative to the size of the protein (usually 1.4g SDS per gram of protein), so that the resulting denatured proteins have an overall negative charge, and all the proteins have a similar charge to mass ratio. Since denatured proteins act like long rods instead of having a complex tertiary shape, the rate at which the resulting SDS coated proteins migrate in the gel is relative only to its size and not its charge or shape.

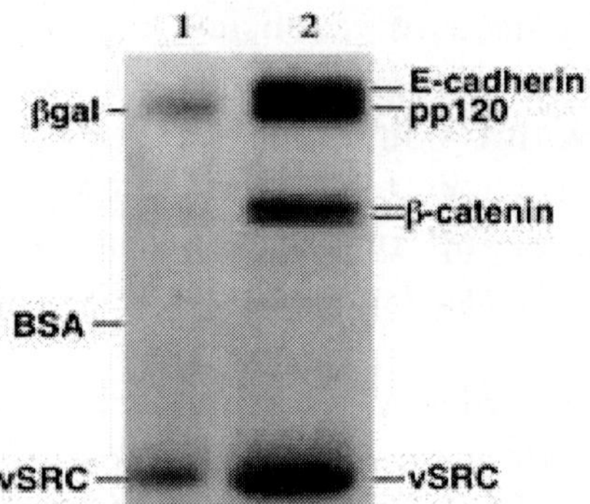

Fig. 2.2: *SDS-PAGE autoradiography*—The indicated proteins are present in different concentrations in the two samples.

Proteins are usually analyzed by sodium dodecyl sulfate polyacrylamide gel electrophoresis (SDS-PAGE), by native gel electrophoresis, by quantitative preparative native continuous polyacrylamide gel electrophoresis (QPNC-PAGE), or by 2-D electrophoresis.

History

- 1930s—first reports of the use of sucrose for gel electrophoresis
- 1955—introduction of starch gels, mediocre separation
- 1959—introduction of acrylamide gels (Raymond and Weintraub); accurate control of parameters such as pore size and stability
- 1964—disc gel electrophoresis (Ornstein and Davis)
- 1969—introduction of denaturing agents especially SDS separation of protein subunit (Beber and Osborn)
- 1970—Laemmli separated 28 components of T4 phage using a stacking gel and SDS
- 1975—2-dimensional gels (O'Farrell); isoelectric focusing then SDS gel electrophoresis
- 1977—sequencing gels
- late 1970s—agarose gels
- 1983—pulsed field gel electrophoresis enables separation of large DNA moleucles
- 1983—introduction of capillary electrophoresis

A 1959 book on electrophoresis by Milan Bier cites references from the 1800s. However, Oliver Smithies made significant contributions. Bier states: "*The method of Smithies... is finding wide application because of its unique separatory power.*" Taken in context, Bier clearly implies that Smithies' method is an improvement.

2.5 Protein Electrophoresis

In chemistry and medicine, *protein electrophoresis* (a.k.a. Immunoelectrophoresis) is a method of analysing a mixture of proteins by means of gel electrophoresis, mainly in blood serum (blood plasma is not suitable). Before the widespread use of electrophoresis gels, protein electrophoresis was performed on paper (sometimes in combination with paper chromotography).

Interpretation

There are two classes of blood proteins: serum albumin and globulin. They are generally equal in proportion, but albumin is much smaller and lightly negatively charged, leading to an accumulation of albumin on the electrophoretic gel. A small band before albumin represents transthyretin (also named prealbumin). Some forms of medication or body chemicals can cause their own band, usually small. Abnormal bands (spikes) are seen in monoclonal gammopathy of undetermined significance and multiple myeloma, and are useful in the diagnosis of these conditions.

The globulins are classified by their banding pattern (with their main representatives):

- The *alpha* (á) band consists of two parts, 1 and 2:
  - α_1—α_1-antitrypsin, α_1-acid glycoprotein.
  - α_2—haptoglobin, α_2-macroglobulin, α_2-antiplasmin, ceruloplasmin.
- The *beta* (β) band—transferrin, LDL, complement
- The *gamma* (γ) band—immunoglobulin (IgA, IgD, IgE, IgG and IgM). Paraproteins (in multiple myeloma) appear in this band.

2.6 Serum Protein Electrophoresis

Serum protein electrophoresis is a laboratory test that examines specific proteins in the blood called globulins. Blood must first be collected, usually into an airtight vial or syringe. Electrophoresis is a laboratory technique where the blood serum (the fluid portion of the blood after the blood has clot) is placed on special paper treated with agarose gel and exposed to an electric current to separate the serum protein components into five classifications by size and electrical charge, those being serum albumin, alpha-1-globulins, alpha-2-globulins, beta globulins, and gamma globulins.

Albumin

A fall of 30 per cent is necessary before the decrease shows on electrophoresis. Usually a single band is seen. Heterozygous individuals may produce bisalbuminaemia—two equally staining bands, the product of two genes. Some variants give rise to a wide band or two bands of unequal intensity but none of these variants is associated with disease. Increased mobility results from the binding of bilirubin, nonesterified fatty acids, penicillin and acetylsalicylic acid, and occasionally from tryptic digestion in acute pancreatitis.

Absence of albumin is rare. This condition is called analbuminaemia. A decreased level of albumin, however, is common in many diseases and is especially important in liver disease.

Albumin-Alpha-1 Interzone

Even staining in this zone is due to alpha-1 lipoprotein (High density lipoprotein—HDL). Decrease occurs in severe inflammation, acute hepatitis, and cirrhosis. Also, nephrotic syndrome can lead to decrease in albumin level; due to its loss in the urine through a damaged leaky glomerulus. An increase appears in severe alcoholics and in women during pregnancy and in puberty.

The high levels of AFP that may occur in hepatomas may result in a sharp band between the albumin and the alpha-1 zone.

Alpha-1 Zone

Orosmucoid and antitrypsin migrate together but orosmucoid stains poorly so alpha-1 antitrypsin constitutes most of the alpha-1 band. Alpha-1 antitrypsin has an SG group and thiol compounds may be bound to the protein altering their mobility. A decreased band is seen in the deficiency state. It is decreased in the nephrotic syndrome and absence could indicate possible Alpha 1-antitrypsin deficiency. This eventually leads to emphysema from unregulated lung elastase breakdown by neutrophils in the lung tissue.

Bence Jones protein may bind to and retard the alpha-1 band.

Alpha-1—Alpha-2 Interzone

Two faint bands may be seen representing alpha-1 antichymotrypsin and vitamin D binding protein. These bands fuse and intensify in early inflammation due to an increase in alpha-1 antichymotrypsin, an acute phase protein.

Alpha-2 Zone

This zone consists principally of alpha-2 macroglobulin and haptoglobin. There are typically low levels in haemolytic anaemia (haptoglobin is a suicide molecule which binds with free haemoglobin released from red blood cells and these complexes are rapidly removed by phagocytes). Haptoglobin is raised as part of the acute phase response and therefore there is a typical increase in alpha-2 inflammation. A normal alpha-2 and a raised alpha-1 is typical in hepatic metastases and cirrhosis.

Haptoglobin/haemaglobin complexes migrate more cathodally than haptoglobin as seen in the alpha-2—beta interzone. This is typically seen as a broadening of the alpha-2 zone.

Alpha-2 macroglobulin is raised in children and the elderly. This is seen as a sharp front to the alpha-2 band. It is of little diagnostic significance but is markedly raised in association with glomerular protein loss—the characteristic increased alpha-2 of the nephrotic syndrome, due to its large size, cannot pass through glomeruli.

Alpha-2 globulins can be raised in cirrhosis, diabetes, and malignancy.

Alpha-2—Beta Interzone

Cold insoluble globulin forms a band here which is not seen in plasma because it is precipitated by heparin. There are low levels in inflammation and high levels in pregnancy.

B lipoprotein forms an irregular crenated band in this zone. High levels are seen in type II hypercholesterolaemia and in the nephrotic syndrome.

Beta Zone

Transferrin comprises the beta-1 band. Increased beta-1 protein (due to the increased level of free transferrin) is typical of Iron deficiency anemia, pregnancy, and oestrogen therapy.

Beta-2 comprises C3 (Complement protein 3). It is raised in the acute phase response.

Gamma Zone

The immunoglobulins are the only proteins present in the normal gamma region. IgA has the most anodal mobility and migrates in the beta-gamma region. High levels give rise to beta/gamma fusion as in cirrhosis, respiratory infection, skin disease, or rheumatiod arthritis. IgA deficiency occurs in 1:500 of the population, as is suggested by a pallor in the gamma zone.

Commonest causes of hypergammaglobulinaemia detected by electrophoresis are severe infection, chronic liver disease, systemic lupus erythematosus, and rheumatoid arthritis.

Hypogammaglobulinaemia is easily identifiable. It is normal in infants.

Zones or faint bands in the gamma region are often seen as a result of clones of immunoglobulins responding to antigenic challenge, *e.g.* in chronic hepatitis and chronic viral infections.

Creactive protein can be seen as a faint band when the level is grossly abnormal, *e.g.* in tissue damage.

Dense narrow bands composing of immunoglobulins of a single class and containing only one type of light chain are known as paraproteins or M (monoclonal) proteins. IgA paraproteins migrate most anodally and may be confused with beta globulins. Paraproteins can arise from benign or malignant clones of B cells and their recognition and investigation is very important.

Lysozyme may be seen as a band cathodal to the slowest gamma in myelomonocytic leukaemia in which it is released from tumour cells.

Myeloma is the commonest cause of IgA and IgG paraproteinaemias but chronic lymphatic leukaemia and lymphosarcoma are not uncommon and usually give rise to IgM paraproteins. Waldenstrom's macroglobulinaemia, a malignant lymphoma, also gives rise to an IgM paraprotein.

Benign paraproteins are usually faint and do not show immunoparesis. Faint bands seen in the gamma region may be due to light chain disease, particularly if there is immune peresis.

Fibrinogen from plasma samples will be seen in the fast gamma region.

3

Applied Environmental Chemistry

3.1 Chemistry

At one time it was easy to define chemistry. The traditional definition goes something like this: Chemistry is the study of the nature, properties, and composition of matter, and how these undergo changes. That served as a perfectly adequate definition as late as the 1930s, when natural science (the systematic knowledge of nature) seemed quite clearly divisible into the physical and biological sciences, with the former being comprised of physics, chemistry, geology and astronomy and the latter consisting of botany and zoology. This classification is still used, but the emergence of important important fields to study such as oceanography, paleobotany, meteorology, pharmacy and biochemistry, for example, have made it increasingly clear that the dividing lines between the sciences are no longer at all sharp. Chemistry, for instance, now overlaps so much with geology (thus we have geochemistry), astronomy (astrochemistry), and physics (physical and analytical chemistry) that it is probably impossible to devise a really good modern definition of chemistry, except, perhaps, to fall back on the operational definition: Chemistry is what chemists do!

Chemistry plays an important part in all of the other natural sciences, basic and applied. Plant growth and metabolism, the formation of igneous rocks, the role played by ozone in the atmosphere, the degradation of environmental pollutants, the properties of lunar soil, the medical action of drugs, establishment of forensic evidence: none of these can be understood without the knowledge and perspective provided by chemistry. Indeed, many people study chemistry so that they can apply it to their own particular field of interest. Of course, chemistry itself is the field of interest for many people, too. Many study chemistry not to apply it to another field, but simply to learn more about the physical world and the behaviour of matter from a chemical viewpoint. Some simply like "what chemists do" and so decide to "do it" themselves.

Chemistry is a way of studying matter. What is matter? As is true with many of those words which are really basic to science, matter is hard to define. It is often said that matter is anything which has mass and occupies space. But then what are "mass" and "space"? Although we can define these, the process yields very little insight into what matter is. So let us just say that matter is anything which has real physical existence; in a word matter is just stuff. Iron, air, wool, gold, milk, aspirin, monkeys, rubber, and pizza—these are all matter. Some things which are not matter are heat, cold, colours, dreams hopes, ideas, sunlight, beauty, fear, and x-rays. None of these is "stuff"; none is matter.

A sample of matter can be either a pure substance or a mixture. A pure substance has a fixed, characteristic composition an a fixed, definite set of properties. Pure substances are for example copper, salt, diamond, water, table sugar, oxygen, mercury, vitamin C, and ozone. A pure substance may be a single element, such as copper or oxygen, or a compound of two or more elements in a fixed ratio, such as salt (39.34 % sodium and 60.66 % chlorine) or table sugar (42.11 % carbon, 6.48 % hydrogen, and 51.41 % oxygen).

A mixture is a collection of pure substances simply mixed together. Its composition is variable, as are its properties. Examples of mixtures are milk, wood, concrete, saltwater, air, granite, motor oil, chocolate, and elephants.

A pure substance can be a solid, a liquid, or a gas; these are the three states of matter A solid maintains its volume and shape; a liquid, its volume only; and a gas, neither. Solids tend to be hard and unyielding; liquids maintain their volumes and flow to adopt the shapes of their containers. The ability to flow is called fluidity, and so gases and liquids are called fluids.

One of the goals of chemistry is to be able to describe the properties of matter in terms of its internal structure, the arrangement and interrelationship of its parts. This word, structure, sometimes refers to the physical arrangement of particles, such as atoms or molecules in space. At other times it is used to indicate some other arrangement, such as the arrangement of energy levels of an electron in an atom. The structure of matter determines its properties. Properties can be classed as either physical or chemical. A physical property of a substance can be characterized without specific reference to any other substance and usually describes the response of the substance to some external influence, such as heat, light, force, electricity, etc. Physical properties include boiling point, melting point, thermal (heat) conductivity, colour, refractive index, viscosity, reflectivity, hardness, tensile strength, and electrical conductivity.

A chemical property, on the other hand, describes a chemical change: the interaction of one substance with another, or the change of one substance into another. Iron rusts in a moist environment, unrefrigerated milk turns sour, wood burns in air, photographs bleach when exposed to sunlight for a long time, dynamite explodes—each of these is a chemical property because each involves chemical change. During chemical changes, substances are actually changed into other substances. The simultaneous disappearance of some substances (called the reactants) and appearance of others (the products) is characteristic in chemical change (chemical reaction). Chemical changes are generally characterized by pronounced internal structural rearrangements.

Physical changes are not characterized by the transformation of one substance into another, but rather by the change of the form of a given substance. The bending of a piece of copper wire fails to change the property of copper into another substance; crushing a block of ice leaves only crushed ice; melting an iron nail yields a substance still called iron: These are all usually accepted as physical changes.

Properties of matter may also be categorized as either macroscopic or microscopic. A macroscopic property describes characteristics or behaviour of a sample which is large enough to see, handle, manipulate, weigh, etc. A microscopic property describes the behaviour of a much smaller sample of matter, an atom or molecule for instance. Macroscopic and microscopic properties are often different. A banana is yellow, but we do not use colour to describe an atom. Some properties, on the other hand, can be either microscopic or macroscopic; mass is one of these.

Another word that is often used is system. A system is a portion of the universe which we wish to observe or consider. The size of the portion is usually small and a system may be a real one (in a test tube or flask, for example), or an imaginary one which this text is just referring to.

2D CH_3

N

O

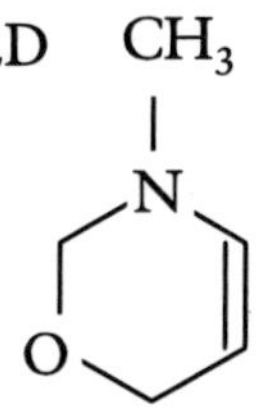

Viewed from an historical point of view, it is clear that scientific knowledge has been obtained and that therefore science has "advanced" in a series of fairly logical steps. On the other hand, counterparts to these steps are difficult to identify in the day-to-day professional activities of a scientist. The way in which science and in particular chemistry advances can be describes in terms of a series of cycles. Observations and data (and laws) lead to the proposal of theories that, in turn, suggest predictions which can be tested by designing new experiments, and the whole process starts all over again.

3.2 Green Chemistry

Green chemistry, also called sustainable chemistry, is a chemical philosophy encouraging the design of products and processes that reduce or eliminate the use and generation of hazardous substances. Whereas environmental chemistry is the chemistry of the natural environment, and of pollutant chemicals in nature, green chemistry seeks to reduce and prevent pollution at its source. In 1990 the Pollution Prevention Act was passed in the United States. This act helped create a *modus operandi* for dealing with pollution in an original and innovative way.

As a chemical philosophy, green chemistry derives from organic chemistry, inorganic chemistry, biochemistry, analytical chemistry, and even physical chemistry. However, the philosophy of green chemistry tends to focus on industrial applications. Contrast this with click chemistry which tends to favor academic applications, although industrial applications are possible. The focus is on minimizing the hazard and maximizing the efficiency of any chemical choice. It is distinct from environmental chemistry which focuses on chemical phenomena in the environment.

In 2005 Ryoji Noyori identified three key developments in green chemistry: use of supercritical carbon dioxide as green solvent, aqueous hydrogen peroxide for clean oxidations and the use of hydrogen in asymmetric synthesis. Examples of applied green chemistry are supercritical water oxidation, on water reactions and dry media reactions.

Bioengineering is also seen as a promising technique for achieving green chemistry goals. A number of important process chemicals can be synthesized in engineered organisms, such as shikimate, a Tamiflu precursor which is fermented by Roche in bacteria.

Principles

Paul Anastas, then of the United States Environmental Protection Agency, and John C. Warner developed 12 principles of green chemistry, which help to explain what the definition means in practice. The principles cover such concepts as:

- The design of processes to maximize the amount of raw material that ends up in the product;

- The use of safe, environment-benign substances, including solvents, whenever possible;
- The design of energy efficient processes; and
- The best form of waste disposal: do not create it in the first place.

The 12 principles are:

1. *Prevent waste:* Design chemical syntheses to prevent waste, leaving no waste to treat or clean up.
2. *Design safer chemicals and products:* Design chemical products to be fully effective, yet have little or no toxicity.
3. *Design less hazardous chemical syntheses:* Design syntheses to use and generate substances with little or no toxicity to humans and the environment.
4. *Use renewable feedstock:* Use raw materials and feedstock that are renewable rather than depleting. Renewable feedstock are often made from agricultural products or are the wastes of other processes; depleting feedstock are made from fossil fuels (petroleum, natural gas, or coal) or are mined.
5. *Use catalysts, not stoichiometric reagents:* Minimize waste by using catalytic reactions. Catalysts are used in small amounts and can carry out a single reaction many times. They are preferable to stoichiometric reagents, which are used in excess and work only once.
6. *Avoid chemical derivatives:* Avoid using blocking or protecting groups or any temporary modifications if possible. Derivatives use additional reagents and generate waste.
7. *Maximize atom economy:* Design syntheses so that the final product contains the maximum proportion of the starting materials. There should be few, if any, wasted atoms.
8. *Use safer solvents and reaction conditions:* Avoid using solvents, separation agents, or other auxiliary chemicals. If these chemicals are necessary, use innocuous chemicals. If a solvent is necessary, water is a good medium as well as certain eco-friendly solvents that do not contribute to smog formation or destroy the ozone.
9. *Increase energy efficiency:* Run chemical reactions at ambient temperature and pressure whenever possible.
10. *Design chemicals and products to degrade after use:* Design chemical products to break down to innocuous substances after use so that they do not accumulate in the environment.

11. *Analyze in real time to prevent pollution:* Include in-process real-time monitoring and control during syntheses to minimize or eliminate the formation of byproducts.
12. *Minimize the potential for accidents:* Design chemicals and their forms (solid, liquid, or gas) to minimize the potential for chemical accidents including explosions, fires, and releases to the environment.

Presidential Green Chemistry Challenge Awards

The Presidential Green Chemistry Challenge Awards began in 1995 as an effort to recognize individuals and businesses for innovations in green chemistry. Typically five awards are given each year, one in each of five categories: Academic, Small Business, Greener Synthetic Pathways, Greener Reaction Conditions, and Designing Greener Chemicals. Nominations are accepted the prior year, and evaluated by an independent panel of chemists convened by the American Chemical Society. Through 2006, a total of 57 technologies have been recognized for the award, and over 1000 nominations have been submitted.

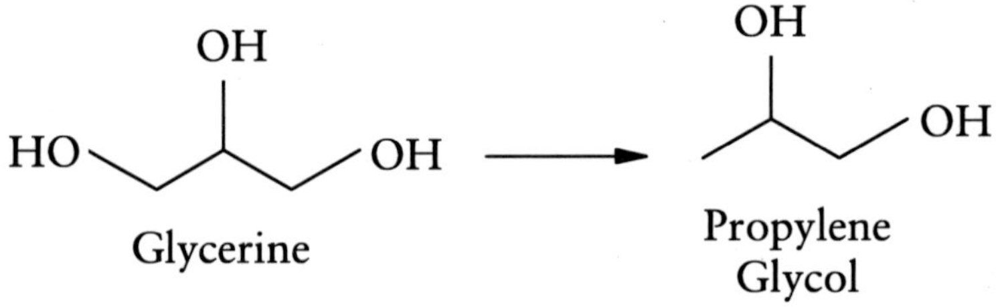

OH
HO OH ⟶ OH OH

Glycerine → Propylene Glycol

Glycerine to Propylene Glycol

In 2006, Professor Galen J. Suppes, from the University of Missouri—Columbia, was awarded the Academic Award for his system of converting waste glycerin from biodiesel production to propylene glycol. Through the use of a copper-chromite catalyst, Professor Suppes was able to lower the required temperature of conversion while raising the efficiency of the distillation reaction. Propylene glycol produced in this way will be cheap enough to replace the more toxic ethylene glycol that is the primary ingredient in automobile antifreeze.

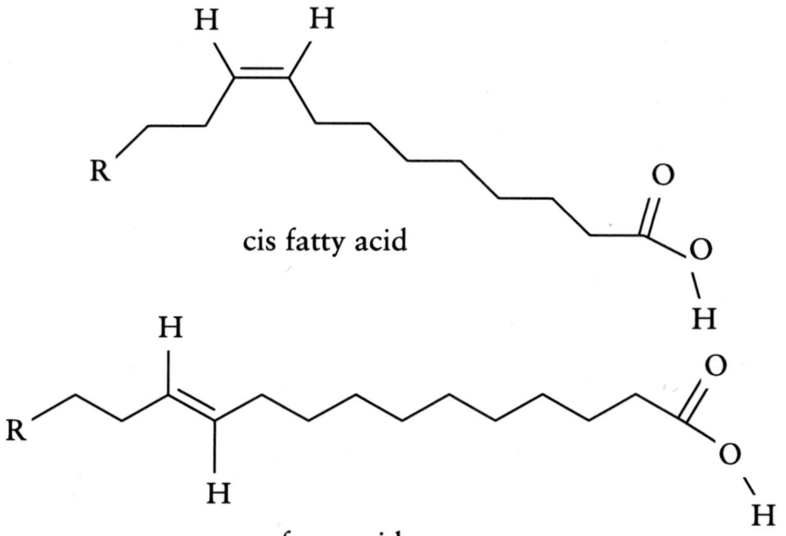

cis fatty acid

trans fatty acid

Trans and CIS Fatty Acids

In 2005, Archer Daniels Midland (ADM) and Novozymes N.A. won the Greener Synthetic Pathways Award for their enzyme interesterification process. In response to the FDA mandated labeling of *trans*-fats on nutritional information by January 1, 2006, Novozymes and ADM worked together to develop a clean, enzymatic process for the interesterification of oils and fats by interchanging saturated and unsaturated fatty acids. The result is commercially viable products without *trans*-fats. In addition to the human health benefits of eliminating *trans*-fats, the process has reduced the use of toxic chemicals and water, prevents vast amounts of byproducts, and reduces the amount of fats and oils wasted.

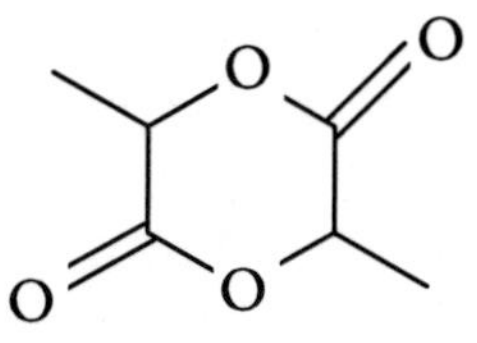

Lactide

❖ In 2002, Cargill Dow (now NatureWorks) won the Greener Reaction Conditions Award for their improved polylactic acid polymerization process. Lactic acid is produced by fermenting corn and converted to lactide, the cyclic dimer ester of lactic acid using an efficient, tin-catalyzed cyclization. The L, L-lactide enantiomer is isolated by distillation and polymerized in the melt to make a crystallizable polymer, which has use in many applications including textiles and apparel, cutlery, and food packaging. Wal-Mart has announced that it is using/will use PLA for its produce packaging. The NatureWorks PLA process substitutes renewable materials for petroleum feedstocks, doesn't require the use of hazardous organic solvents typical in other PLA processes, and results in a high-quality polymer that is recyclable and compostable.

- ❖ In 1996, Dow Chemical won the 1996 Greener Reaction Conditions award for their 100 per cent carbon dioxide blowing agent for polystyrene foam production. Polystyrene foam is a common material used in packing and food transportation. Seven hundred million pounds are produced each year in the United States alone. Traditionally, CFC and other ozone-depleting chemicals were used in the production process of the foam sheets, presenting a serious environmental hazard. Flammable, explosive, and, in some cases toxic hydrocarbons have also been used as CFC replacements, but they present their own problems. Dow Chemical discovered that supercritical carbon dioxide works equally as well as a blowing agent, without the need for hazardous substances, allowing the polystyrene to be more easily recycled. The CO_2 used in the process is reused from other industries, so the net carbon released from the process is zero.
- ❖ In 2003 Shaw Industries was recognized with the Designing Greener Chemicals Award for developing EcoWorx Carpet Tile. Historically, carpet tile backings have been manufactured using bitumen, polyvinyl chloride (PVC), or polyurethane (PU). While these backing systems have performed satisfactorily, there are several inherently negative attributes due to their feedstocks or their ability to be recycled. Shaw selected a combination of polyolefin resins as the base polymer of choice for EcoWorx due to the low toxicity of its feedstocks, superior adhesion properties, dimensional stability, and its ability to be recycled. The EcoWorx compound also had to be designed to be compatible with nylon carpet fiber. Although EcoWorx may be recovered from any fiber type, nylon-6 provides a significant advantage. Polyolefins are compatible with known nylon-6 depolymerization methods. PVC interferes with those processes. Nylon-6 chemistry is well-known and not addressed in first-generation production. From its inception, EcoWorx met all of the design criteria necessary to satisfy the needs

of the marketplace from a performance, health, and environmental standpoint. Research indicated that separation of the fiber and backing through elutriation, grinding, and air separation proved to be the best way to recover the face and backing components, but an infrastructure for returning postconsumer EcoWorx to the elutriation process was necessary. Research also indicated that the postconsumer carpet tile had a positive economic value at the end of its useful life. EcoWorx is recognized by MBDC as a certified Cradle to Cradle design.

Other Awards

The Royal Australian Chemical Institute (RACI) presents Australia's Green Chemistry Challenge Awards. This awards program is similar to that at the EPA, although the Institute has included a category for Green Chemistry education as well as Small Business and Academic or Government.

The Canadian Green Chemistry Medal is an annual award given to an individual or group for promotion and development of green chemistry in Canada and internationally. The winner is presented with a citation recognizing the achievements together with a sculpture.

Green Chemistry activities in Italy center around an inter-university consortium known as INCA. Beginning in 1999, the INCA has given three awards annually to industry for applications of green chemistry. The winners receive a plaque at the annual INCA meeting.

In Japan, The Green & Sustainable Chemistry Network (GSCN), formed in 1999, is an organization consisting of representatives from chemical manufacturers and researchers. In 2001, the organization began an awards program. GSC Awards are to be granted to individuals, groups or companies who greatly contributed to green chemistry through their research, development and their industrialization. The achievements are awarded by Ministers of related government agencies.

In the United Kingdom, the Crystal Faraday Partnership, a non-profit group founded in 2001, awards businesses annually for incorporation of green chemistry. The Green Chemical Technology Awards have been given by Crystal Faraday since 2004; the awards were presented by the Royal Society of Chemistry prior to that time. The award is given only to a single researcher or business, while other notable entries are given recognition as well.

The Nobel Prize Committee recognized the importance of green chemistry in 2005 by awarding Yves Chauvin, Robert H. Grubbs, and Richard R.

Schrock the Nobel Prize for Chemistry for "the development of the metathesis method in organic synthesis." The Nobel Prize Committee states, "this represents a great step forward for 'green chemistry', reducing potentially hazardous waste through smarter production. Metathesis is an example of how important basic science has been applied for the benefit of man, society and the environment."

Trends

Attempts are being made not only to quantify the *greenness* of a chemical process but also to factor in other variables such as chemical yield, the price of reaction components, safety in handling chemicals, hardware demands, energy profile and ease of product workup and purification. In one quantitative study, the reduction of nitrobenzene to aniline receives 64 points out of 100 marking it as an acceptable synthesis overall whereas a synthesis of an amide using HMDS is only described as adequate with a combined 32 points.

Green chemistry is increasingly seen as a powerful tool that researchers must use to evaluate the environmental impact of nanotechnology. As nanomaterials are developed, the environmental and human health impacts of both the products themselves and the processes to make them must be considered to ensure their long-term economic viability.

Examples

Supramolecular Chemistry

Research is currently ongoing in the area of supramolecular chemistry to develop reactions which can proceed in the solid state without the use of solvents. The cycloaddition of *trans*-1, 2-bis(4-pyridyl) ethylene is directed by resorcinol in the solid state. This solid-state reaction proceeds in the presence of UV light in 100 per cent yield.

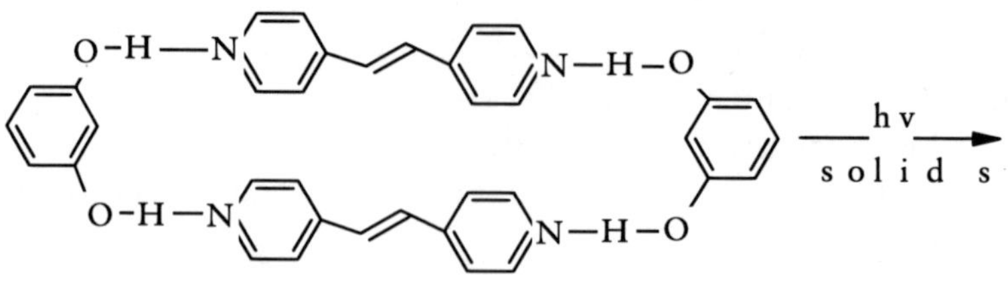

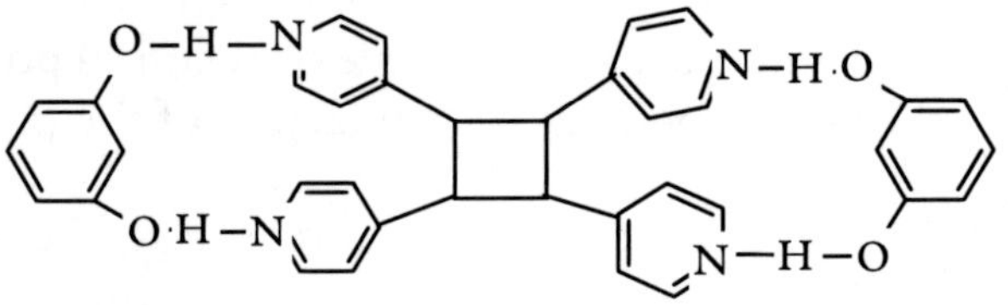

Cycloaddition of *trans*-1, 2-bis(4-pyridyl) ethylene

Reducing Market Barriers

In March 2006, the University of California published a landmark report by Dr. Michael P. Wilson and colleagues, Daniel A. Chia and Bryan C. Ehlers, on green chemistry and chemicals policy for the California Legislature entitled, Green Chemistry in California: A Framework for Leadership in Chemicals Policy and Innovation (http://coeh.berkeley.edu/news/06_wilson_policy.htm). The report finds that long-standing weaknesses in the U.S. chemical management program, notably the Toxic Substances Control Act (TSCA) of 1976, have produced a chemicals market in the U.S. that discounts the hazardous properties of chemicals relative to their function, price, and performance. The report concludes that these market conditions represent a key barrier to the scientific, technical, and commercial success of green chemistry in the U.S., and that fundamental policy changes are needed to correct these weaknesses.

The report describes three primary U.S. policy weaknesses:

(1) *The Data Gap*: TSCA does not require chemical testing prior to placing them on the market. However, any test results on the properties or hazards of the chemical that are in the possession of the submitter must be submitted to the U.S. Environmental Protection Agency. Even if test results are submitted, they may be claimed as confidential and cannot be disclosed. As a consequence, industrial buyers, workers, and consumers may not have the information they need to make informed decision about the chemicals they use. This data gap allows potentially hazardous chemicals to remain competitive in the market, and may undermine the commercial success of less hazardous products;

(2) *The Safety Gap*: Public agencies are overly constrained in their capacity to assess chemical risks and control those of greatest concern to public and environmental health; and

(3) *The Technology Gap*: Together, the Data and Safety Gap have produced market conditions in the U.S. that have damped the

> motivation of the private sector to invest in green chemistry at a level commensurate with the pace and scale of chemical production; green chemistry therefore operates at the margins of the industrial system.

The UC report calls for a modern, comprehensive chemicals policy to motivate new investment in green chemistry by improving transparency and accountability in the chemicals market. The report argues that these changes are needed soon, given the growing body of scientific information on the health and environmental effects of many chemicals, and the expected doubling of global chemical production over the next 25 years. Recommendations include (1) regulations to improve the generation and flow of information on the health and environmental effects of chemicals; (2) enhancing the capacity of public agencies to assess chemical risks and control those of greatest concern; and (3) increasing public investments in green chemistry research, education, and technology diffusion. The report argues that by taking these steps, California can position itself to become a global leader in green chemistry innovation, and that doing so will address a growing set of health and environmental problems related to chemicals and will open new possibilities for investment, employment, and productive capacity in California in green chemistry.

In January 2008, the University of California at Berkeley and Los Angeles produced a second report, Green Chemistry: Cornerstone to a Sustainable California (http://coeh.berkeley.edu/greenchemistry/briefing/). Commissioned by California EPA and endorsed by 127 UC faculty members from seven UC campuses and the UC national labs, the Cornerstone Report builds on the findings of the 2006 UC report to the California Legislature (noted above) and presents new cost estimates of the health and environmental consequences of existing chemical and product management approaches. The Cornerstone Report proposes policy strategies to stimulate investment in green chemistry by steadily closing the data, safety and technology gaps. California EPA Secretary Linda Adams released a statement noting that "the University of California's green chemistry report supports and reinforces the significance of the state's new environmental protection initiative on green chemistry. The University of California is an important partner in our efforts to establish a first-of-its-kind comprehensive policy for managing toxic chemicals in products."

3.3 Green Chemistry Network (GCN)

The Green Chemistry Network (GCN) was launched by the Royal Society of

Chemistry with initial core funding from the RSC for the first five years (1998—2003). The GCN is based within the Department of Chemistry at the University of York and is now a self-funding organisation.

The GCN* document gives information about ways of getting more involved in the Network.

The main aim of the GCN is to promote awareness and facilitate education, training and practice of Green Chemistry in industry, commerce, academia and schools. This will be achieved by:

- Providing links to other organisations and government departments.
- Organising conferences/workshops and training courses.
- Providing educational material for universities & schools.
- Newsletters and books with close links to the Green Chemistry journal.
- Providing prizes and awards for companies & university researchers.
- Running specific-themed projects targeting key areas and groups.

Green chemistry includes such concepts as waste minimisation, solvent selection, atom utilisation, intensive processing and alternative synthetic routes from sustainable resources. The challenge for chemists is to develop products, processes and services in a sustainable manner to improve quality of life, the natural environment and industry competitiveness.

Green Chemistry issues are here to stay. The most successful chemical companies of the future will be those who exploit its opportunities to their competitive advantage, and the most successful chemists of the future will be those who use Green Chemistry concepts in R & D, innovation and education.

The Green Chemistry Network aims to help these chemical companies and chemists by sharing best practice, promoting green technology transfer and providing data to show that adoption of green practices can also provide cost benefits for industry. In addition, we aim to make it possible for chemists of the future to grow up in an environment where green issues are not taught in isolation but form the underlying principles of all courses.

The 12 Principles of Green Chemistry according to the American Chemical Society and the US Environmental Protection Agency can be found on the ACS web site

3.4 Sonochemistry

In chemistry, the study of sonochemistry is concerned with understanding the effect of sonic waves and wave properties on chemical systems. Since and molecular chemistry is unique as well. Often these effects are most apparent in ultrasonic systems. This is demonstrated in phenomena such as ultrasound, sonication, sonoluminescence, and sonic cavitation.

The influece of sonic waves traveling through liquids was first reported by Robert Williams Wood (1868-1955) and Alfred Lee Loomis (1887-1975) in 1927, but the article was left mostly unnoticed. Sonochemistry experienced a renaissance in the 1980s with the advent of inexpensive and reliable generators of high-intensity ultrasound.

Upon irradiation with high intensity sound or ultrasound, acoustic cavitation usually occurs. Cavitation— the formation, growth, and implosive collapse of bubbles irradiated with sound—is the impetus for sonochemistry and sonoluminessence. Bubble collapse in liquids produces enormous amounts of energy from the conversion of kinetic energy of the liquid motion into heating the contents of the bubble. The compression of the bubbles during cavitation is more rapid than thermal transport, which generates a short-lived localized hot-spot. Experimental results have shown that these bubbles have temperatures around 5000 K, pressures of roughly 1000 atm, and heating and cooling rates above 10^{10} K/s. These cavitations can create extreme physical and chemical conditions in otherwise cold liquids.

With liquids containing solids, similar phenomena may occur with exposure to ultrasound. Once cavitation occurs near an extended solid surface, cavity collapse is nonsphereical and drives high-speed jets of liquid to the surface. These jets and associated shock waves can damage the now highly heated surface. Liquid-powder suspensions produce high velocity interparticle collisions. These collisions can change the surface morphology, composition, and reactivity.

Three classes of sonochemical reations exist: homogeneous sonochemistry of liquids, heterogeneous sonochemistry of liquid-liquid or solid-liquid systmes, and, overlapping with the aforementioned, sonocatalysis. Sonoluminessence is typically regarded as a special case of homogeneous sonochemistry. The chemical enhancement of reactions by ultrasound has been explored and has beneficial applications in mixed phase synthesis, materials chemistry, and biomedical uses. Because cavitation can only occur in liquids, chemical reactions are not seen in the ultrasonic irradiation of solids or solid-gas systems.

For example, in chemical kinetics, it has been observed that ultrasound can greatly enhance chemical reactivity in a number of systems by as much as a million-fold; effectively acting as a catalyst by exciting the atomic and molecular modes of the system (such as the vibrational, rotational, and translational modes). In addition, in reactions that use solids, ultrasound breaks up the solid pieces from the energy released from the bubbles created by cavitation collapsing through them. This gives the solid reactant a larger surface area for the reaction to proceed over, increasing the observed rate of reaction.

While the application of ultrasound often generates mixtures of products, a paper published in 2007 in the journal *Nature* described the use of ultrasound to selectively effect a certain cyclobutane ring-opening reaction.

Sonochemistry can be performed by using a bath (usually used for ultrasonic cleaning) or with a high power probe.

3.5 Ultrasonic

Ultrasonics is a trade term coined by the Ultrasonic Manufacturers Association and used by its successor, the Ultrasonic Industry Association, to refer to the use of high-intensity acoustic energy to change materials. This usage is contrasted to *ultrasound*, which is generally reserved for imaging, as in sonar, materials examination (NDI), and diagnostics (mammography, doppler bloodflow, etc.). However, in spite of this distinction, much technical material on ultrasound imaging actually uses the term *ultrasonics*, for example:

- *Ultrasonic Flaw Detection for Technicians*, 3rd ed., 2004 by J. C. Drury.
- *Ultrasonic nondestructive evaluation: engineering and biological material characterization*, Boca Raton, FL: CRC Press, 2004, by Tribikram Kundu.

Ultrasonication offers great potential in the processing of liquids and slurries, by improving the mixing and chemical reactions in various applications and industries. Ultrasonication generates alternating low-pressure and high-pressure waves in liquids, leading to the formation and violent collapse of small vacuum bubbles. This phenomenon is termed cavitation and causes high speed impinging liquid jets and strong hydrodynamic shear-forces. These effects are used for the deagglomeration and milling of micrometre and nanometre-size materials as well as for the disintegration of cells or the mixing of reactants. In this aspect, ultrasonication is an alternative to high-speed mixers and agitator bead mills. Ultrasonic foils under the moving wire in a paper machine will use the shock waves from the imploding bubbles to distribute the cellulose fibres more uniform in the produced paper web, which will make a stronger paper with more even surfaces, see more on Ultra Technology. Furthermore, chemical reactions benefit from the free radicals created by the cavitation as well as from the energy input and the material transfer through boundary layers. For many processes, this sonochemical (see sonochemistry) effect leads to a substantial reduction the reaction time, like in the transesterification of oil into biodiesel. Ultrasonication can easily be tested in lab scale for its effect on various liquid formulations. Within the past five years equipment manufacturers like Hielscher developed a number of larger ultrasonic processors of up to 16 kW power.

3.6 Laser-Based Detectors

Detectors that utilize lasers have been developing and some laser-based detectors are available commercially, but they are not widely used in routine chromatographic practice. Lasers offer several beneficial properties relevant to the chromatographer. In most cases, they offer improved sensitivity and selectivity than other detectors. In other cases, improved signal-to-noise ratios (S/N) and deletion of background disturbances are observed. Lasers are capable of better resolution than most liquid chromatographic needs, and they are well suited for pulsed—detection. On the other hand, they are not only costly, but the instrumentation can be complex. High energies of lasers can cause thermal distortions, and sensitivity can decrease due to scattering at the optical sections of the system. Listed below are some of the types of laser-based detectors used:

1. *Laser-Induced Absorption:* In most cases, this system is limited to trace concentrations due to difficulties in comparing the laser source signal from the altered signal. *Laser-induced photoacoustic (PA)* detection has been shown where a 488 nm argon laser source is emitted through an effluent cell. A PA detector measures pressure changes resulting from absorbed radiation. The data is sensed by a piezoelectric transducer (PZT) positioned posterior to a foil-covered slit. The signal is subsequently detected by an amplifier. Compared to a typical UV detector, the PA offers 25 times more sensitivity (Green, 1983).
2. *Laser-Induced Fluorescence*: Offers increased sensitivity but Rayleigh light scattering at optical sections and in the elements can present a problem to certain detectors. Detection can proceed through one—or two-photon excited fluorescence (OPEF or TPEF). TPEF results in absorption of a pair of photons that occupy a certain molecular energy level; it is feasible due to availability of high-powered lasers and absence of noise levels often seen in OPEF (Green, 1983). There are

several detector types: A) *Flowing Droplet Cell*: Four µl of eluent is kept between the end of an HPLC column and a solid probe. A laser is emitted perpendicular to this flow channel. The fluorescence is measured using an amplifier. Rayleigh scattering of the radiation source is mostly avoided B) *Laminar-Flow Cell*: Effluent is confined in the cell by one—way, laminar flow. The cell utilizes a sub-microliter flow-through cuvette whose windows are 5 mm from the stream which minimizes scattering. The laser beam and effluent are perpendicular to the optical component C) *Optical-Fiber Cell*: A capillary tube cell is connected to an optical fiber, which conducts light rays to the detector flow. The fluorescence collection optics are placed at a 30 degree angle in relation to the capillary in order to minimize scattering. (Green, 1983).

3. Laser-Based Refractive Index Detector: Differences in RI's are accomplished by using interferometry. A monochromatic laser measures changes in RI for a sample contained in a Fabry-Perot interferometer. A photo amplifier senses the light when the interferometer is scanned. A computer calculates the likely position of maximum interference and converts this data into a signal that represents a change in RI.
4. *Laser Light Scattering*: Considered the most "mature" of the laser-based techniques, this process involves the detection of Rayleigh scattering (Green, 1983). This approach utilizes a low-angle laser light scattering (LALLS) photometer and a gel permeation chromatograph (GPC). GPC, otherwise known as size-exclusion chromatography (SEC), separates compounds based on molecular weight. The LALLS focuses a laser beam on the sample placed between two quartz windows separated by a Teflon spacer that reduces background noise: (Green, 1983). The scattered light can be measured at scattering angles as little as 2 degrees. In series, the GPC/LALLS can provide on-line molecular weight distributions; consequently, this technique is well-suited for polymer characterization (Green, 1983).5) *Laser-Induced Raman Scattering*: Raman scattering is directly proportional to intensity of the source. Visible or UV lasers can therefore be useful in this technique. When the laser wavelength corresponds to the energetic vibrational level in an "excited" state, a phenomenon known as resonance Raman scattering (RRS) occurs (Green, 1983). Resonance enhancement of the normal Raman scattering (NRS) results in increased intensity of RRS bands, as well as improved efficiency since an increase is observed in the number of scattered photons-to-number of incident photons ratio.)

Laser-Based Optical Activity: By coupling a laser-based micropolarimeter with a HPLC, a system is devised which is both selective and it improves the extinction coefficient of the polarimeter up to 4 orders of magnitude (Green, 1983). The important components of this system include the Glan prism, the air-based Faraday rotators, and the cell-window. Also important to note is that many of the mobile phase eluents are not optically active and this, in turn, helps facilitate the detection of optically active samples. Green, R.B. *Analytical Chemistry*, 1983, Vol. 55, pp. 27.

3.7 Time-Resolved Microscopy of Laser Microchemistry

Some of the most important practical applications of laser chemistry are found in the lithography and graphic arts industry. Recently a great deal of attention has been devoted to utilizing laser microchemical processes in high resolution graphic arts applications. In the last few years, the Dlott group has become a leader in developing a fundamental understanding of the mechanisms of laser photothermal imaging processes used in the graphic arts industry, thanks to financial support from Presstek, Inc., Optodot, Inc., and the US Army Research Office.

> "Presstek, Inc., founded in 1987, has developed a patented proprietary non-photographic, toxic-free, digital imaging and printing plate technology for the printing and graphic arts industries on a worldwide basis. PEARL's thermal laser diode system is capable of imaging various types of Presstek printing plates either off-press or on-press to produce high quality, full-color lithographic printed materials. Presstek is also engaged in the development of additional products and applications that incorporate its proprietary PEARL technologies and consumables, including both computer-to-plate and other direct-to-press applications."
>
> —*From a Presstek press release*

> "Pioneering research at Optodot, by leading experts in organic semiconductor and microporous materials, is producing scientific discoveries that meet the needs of a progressively interdependent."

4

Environmental Methods for Chemical Analysis

4.1 Quantitative Analysis

In chemistry, *quantitative analysis* is the determination of the absolute or relative abundance (often expressed as a concentration) of one, several or all particular substance(s) present in a sample.

Once the presence of certain substance(s) in a sample is known, the study of their absolute or relative abundance can help in determining specific properties.

For example, quantitative analysis performed by mass spectrometry on biological samples can determine, by the relative abundance ratio of specific proteins, indications of certain diseases, like cancer.

Quantitative vs. Qualitative

The term "quantitative analysis" is often used in comparison (or contrast) with "qualitative analysis", which seeks information about the identity or form of substance present. For instance, a chemist might be given an unknown solid sample. He or she will use "qualitative" techniques (perhaps NMR or IR spectroscopy) to identify the compounds present, and then quantitative techniques to determine the amount of each compound in the sample. Careful procedures for recognizing the presence of different metal ions have been developed, although they have largely been replaced by modern instruments; these are collectively known as qualitative inorganic analysis. Similar tests for identifying organic compounds (by testing for different functional groups) are also known.

Many techniques can be used for either qualitative or quantitative measurements. For instance, suppose an indicator solution changes color in the presence of a metal ion. It could be used as a qualitative test: does the indicator solution change color when a drop of sample is added? It could also be used as a quantitative test, by studying the colour of the indicator solution with different concentrations of the metal ion. (This would probably be done using ultraviolet-visible spectroscopy.)

People are curious about the substances contained in material with which they come in contact. Precious metals may make them rich, medications may cure diseases and toxins and poisons may make them ill or worse. Analytical chemistry is the science which helps to satisfy such curiosity. Qualitative Analysis tries to answer the question "What's in this stuff?" and Quantitative Analysis tries to answer "How much?"

In 1894 the renowned chemist Wilhelm Ostwald summed up the matter when he wrote:

> "Analytical chemistry, or the art of recognizing different substances and determining their constituents, takes a prominent position among the applications of science, since the questions which it enables us to answer arise wherever chemical processes are employed for scientific or technical purposes. Its supreme importance has caused it to be assiduously cultivated from a very early period in the history of chemistry, and its records comprise a large part of the quantitative work which is spread over the whole domain of science."

Today in routine medical examinations a single sample of blood can yield quantitative information on several dozen constituents. Quantitative analyses are used in one form or another in the processing of all raw materials, from the determination of carbon, nickel and chromium to determine the hardness of steel to the analysis of sugar content of grapes hour by hour near the time of harvest so as to give the vintner the best quality grape for fermentation into wine.

This class will offer you some experiences with four methods of quantitative analysis: gravimetric, acid-base volumetric, iodometric volumetric and spectroscopic. Some other methods, certainly equally as important as the ones with which you will come in contact ought at least to be mentioned: mass spectrometry for the determination of the masses of molecular fragments and their quantities, radioactivity for the determination of the abundance of certain isotopes, heat and rate of reaction, thermal conductivity, optical activity and refractive index.

Doing an Analysis

The process of doing an analysis involves a number of steps which might not be done in the sequence shown, but are important points to consider and usually to execute in any analysis:

- Deciding on an analytical method,
- Collecting and preparing the bulk sample(s),
- Preparing the lab sample,
- Defining samples to be analyzed (replicate samples),
- Preparing solutions of a sample,
- Paying attention to properties of the sample matrix so as to be able to eliminate interferences,
- Calibration, measurement and calculating results, and
- Evaluation of results and estimation of reliability.

1. The field of analytical chemistry is so very important in commerce that money is often involved and cost and profits are routinely considered. The cost of an analysis usually goes up with the level of accuracy demanded so it is not surprising that the method may be a compromise between accuracy and economics. The method to be used also is a function of the number of components in the sample which may interfere with a certain analysis.
2. The decision of how to collect and to prepare your bulk samples depends sometimes on what it is that one wants to demonstrate. *Pfiesteria piscicida*, a single-celled microorganism which lives in the Chesapeake Bay and other bodies of water near the Virginia and North Carolina coasts, is known to thrive where there is an overabundance of nutrients such as nitrogen and phosphorus in the water. These include nutrient-rich human sewage, fertilizers, certain industrial by-products and animal wastes from swine and poultry. Such an environment allows algae to proliferate and the algae provide a rich food source for *Pfiesteria*. In 1991 a billion fish died in the Neuse Estuary of North Carolina as the result of succumbing to an agent or an effect connected with an organism in the food chain of which *Pfiesteria* is one. Some say *Pfiesteria* produces a powerful toxin which kill the fish. Others suggest that the fish become part of the food chain and are eaten by the microorganisms. The matter is further complicated by evidence which suggests that some but not all varieties of a species of algae produce toxins and studies of the life cycle of

Pfiesteria have produced ambiguous results. Proponents of one hypothesis will more likely than not defend their position vigorously in the face of evidence which other investigators find to be less than convincing. Periods of "red tide" along much of the coastline of the U.S. in which shellfish cannot be eaten because of their toxin content further complicate finding the source of natural products which are toxic to humans; moreover, one researcher claims that agricultural and industrial wastes are not the cause of algal blooms but the fault lies in red sand blown across the Atlantic Ocean from the Sahara Desert which triggers the algal blooms in the mid-Atlantic. As for the *Pfiesteria* controversy, more than a decade has passed and the matter seems to be nowhere near resolution. One would want to choose carefully how samples are to be taken based on whatever hypothesis it was to be tested. If the nutrient—algae-*Pfiesteria* link was a hypothesis to be supported, then water samples would have to be tied to a program of culture growth of the microorganism. If the link was felt to be well-established and the analysis was a part of an ongoing monitoring program then one might concentrate on regions where high nutrients might be expected to be found throughout the year. If the purpose of the analysis was a part of a program to protect the fish population then the samples might be taken during only certain months.

If a large quantity of precious metal ore is to be analyzed for the metal before a price can be agreed upon, one would definitely want to take samples from locations within the shipment at points of some wide separation so as to be able better to determine the variability of the metal within the shipment. Ores, which are commonly heterogeneous, require prudent choices to obtain representative samples.

3. Preparing the lab sample. Homogeneity of each sample taken from the larger bulk sample must be achieved at this point. Grinding and drying may be involved at this step. The determination of the amount of H_2O may be required as a part of the overall analysis. The grinding process not only achieves homogeneity but gives the technician a sample with finely divided particles for later work in producing a solution.
4. Defining the samples to be analyzed (replicate samples). Replication improves reliability because the variance between individual samples of the determined quantity of analyte establishes a level of reliability of the method used. Often individual quantities are weighed and those

quantities become the laboratory samples; the process of analysis starts after this weighing. Sometimes the grinding operation is found not to produce a bulk sample sufficiently homogeneous to use this method. In this case then a larger sample may be weighed and that sample then dissolved in a measured volume of appropriate liquid. This process assures homogeneity of the analyte and measured volumes of the solution are then taken as the lab samples.

5. Preparing solutions of a sample. Most analyses are done on solutions because the analytical reactions go at greatest speed when the analyte is divided down to the atomic, ionic or molecular level. All of the sample must be dissolved so as to assure that none of the analyte has occluded onto the surface or within the crystal lattice of any insoluble residue. Unfortunately, this step may be time-consuming but must be followed so as to exceed any reasonable doubt about the state of the analyte.
6. Paying attention to the properties of the sample matrix so as to be able to eliminate interferences. The families of elements found in the Periodic Table attest to there being few elements with unique properties. Thus, in practically all analyses there can be many interferences. Lead precipitates as the sulfide, but so does copper, mercury, nickel and silver. The oxidized forms of manganese (permanganate) and chromium (dichromate) have useful absorption spectra in the visible region. Both of these may be present in small amounts in steel. But iron which is present in the largest amount in steel oxidizes to Fe^{3+} and has an absorption spectrum in the same region. Citrus fruits contain a large number of natural acids which would preclude the analysis for any one of them using simple acid-base titrimetry. It is often necessary that subtle differences in the chemistry of elements within families be exploited so as to eliminate all interferences from any analysis.
7. Calibration, measurement and calculating results. The measurement which leads to the final result is usually directly proportional to that result. In the gravimetric determination of sulfate, all of the sulfate in the sample precipitates as barium sulfate, $BaSO_4$. The sulfate, the SO_4^{2-}, is the amount one wishes to calculate, but the weight of $BaSO_4$ is the measured quantity. In many cases for quantitative analyses, the calculation takes on the simple form:

$$C_A = k \times X$$

C_A is the desired quantity, the mass of SO_4^{2-} in 100 g of sample, but

X, the weight of $BaSO_4$ is that which is measured. k is the proportionality constant which relates one to the other, in this case the ratio of the atomic weight sum for SO_4^{2-} divided by the atomic weight sum for $BaSO_4$. Often one cannot so easily calculate k; in the case of the colorimetric determination of manganese in steel, the absorbance of a known concentration of manganese must be measured. One then has and k can be calculated from the absorbance of the known concentration by the quotient of C_A and X. The assumption is that the proportionality holds for all solutions of Mn prepared in the same way, both known and unknown. A value of X for an unknown sample will yield a calculated value for the concentration of Mn in that sample.

$$C_{A_{inswn}} = K \times X_{measured}$$

8. Evaluating results and estimating reliability. The announcement of an experimentally determined value alone without some indication of its reliability has no scientific worth. In experimental science we have a means for implying some level of reliability: significant figures. As imperfect as is the use of significant figures the method does at least set some rather broad boundaries of reliability. Reporting a value of 2.50 per cent Cr in a sample of stainless steel might suggest a maximum experimental deviation anywhere between ±0.01 per cent and ±0.09 per cent. If the experimental deviation were found to be between ±0.06 per cent and ±0.09 per cent, then one might consider reporting the value as 2.5 ±0.1 per cent Cr. If *nothing* were known about the reliability of a quantity determined experimentally there would be no justification for reporting any value of that quantity to any number of significant figures. Any report of the measurement would be worthless. It is fortunate that rarely is *nothing* known about the reliability of any experimental measurement. More often than not the problem is one of data being taken poorly with bad estimates of their reliability, leading to results the validity of which is open to withering criticism.

4.2 Basic Concepts

2-1 Système International or SI

A system of units and prefixes established to simplify and to unify communication between scientists.

Seven basic quantities	Unit	Abbreviation
Mass	kilogram	kg
Length	meter	m
Time	second	s
Temperature	kelvin	K
Amount of substance	mole	mol
Electric current	ampère	A
Luminous intensity	candela	cd

Definitions of the Seven Fundamental Units

Mass—The Kilogram (kg)

The *kilogram* of a particular cylinder of platinum-iridium alloy, called the International Prototype Kilogram, which is preserved in a vault at Sèvres, France, by the International Bureau of Weights and Measures.

Length—The Meter (m)

The *meter* is the distance that light travels in a vacuum in 1/299 792 458 of a second.

Time—The Second (s)

The *second* is the unit of time of the International System of Units. The definition adopted at the October 13, 1967 meeting of the 13th General Conference on Weights and Measures is: "The second is the duration of 9,192,631,770 periods of the radiation corresponding to the transition between the two hyperfine levels of the fundamental state of the atom of cesium 133." The frequency (9,192,631,770 Hz) which the definition assigns to the cesium radiation was carefully chosen to make it impossible, by any existing experimental evidence, to distinguish the new second from the "ephemeris second" based on the earth's motion. Therefore no changes need to be made in data stated in terms of the old standard in order to convert them to the new one. The atomic definition has two important advantages over the previous definition: (1) it can be realized (*i.e.*, generated by a suitable clock) with sufficient precision, ± 1 part per hundred billion (10^{11}) or better, to meet the most exacting demands of modern metrology; and (2) it is available to anyone who has access to or who can build an atomic clock controlled by the specified cesium radiation. (A description of such clocks is given in "Atomic Frequency Standards," *NBS Tech. News Bull.* 45, 8-11 (Jan., 1961). For more recent developments and technical details, see R. E. Beehler, R. C.

Mockler, and J. M. Richardson, "Cesium Beam Atomic Time and Frequency Standards," *Metrologia* 1, 114-131 (July, 1965)). In addition one can compare other high-precision clocks directly with such a standard in a relatively short time—an hour or so compared against years with the astronomical standard. Laboratory-type atomic clocks are complex and expensive, so that most clocks and frequency generators will continue to be calibrated against a standard such as the National Institutes of Standards and Technology (NIST) Frequency Standard, controlled by a cesium atomic beam, at the Radio Standards Laboratory in Boulder, Colorado. In most cases the comparison will be by way of the standard-frequency and time-interval signals broadcast by NBS radio stations WWV, WWVH, WWVB, and WWVL. There has been a recent proliferation of desk and wall clocks which tune into these stations a couple of times a day and reset themselves. Often they operate on an AA battery. When they first arrived on the mass market they cost around $400, but as in all things "hi-tech" the prices have fallen. Desk models can now be purchased for under $30. Come in to your instructor's office and see the wall-mount version.

Temperature—The Kelvin (K)

The *Kelvin*, the unit of thermodynamic temperature, is the fraction 1/273.16 of the thermodynamic temperature of the triple point of water. The decision was made at the 13th General Conference on Weights and Measures on October 13, 1967 that the name of the unit of thermodynamic temperature would be changed from *degree Kelvin* (symbol: °*K*) to ***kelvin*** (symbol: *K*). The name (***kelvin***) and symbol (*K*) are to be used for expressing temperature intervals. The former convention which expressed a temperature interval in *degrees Kelvin* or, abbreviated, *deg. K* is dropped. However, the old designations are acceptable temporarily as alternatives to the new ones. One may also express temperature intervals in *degrees Celsius*.

Amount of Substance—The Mole (mol)

The *mole* is the number of atoms of C^{12} in 12.0000 g of C^{12}.

Electrical Current—The Ampère (A).

The *ampère* (unit of electric current) is the constant current which, if maintained in two straight parallel conductors of infinite length, of negligible circular sections, and placed 1 meter apart in a vacuum, will produce between these conductors a force equal to 2×10^{-7} Newton per meter of length.

Luminous Intensity—The Candela (cd).

The *candela* is the luminous intensity, in the direction of the normal, of a black body surface 1/600,000 square meter in area, at the temperature of solidification of platinum under a pressure of 101,325 Newton per square meter.

Prefixes vary from yotta (factor of 10^{24}) to yocto (factor of 10^{-24})

Exponent Prefixes

Factor	*Prefix*	*Symbol*
10^{24}	yotta	Y
10^{21}	zetta	Z
10^{18}	exa	E
10^{15}	peta	P
10^{12}	tera	T
10^{9}	giga	G
10^{6}	mega	M
10^{3}	kilo	k
10^{2}	hecto	h
10^{1}	deka	da
10^{-1}	deci	d
10^{-2}	centi	c
10^{-3}	milli	m
10^{-6}	micro	μ
10^{-9}	nano	n
10^{-12}	pico	p
10^{-15}	femto	f
10^{-18}	atto	a
10^{-21}	zepto	z
10^{-24}	yocto	y

These are the seven *fundamental base units* and accompanying prefixes. All of the other units may be derived from these—they are called *derived units.*

Thus, 1 mL = 1 cm^3 and 1 Newton = 1 (kg m)/s^2

The Mole

A *mole* of molecules is Avogadro's number of molecules. Avogadro's number is the number of ^{12}C atoms in 12.0000 g of ^{12}C and has been experimentally determined to equal 6.022×10^{23}.

The *molar mass* is the mass of one mol of a substance.

Solutions and Their Concentrations

The following concentrations and concentration relationships are of importance to and will often be found in studies involving the quantitative analyses of chemical substances:

(a) Molar concentration.
(b) Analytical morality.
(c) Equilibrium molarity of a particular species.
(d) Percent concentration.
(e) Parts per million/billion (ppm, ppb).
(f) Volume ratios for dilution procedures.
(g) p-functions.

Percent Concentration

Percent concentration may be thought of as parts per hundred.

There are three forms of percent concentration which may be encountered:

1. wt. % (w/w) = (mass solute) ÷ (mass solution) × 100%
2. volume % (v/v) = (volume solute) ÷ (volume solution) × 100%
3. wt/volume % (w/v) = (mass solute) ÷ (volume solution, mL) × 100%

Weight percent is often used to express the concentration of commercial reagent grade acids. Concentrated aqueous ammonia is sold as 28% (w/w) NH_3.

Volume % often is used where one liquid is diluted by another:

10 per cent aqueous butanol, for example, would be a solution in which 10 mL pure butanol is diluted to give 100 mL solution.

Wt/vol solutions are often dilute aqueous solutions in which the calculation involved takes advantage of the fact that the density of water is close to 1 g/mL. Dilute w/v solutions have concentrations very close to the values which would be reported for their w/w concentrations. At higher concentrations there is a greater divergence. For example, 50% w/w NaOH is 76.3% (w/v) NaOH because the density of the solution rises significantly above 1.00 g/mL

The use of (w/w)% for (weight/weight)% is an historical artifact which is still seen from time to time. A better designation would be (m/m)% for (mass/mass)% but because the abbreviation "m" is that which is used for the metric unit of length, (w/w) has stuck around far longer than is justifiable. To get around the use of (w/w), companies which produce reagent grade acids simply say "Assay". That is, on a bottle of Spectrum reagent grade sulfuric acid one finds the designation "Assay (H_2SO_4)....... 95.0—98.0 per cent". That designation very definitely describes a per cent concentration found by the operation

4.3 Gravimetry

Gravimetric methods of analysis are used where weights of reactants and products of chemical reactions are reproducible, stable and reflect the presence of constituents which are important in the establishment of identity.

Two important methods deal with the trapping and weighing of products in the solid and gaseous phases. The first of these falls into the category of a precipitation method.

Precipitation Methods

Many metallic elements in their ionic forms react with negative counter ions to produce stable precipitates. Silver ions form stable and highly insoluble salts with chloride, bromide and iodide. Calcium precipitates quantitatively with oxalate and can be measured reproducibly at any of three temperature dependent plateaus as the oxalate, the carbonate and the oxide. Barium precipitates quantitatively as the sulfate. The reactions often follow the same patterns:

$$Ca^{2+}(aq) + C_2O_4^{2-}(aq) \rightarrow CaC_2O_4(s)$$

$$Ag^{+}(aq) + Cl^{-}(aq) \rightarrow AgCl(s)$$

Positive and negative ions in an aqueous solution, otherwise soluble with the counter ions in their environment, produce highly insoluble precipitates with certain added reagents.

Volatilization Methods

An interesting volatilization method which is entirely gravimetric is the one shown by the equations below.

The analyte can be bicarbonate as shown or a mixture of carbonate and bicarbonate. The total amount of carbonate in whatever form is found by placing the analyte in a solution containing an excess of H_2SO_4. This solution is in a flask connected to incoming nitrogen gas gently bubbled through the solution and an exit tube first to a drying agent to absorb aerosolized water and water vapor and then to a mixture of NaOH and drying agent to absorb the CO_2 and

$$NaCO_3\,(aq) + H_2SO_4 \rightarrow CO_2(g) + H2O(l)\ NaHSO_4(aq)$$

water subsequently produced by the absorption by NaOH:

$$CO_2(g) + 2\,NaOH(s) \rightarrow Na_2SO_4(s) + H_2O(l)$$

The apparatus is shown below. The tube containing the NaOH on asbestos and the $CaSO_4$ to absorb the final water product is pre- and post-weighed to given the total amount of carbonate in the sample. Note that the nitrogen gas acts only as a carrier and does not take part in any reaction.

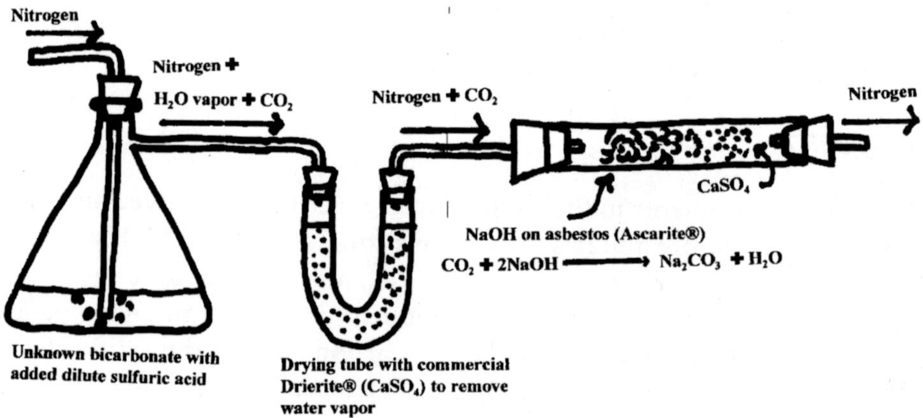

Considerations for the Isolation of Precipitates

Precipitates ought to be easy to wash free of contaminants without loss of the precipitate either in solution or through the filter. The particle size of the precipitate ought to be large enough not to escape through the filter pores. That the precipitate has a low solubility is paramount. The precipitate ought not to react with the atmosphere and it must have a known composition which remains stable after ignition.

Substances of low solubility have the nasty habit of forming colloidal suspensions. Colloidal particles have diameters from 10^{-7} cm to 10^{-4} cm. That is on the order of from 10 atomic diameters to 10,000 atomic diameters. Particles in this size range are still sufficiently jostled about by thermal molecular motion to remain in suspension. Where they are the result of a process of precipitation brought about by the addition of ionic species, the particles are surrounded by the excess ionic species. If Ba^{2+} is added in excess to SO_4^{2-} the $BaSO_4$ precipitate which is formed is considered to be surrounded by Ba^{2+} ions. If the opposite procedure were being followed, the precipitate would be surrounded by SO_4^{2-} ions. That these particles all have like charge and therefore repel each other suggests that your technique must favor the formation of large rather than small precipitate particles and to offer ways of encouraging the coagulation of particles after they have formed.

This can be done by carrying out the precipitation at a temperature close to the boiling point of water, in a dilute solution of your analyte and with constant stirring for the reasons given below.

Although analytical chemists still have some disagreement as to the mechanism of precipitation, there is wide agreement that a quantity called the *relative supersaturation* affects the particle size. Relative supersaturation is given as:

$$\frac{Q-S}{S}$$

where Q is the instantaneous concentration of the species added to effect precipitation and S is the equilibrium solubility of the substance which precipitates. Particle size seems to be inversely proportional to Relative Supersaturation because a high concentration of added reagent increases the probability that oppositely charged ions will begin the precipitation process at late as well as early stages of the addition and the resulting particles will be on the order of atomic dimensions, whereas the maintenance of a value of Q just slightly above S lowers that probability but offers in any case a layer of the added reagent ions around existing particles for their further growth.

4.4 The Electric Double Layer

If a particle of precipitate is surrounded by the ion in excess, say Ba^{2+} in the case of the determination of SO_4^{2-}, any negative ions in the immediate surroundings will be attracted to that primary positive layer. In the case of the addition of a $BaCl_2$ solution to a Na_2SO_4 solution, the ions Cl^- and SO_4^{2-} are available. As the sulfate is used up in the precipitation process it is the chloride which is left and which forms the second layer. Thus we have an *electric double layer*, made up first of barium ions then of chloride ions. This double layer keeps the colloidal precipitate particles from coming into contact with each other for further coagulation.

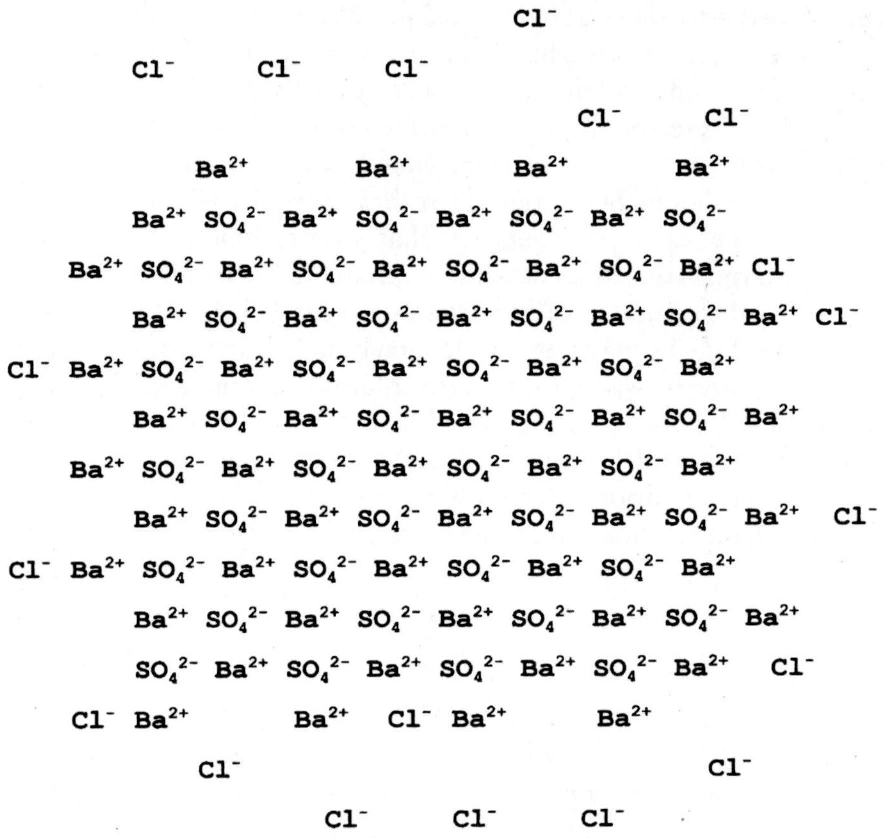

Cl^-

$Cl^- \quad Cl^- \quad Cl^-$

$Cl^- \quad Cl^-$

$Ba^{2+} \quad Ba^{2+} \quad Ba^{2+} \quad Ba^{2+}$

$Ba^{2+} \; SO_4^{2-} \; Ba^{2+} \; SO_4^{2-} \; Ba^{2+} \; SO_4^{2-} \; Ba^{2+} \; SO_4^{2-}$

$Ba^{2+} \; SO_4^{2-} \; Ba^{2+} \; SO_4^{2-} \; Ba^{2+} \; SO_4^{2-} \; Ba^{2+} \; SO_4^{2-} \; Ba^{2+} \; Cl^-$

$Ba^{2+} \; SO_4^{2-} \; Ba^{2+} \; SO_4^{2-} \; Ba^{2+} \; SO_4^{2-} \; Ba^{2+} \; SO_4^{2-} \; Ba^{2+} \; Cl^-$

$Cl^- \; Ba^{2+} \; SO_4^{2-} \; Ba^{2+} \; SO_4^{2-} \; Ba^{2+} \; SO_4^{2-} \; Ba^{2+} \; SO_4^{2-} \; Ba^{2+}$

$Ba^{2+} \; SO_4^{2-} \; Ba^{2+} \; SO_4^{2-} \; Ba^{2+} \; SO_4^{2-} \; Ba^{2+} \; SO_4^{2-} \; Ba^{2+}$

$Ba^{2+} \; SO_4^{2-} \; Ba^{2+} \; SO_4^{2-} \; Ba^{2+} \; SO_4^{2-} \; Ba^{2+} \; SO_4^{2-} \; Ba^{2+}$

$Ba^{2+} \; SO_4^{2-} \; Ba^{2+} \; SO_4^{2-} \; Ba^{2+} \; SO_4^{2-} \; Ba^{2+} \; SO_4^{2-} \; Ba^{2+} \quad Cl^-$

$Cl^- \; Ba^{2+} \; SO_4^{2-} \; Ba^{2+} \; SO_4^{2-} \; Ba^{2+} \; SO_4^{2-} \; Ba^{2+} \; SO_4^{2-} \; Ba^{2+}$

$Ba^{2+} \; SO_4^{2-} \; Ba^{2+} \; SO_4^{2-} \; Ba^{2+} \; SO_4^{2-} \; Ba^{2+} \; SO_4^{2-} \; Ba^{2+}$

$SO_4^{2-} \; Ba^{2+} \; SO_4^{2-} \; Ba^{2+} \; SO_4^{2-} \; Ba^{2+} \; SO_4^{2-} \; Ba^{2+} \quad Cl^-$

$Cl^- \; Ba^{2+} \quad Ba^{2+} \; Cl^- \; Ba^{2+} \quad Ba^{2+}$

$Cl^- \quad Cl^-$

$Cl^- \quad Cl^- \quad Cl^-$

Fig. 4.1: The electric double layer.

There are two ways to bring the particles closer together and to increase the probability of coagulation: (1) heating increases overall thermal motion, affecting both the mobility of adsorbed ions and of the colloidal precipitate particles themselves. The summary effect is that there are collisions of particles which result in the increase in particle size due to increased coagulation; (2) increasing the electrolyte concentration of the solution, for reasons not entirely clear, results in a decrease in the mean radius of the electric double layer and encourages further coagulation. Carrying out both of these operations results in *digestion* of the precipitate, an unfortunate term because biological digestion usually refers to the dissolving of food and absorption at the molecular level through the wall of the intestine. *Digestion* in quantitative analysis refers to the coagulation of a precipitate into a filterable form. Unfortunately after successful digestion, some of the primary electric layer is made up of Na^+ ions which must be washed away ultimately for quantitative results to be achieved. The Ba^{2+} ions as well will give a positive error if not removed and end up being dried with the precipitate as excess $BaCl_2$. Many coagulated precipitates do not respond well to washing with distilled water because as the second electric layer is removed (excess Cl^- for example) the first remains on all particles with an electric charge of the same sign. The result is that there is a return to the repulsive state and an effective increase in the radius of the particles which then begin once again to separate as colloidal particles. The process is called *peptization* and is to be avoided if some of the precipitate is not to be lost. One way around this for many precipitates is to encourage digestion by heating and also by increasing the electrolyte concentration by washing with a reagent which will go off as a gas during the drying process. Dilute nitric acid, HNO_3, is effective for washing excess ions from AgCl. In choosing such a wash, it is imperative that the procedure has been carried out and has been shown to yield reproducible, quantitative results. Unexpected side reactions, complex formation and changes in solubility with added reagents are sufficiently unpredictable to make intuition in the absence of experience unacceptable.

Other Demons which can Plague Quantitative Precipitate Isolation

During the precipitation procedure a number of other problems can arise to give erroneous positive or negative results. Among these are surface adsorption, mixed crystal formation, occlusion and mechanical entrapment.

Any ions may be carried down during a precipitation as the result of *surface adsorption.* Na^+, or Cl^- in the case of the determination of SO_4^{2-} by the addition of dilute $BaCl_2$ solution to a $NaSO_4$ solution. Both Ba^{2+} and Na^+ can compete for lattice positions as the particles form. Likewise, the ions Cl^- and SO_4^{2-} can have the same effect. In the quantitative determination of some transition metals, iron for example as $Fe(OH)_3$, zinc, cadmium and manganese may be present as impurities and all three form sparingly soluble hydroxides as well, though each with greater solubility than the hydroxide of iron:

Compound	*Solubility Product*
$Fe(OH)_3$	4×10^{-38}
$Cd(OH)_2$	2.5×10^{-14}
$Mn(OH)_2$	1.9×10^{-9}
$Zn(OH)_2$	1.2×10^{-17}

Mixed crystal formation can occur if two ions have the same charge, if their ionic diameters are sufficiently close to fit into the same crystal lattice.

Ions which commonly interfere with each other are shown in the table below with their ionic diameters in picometers given after each.

Interfering	*Ions*
K^+, 133 pm	NH_4^+, 148 pm
Sr^{2+}, 113 pm	Ba^{2+}, 135 pm
Mn^{2+}, 80 pm	Cd^{2+}, 97 pm

In cases where one has a known interference of one ion with the other it is necessary to find methods of removing one before carrying out a precipitation of the other, or using a precipitating reagent in which there is no interference.

Occlusion and mechanical entrapment. If a precipitation procedure is carried out too quickly, pockets of solvent and spectator ions can form, trapping them within the precipitate particles and dashing one's hope of removing them during the washing procedure. This is another reason why the relative supersaturation must be kept as low as possible so that in principle at least, all precipitation occurs only at the surface of a growing solid particle, devoid of solvent pockets.

All of these problems of coprecipitation of unwanted ions can lead to positive or negative errors. In the example above where it pointed out that Na^+ or Cl^- may coprecipitate in the SO_4^{2-} determination, surface adsorption will produce a positive error. In the case of mixed crystal formation, the direction of the error depends on the relative atomic weight of the ion which replaces that which is desired in the precipitate. In the case of the precipitation of zinc hydroxide, mixed crystal formation with manganese would produce a negative error but with cadmium or zinc a positive error.

		Desired precipitate	
Compound	$Mn(OH)_2$	$Zn(OH)_2$	$Cd(OH)_2$
At. Wt. of M^{2+}	54.94	65.39	112.41
Direction of error	negative	—	positive

The use of the technique of homogeneous solutions to effect precipitation. A solution containing a reagent which produces a desired ion to effect precipitation, often by gentle heating of the solution, offers an exquisite means for obtaining well-formed large crystal particles which lend themselves splendidly to the technique of filtration.

The model we use to explain why this happens also uses the concept of relative supersaturation. The initial nucleation of sparingly soluble particles offers a surface *template* which favors "locking" onto ions in the vicinity which by the luck of the draw (and the kinetic molecular theory) find themselves at the right energy and orientation to enter the crystal lattice. Ions isolated from a growing crystal are not favored to enter this process because at least two are required, both at the right energy and orientation to start the growth of a new crystal. If the concentration of one ion of a sparingly soluble salt increases gradually by slow homogeneous synthesis in a solution, then as its concentration reaches the threshold of supersaturation for the ion pair, a relatively small number of nucleated particles grows to larger size (because the probability of finding a place in an existing crystal lattice for any single ion is greater than that of a spontaneous creation a new crystal from dissolved and randomly arranged ions) rather than a large number of nucleated particles growing in constant competition with the rest and thus remaining small. The result for the latter is a non-filterable precipitate, but one in the former which filters quite well.

Here is a table of common reagents useful for the preparation of ions often needed for precipitation processes.

Reagent	*Precipitating species*	*Precipitation reaction*	*Elements which yield to this reaction*
Urea	OH^-	$(NH_2)_2CO + 3H_2O \rightarrow CO_2 + 2NH_4^+ + 2OH^-$	Al, Ga, Th, Bi, Fe, Sn
Trimethyl phosphate	PO_4^{3-}	$(CH_3O)_3PO + 3H_2O \rightarrow 3CH_3OH + H_3PO_4$	Zr, Hf
Ethyl oxalate	$C_2O_4^{2-}$	$(C_2H_5)_2C_2O_4 + 2H_2O \rightarrow 2C_2H_5OH + H_2C_2O_4$	Mg, Zn, Ca
Dimethyl sulfate	SO_4^{2-}	$(CH_3O)_2SO_2 + 4H_2O \rightarrow 2CH_3OH + SO_4^{2-} + 2H_3O^+$	Ba, Ca, Sr, Pb
Trichloroacetic acid	CO_3^{2-}	$Cl_3CCOOH + 2OH^- \rightarrow CHCl_3 + CO_3^{2-} + H_2O$	La, Ba, Ra
Thioacetamide	H_2S	$CH_3CSNH_2 + H_2O \rightarrow CH_3CONH_2 + H_2S$	Sb, Mo, Cu, Cd
Dimethyl glyoxime	$CH_3(CNOH)_2CH_3$	$CH_3COCOCH_3 + 2H_2NOH \rightarrow DMG + 2H_2O$	Ni
8-Acetoxyquinoline	C_9H_6NOH	$CH_3COOQ + H_2O \rightarrow CH_3COOH + HOQ$	Al, U, Mg, Zn

Preparation of a Dry Weight of Your Precipitate

The resulting precipitate must be heated until a stable dry state is reached. Some understanding of typical precipitate properties is mandatory for repeatable results to be achieved.

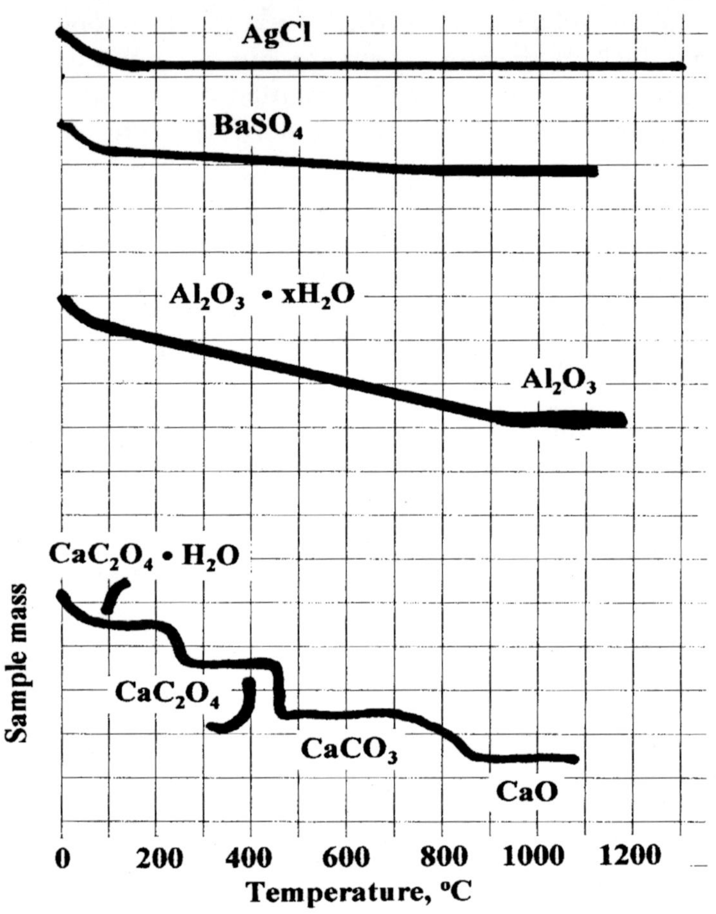

Note in the figure at the right that whereas AgCl achieves a stable dry weight just above 100°C, $BaSO_4$ does not do so until it reaches a temperature in the vicinity of 700°C Aluminum oxide, Al_2O_3, loses water slowly as the temperature rises to 1000°C at which point it achieves stability. Some compounds decompose in several stages, reaching stable plateaus. Calcium oxalate, CaC_2O_4 H_2O, loses all its water at around 200°C and remains stable as CaC_2O_4 until just above 400°C at which point it decomposes to calcium carbonate, $CaCO_3$ where it remains stable up to 700°C. Between 700°C and 850°C it slowly decomposes to CaO where it remains stable until its melting point at 2614°C.

A device not seen often in analytical laboratories but useful for producing automatic plots of mass of sample vs. temperature such as those at the right is the thermobalance (below). Region A includes the heating circuit, a temperature sensor, sample cup and counter weight resting on one end of the balance arm. B. A light wave is partially attenuated at C, giving a negative feedback to the amplifier circuit at D, designed to yield an output voltage which increases with the force necessary to keep the balance in equilibrium (and the attenuation at a constant value). One can adjust the baseline voltage at E, the tare adjuster, so as to produce the graph at the chart recorder, F.

Example 3-1: A 0.3427 g sample of bronze-age jewelry is analyzed for silver content by first dissolving it in concentrated nitric acid and precipitating it as AgCl. The precipitate is transferred to a dry sintered glass filter weighing 12.2347 g where it is separated from the filtrate and washed with dilute nitric acid. The filter and precipitate are dried at 150°C, cooled, weighed and found to weigh 12.4373 g. Calculate the %Ag in the jewelry.

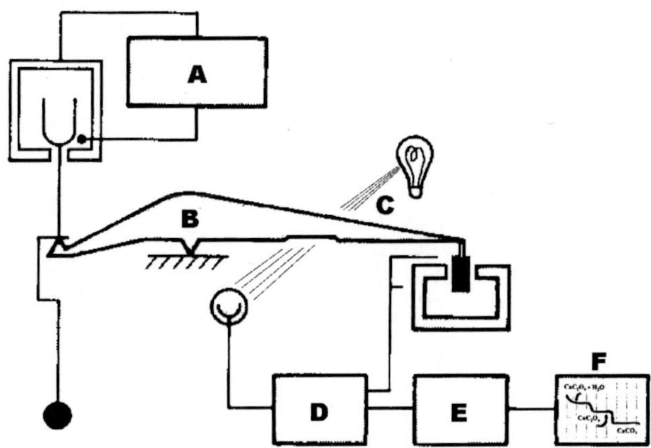

Example 3-2: A sample of iron ore weighing 0.4275 g is dissolved in 12M HCl. The resulting solution is slowly made basic with NaOH until the first hint of a turbid solution is detected. 10.0 g urea are dissolved and the solution heated just to the boiling point for 4 hours. The precipitate, Fe_2O_3 x H_2O, is trapped using ash less filter paper. The precipitate and filter paper are fired in a porcelain crucible of empty weight 12.2837g until the ash less filter paper is completely incinerated and anhydrous Fe_2O_3 is left. The resulting weight of crucible and precipitate is 12.4274 g. Determine the %Fe, the $\%Fe_2O_3$ and the $\%Fe_3O_4$.

Example 3-3: A sample known to contain only KCl and NaCl and weighing 0.4263 g is dissolved in water and treated to an excess of $AgNO_3$, using standard methods of precipitation. The AgCl precipitate is caught on a Gooch Crucible of original dry weight of 15.2748 g. The AgCl precipitate is dried at 150˚C, cooled and the crucible and precipitate are found to weigh 16.2872 g. Determine the %KCl and the %NaCl in this sample.

Solution is based on the difference in the %Cl in the pure salts:

Salt	*%Cl*
NaCl	60.66
KCl	47.55

Preferred methods of gravimetric analysis. Most inorganic ions have yielded to gravimetric analytical techniques, but one finds many interfering ions. The table below illustrates both the abundance of reagents available for use as well as the problems which can be encountered by interfering ions:

Analyte	*Precipitate*	*Measured form*	*Interferences*
K^+	$KB(C_6H_5)_4$	$KB(C_6H_5)_4$	NH_4^+, Ag^+, Hg^{2+}, Tl^+, Rb^+, Cs^+
Mg^{2+}	$Mg(NH_4)PO_4.6H_2O$	$Mg_2P_2O_7$	Many metals (none from Na^+ and K^+)
Ca^{2+}	$CaC_2O_4.H_2O$	$CaCO_3$ or CaO	Many metals (none from Mg^{2+}, Na^+ and K^+)
Ba^{2+}	$BaSO_4$	$BaSO_4$	Na^+, K^+, Li^+, Ca^{2+}, Al^{3+}, Cr^{3+}, Fe^{3+}, Sr^{2+}, Pb^{2+}
Ti^{4+}	TiO(5, 7-dibromo-8-hydroxyquinoline)$_2$	TiO(5, 7-dibromo-8-hydroxyquinoline)$_2$	Fe^{3+}, Zr^{4+}, Cu^{2+}, $C_2O_4{}^{2-}$, citrate, HF
VO_4^{3-}	Hg_3VO_4	V_2O_5	Cl^-, Br^-, I^-, SO_4^{2-}, CrO_4^{2-}, AsO_4^{3-}, PO_4^{3-}
Cr^{3+}	$PbCrO_4$	$PbCrO_4$	NH_4^+, Ag^+
Mn^{2+}	$Mn(NH_4)PO_4.H_2O$	$Mn_2P_2O_7$	Interferences from numerous metals
Fe^{3+}	$Fe(HCO_2)_3$	Fe_2O_3	Interferences from numerous metals
Co^{2+}	Co(1-nitroso-2-naphtholate)$_3$	$CoSO_4$ (by reaction with H_2SO_4)	Fe^{3+}, Zr^{4+}, Pd^{2+}
Ni^{2+}	Ni(dimethylglyoximate)$_2$	Ni(dimethylglyoximate) 2	Pd^{2+}, Pt^{2+}, Bi^{3+}, Au^{3+}
Cu^{2+}	CuSCN	CuSCN	NH_4^+, Pb^{2+}, Hg^{2+}, Ag^+
Zn^{2+}	$Zn(NH_4)PO_4.H_2O$	$Zn_2P_2O_7$	Interferences from numerous metals
Ce^{4+}	$Ce(IO_3)_4$	CeO_2	Th^{4+}, Ti^{4+}, Zr^{4+}
Al^{3+}	Al(8-hydroxyquinolate)$_3$	Al(8-hydroxyquinolate)$_3$	Interferences from numerous metals
Sn^{4+}	Sn(cupferron)$_4$	SnO_2	Cu^{2+}, Pb^{2+}, As(III)
Pb^{2+}	$PbSO_4$	$PbSO_4$	Ca^{2+}, Sr^{2+}, Ba^{2+}, Hg^{2+}, Ag^+, HCl, HNO_3
NH_4^+	$NH_4B(C_6H_5)_4$	$NH_4B(C_6H_5)_4$	K^+, Rb^+, Cs^+
Cl^-	AgCl	AgCl	Br^-, I^-, SCN^-, S^{2-}, $S_2O_3^{2-}$, CN^-
Br^-	AgBr	AgBr	Cl^-, I^-, SCN^-, S^{2-}, $S_2O_3^{2-}$, CN^-
I^-	AgI	AgI	Br^-, Cl^-, SCN^-, S^{2-}, $S_2O_3^{2-}$, CN^-
SCN^-	CuSCN	CuSCN	NH_4^+, Pb^{2+}, Hg^{2+}, Ag^+
CN^-	AgCN	AgCN	Cl^-, Br^-, I^-, SCN^-, S^{2-}, $S_2O_3^{2-}$
F^-	$(C_6H_5)_3SnF$	$(C_6H_5)_3SnF$	Except alkali metals, many interferences, and SiO_4^{4-}, CO_3^{2-}
ClO_4^-	$KClO_4$	$KClO_4$	
SO_4^{2-}	$BaSO_4$	$BaSO_4$	Na^+, K^+, Li^+, Ca^{2+}, Al^{3+}, Cr^{3+}, Fe^{3+}, Sr^{2+}, Pb^{2+}
PO_4^{3-}	$Mg(NH_4)PO_4.6H_2O$	$Mg_2P_2O_7$	Many interferences except Na^+, K^+
NO_3^-	Nitron nitrate	Nitron nitrate	ClO_4^-, I^-, SCN^-, CrO_4^{2-}, ClO_3^-, NO_2^-, Br^-, $C_2O_4^{2-}$
CO_3^{2-}	CO_2 (by addition of acid)	CO_2	CO_2 is trapped as Na$_2$CO3 on Ascarite

There are a number of organic functional groups which precipitate with metal ions by one of two routes: (1) chelating agents are organic compounds which "wrap around" a metal ion thanks to cationic side chains which form coordinate covalent bonds with the ion, and (2) a straightforward ion-ion bond which produces a new species that excludes water of solvation and thus precipitates. Good examples of chelating agents include Ethylene Diamine Tetraacetic Acid (EDTA), oxalic acid, glycine, 8-hydroxyquinoline and dimethylglyoxime.

Some common organic precipitating agents:

Compound	*Ions precipitated*
Dimethylglyoxime	Ni^{2+}, Pd^{2+}, Pt^{2+}
EDTA (Ethylenediamine tetraacetic acid)	Zn^{2+}, Cu^{2+}, Pb^{2+}, Ca^{2+}, Ni^{2+}, Fe^{3+}
Cupferron	Fe^{3+}, VO_2^+, Ti^{4+}, Zr^{4+}, Ce^{4+}, Ga^{3+}, Sn^{4+}
8-Hydroxyquinoline	Fe^{3+}, Al^{3+}, Mg^{2+}, Zn^{2+}, Cu^{2+}, Cd^{2+}, Pb^{2+}, Bi^{3+}, Ga^{3+}, Th^{4+}, Zr^{4+}, TiO^{2+}, UO_2^{2+}
Salicylaldoxime	Bi^{3+}, Ni^{2+}, Pd^{2+}, Zn^{2+}, Cu^{2+}, Pb^{2+}
1-Nitroso-2-naphthol	Fe^{3+}, Co^{2+}, Pd^{2+}, Zr^{4+}
Nitron ($C_{20}H_{16}N_4$)	NO_3^-, ClO_4^-, BF_4^-, WO_4^{2-}
Sodium tetraphenylborate	NH_4^+, organic ammonium, Ag^+, Cs^+, Rb^+, K^+
Tetraphenylarsonium chloride	$Cr_2O_7^{2-}$, MnO_4^-, ReO_4^-, MoO_4^{2-}, WO_4^{2-}, ClO_4^-

4.5 Titrimetry

Definitions

Titrimetry refers to that group of analytical techniques which takes advantage of titers or concentrations of solutions. The word "titer" is also used to denote "equivalence" or that amount of a solution required to complete a chemical reaction. In medicine it is often used to describe a definite result in a diagnostic test.

Though in chemistry the term titrimetry often refers to the use of some volume of a solution of known concentration to determine the quantity of analyte, there are still some variations on the use of the term. It is used rather to denote a quantity of some other measurement parameter which relates directly to the quantity of analyte which is to be measured:

Volumetric titrimetry establishes a quantity of analyte using volumes of reagents of known concentrations and the knowledge of the stoichiometry of the reactions between the reagents and the analyte(s).

Gravimetric titrimetry determines the quantity of analyte by a measure of the *mass* of a solution of known concentration.

Coulometric titrimetry arrives at the amount of analyte by measuring the duration of a given electrical current. Since amperes x time = coulombs or total charge, the number of equivalents of analyte can be measured by relating the extent of reaction to the number of moles of electrons (Faradays).

$$1\ ampere = 1\frac{coulomb}{second}$$

$$96500\ coulobms = 1\ Faraday$$

$$1\ Faraday = 1\ mole\ of\ electorns$$

The *equivalence point* is the point at which a volume, a mass or a quantity of charge equivalent to the amount of analyte present in the sample to be measured is reached. It is the point of stoichiometric chemical equivalence.

The *end point* is the point at which some detection technique tells you that chemical equivalence has been reached. The end point may occur before or after the equivalence point, giving a *titration error*. It is for this reason that blank samples are often used. Blank samples are prepared so that you have a measure of the amount that needs always to be added to or subtracted from the end point (the titration error) to achieve the equivalence point.

Standard Solutions

In volumetric titrimetry one uses a *standard solution* the concentration of which is known with great precision and which reacts stoichiometrically with the analyte. Standard solutions are referred to either as *primary standards* or *secondary standards*. Primary standards can be prepared by weighing directly and dissolving to a measured volume the reagent which is to react with the analyte. But primary standards must meet stringent requirements:

1. High purity,
2. Stability in presence of air,
3. Absence of any water of hydration which might vary with changing humidity and temperature,
4. Cheap,
5. Dissolves readily to produce stable solutions in solvent of choice, and
6. A larger rather than smaller molar mass.

These conditions are met by few materials. Anhydrous sodium carbonate, silver nitrate, potassium hydrogen phthalate are a few which do meet these conditions. The National Institute of Standards and Technology (NIST), formerly the National Bureau of Standards publishes lists of and is a reliable source of exhaustively analyzed primary standards. Moreover, its NIST Chemistry Web book is a useful source of data on many inorganic and organic compounds.

Standard Solutions

So as to be useful as a measure of the quantity of analyte in a sample, a standard solution must meet four criteria:

1. The reagent in the solution must have sufficient stability so that its concentration need be determined only once. Any deterioration in strength due to reaction with components of water or air would make the reagent unacceptable.
2. The reagent in the solution must react rapidly with the analyte. Slow approaches to equilibrium can cause erroneous judgments about reaction completeness.
3. The reaction of the analysis must occur with a completeness easily detectable by an appropriate indicator.
4. The reagent must react with the analyte in a simple and stoichiometrically predictable manner. Any side reactions would render a reagent unacceptable.

Two excellent standard solutions are those of hydrochloric acid, HCl and potassium hydrogen phthalate, $KHC_8H_4O_4$. Three others which have some instability but with appropriate precautions have proven to be excellent standard solutions are sodium thiosulfate, $Na_2S_2O_3$ (light sensitivity, susceptible to bacterial oxidation), silver nitrate, $AgNO_3$ (light sensitivity) and potassium permanganate, $KMnO_4$ (water oxidation catalyzed by light, heat, Mn^{2+} and MnO_2).

Determination of the Concentration of a Standard Solution

The *direct method* of solution standardization is the obvious choice if the reagent is a primary standard which meets all of the criteria described above and whose weight offers a repeatable observation proportional to the number of moles of substance expected. Solutions of sodium carbonate and silver nitrate can be prepared in this manner.

The *method of standardization* can be used if a primary standard reacts quantitatively with the reagent needed in the standard solution. HCl cannot be considered to be a primary standard because of its gaseous form at room temperature, but its solutions may be standardized against anhydrous Na_2CO_3.

For the purposes of this course, concentration in *molarity* will be given the symbol of c and that of *normality* the symbol c_N. The following two relationships will be useful:1

Sygmoidal Titration Curves

For the purpose of demonstrating the origin of sygmoidal titration curves, even the necessity of portraying titration curves as sygmoidal, it is instructive to look at what happens to the hydronium ion concentration during the titration of a strong acid with a strong base. Consider the titration of 50.00 mL 0.1000 M NaOH with 0.1000 M HCl. The hydronium ion concentration remains quite low almost to the equivalence point. As all of the hydroxide ion is used up, the hydronium ion concentration suddenly increases by 4-5 orders of magnitude within a volume change of a few hundredths of a mL. This translates to 4-5 pH units and illustrates why titration curves must be portrayed in a log-linear manner. Consider the following table (the student is invited to fill in the blank cells:

mL 0.1000 M HCl	*mmoles HCl added*	*Total volume (mL)*	*mmoles OH^- remaining*	*mmoles H_3O^+ in excess*	*$[OH^-]$*	*$[H_3O^+]$*	*pH*
0.00	0	50.00	5.000		0.100	$1x10^{-13}$	13
40.91	4.091	90.91			$1x10^{-2}$		
49.01		99.01	0.099		$1x10^{-3}$		
49.90	4.990		.01		$1x10^{-4}$		
49.99	4.999	99.99	.001		$1x10^{-5}$		
50.01	5.001			0.001	$1x10^{-9}$	10^{-5}	
50.10	5.010	100.10			$1x10^{-10}$	10^{-4}	
50.99	5.099	100.99					
59.09	5.909	109.09		0.909	1.25×10^{-12}	$8.3x10^{-3}$	
100.00	10.00	150.00		5.00	$3x10^{-13}$	0.033	

At and near the equivalence point, two phenomena affect the pH: (1) the stoichiometry of the chemical equation and (2) the autoionization or *autoprotolysis* of water. As the amount of HCl added approaches the equivalence point, the total amount of OH^-decreases by virtue of the stoichiometry of the reaction, but as that amount approaches zero molarity, the amount always present due to autoprotolysis becomes more and more important. As the equivalence point is approached in an acid-base titration the concentration of the species in excess continues to decrease. Throughout the titration, on either side of the equivalence point and quite close to it the pH can be estimated with very good precision by simple arithmetic. For example, in the titration of 50 mL of 0.1000 M NaOH with 0.1000 M HCl discussed above, the cells of the table showing various values of amount, volume and concentration can be calculated using only addition, subtraction, multiplication and division. But when the excess OH^- becomes very small, its value approaches that which is produced by protoloysis, that is,

$$2H_2O(l) + H_3O^+ (aq) + OH^- (aq)$$

This process governs the value of the protolysis equation:

$$[H_3O^+] [OH^-] = 1.0 \times 10^{-14}$$

The simple method of estimating the value of the hydronium ion and hydroxide ion concentrations then is complicated by this additional source of acidic and basic species, and the calculation moves into the realm of the quadratic equation.

The general form for a *quadratic equation* can be written as

$$ax^2 + bx + c = 0$$

The *quadratic formula* shows the root, or value of x for the quadratic equation:

$$x = \frac{-b \pm \sqrt{b^2 - 4ac}}{2a}$$

Note in the table above that as HCl is added, the mmoles of OH^- decrease toward zero, but until the equivalence point is approached that value greatly exceeds the mmoles of OH^- which are produced by protolysis. For example, in the table above for the addition of 49.99 mL HCl,

mL 0.1000 M HCl	*mmoles HCl added*	*Total volume (mL)*	*mmoles OH^- remaining*	*mmoles H_3O^+ in excess*	*$[OH^-]$*	*$[H_3O^+]$*	*pH*
49.99	4.999	99.99	0.001	////////	1×10^{-5}	1×10^{-9}	9

all calculations are accomplished by addition, subtraction and division. But look what happens if that process is carried to the following logical absurdity:

mL 0.1000 M HCl	*mmoles HCl added*	*Total volume (mL)*	*mmoles OH^- remaining*	*mmoles H_3O^+ in excess*	*$[OH^-]$*	*$[H_3O^+]$*	*pH*
49.99	4.999	99.99	0.001	////////	$1x10^{-5}$	$1x10^{-9}$	9
49.999	4.9999	99.999	0.0001	////////	$1x10^{-6}$	$1x10^{-8}$	8
49.9999	4.99999	99.9999	0.00001	////////	$1x10^{-7}$	$1x10^{-7}$	7
49.99999	4.999999	99.99999	0.000001	////////	$1x10^{-8}$	$1x10^{-6}$	6

But the result is ridiculous. On the one hand the equivalence point has not yet been reached. The solution ought still to be a little basic but the pH is calculated to be 6. This paradox arises because the protolysis leading to hydroxide and hydronium ions has been ignored. That is, the calculation of excess hydroxide ion by the simple arithmetic process is in error. If we let x = the mmoles of H_3O^+ produced by protolysis we have to admit that x moles of OH^- are also produced by protolysis, because stoichiometrically the ratio of production is 1 : 1, according to the protolysis equation

$$2H_2O(l) = H_3O^+(aq) + OH^-(aq)$$

Then the *total* number of mmoles of OH^- present, for the last row above is $0.000001 + x$

Since the total volume is nearly 100.0 mL, the protolysis equation becomes

$$\frac{x}{100} \times \frac{0.000001 + x}{100} = 1.0 \times 10^{-14}$$

which simplifies to the following quadratic form:

$$x^2 + 0.000001\,x - 10^{-10} = 0$$

The value of x (millimoles of H_3O^+) is found by solving the quadratic formula,

$$x = \frac{-0.000001 \pm \sqrt{10^{-12} + 4 \times 10^{-10}}}{2} = 9.5 \times 10^{-6} \textit{ mmoles } H_3O+ \textit{ (positive root)}$$

to yield $[H_3O^+] = 9.5 \times 10^{-8}$, giving a pH of 7.02, just slightly basic, as one would expect if the titration with HCl hasn't quite reached the equivalence point.

Exercise: Carry out this calculation for mmoles OH^- = 0.0001 and 0.00001.

All this is by way of saying that when one approaches the equivalence point, the pH cannot be estimated without the use of a quadratic equation.

The Henderson-Hasselbalch Equation

A weak acid, symbolized by the formula HA, hydrolyzes (reacts with water) according to an equation showing equilibrium between its acidic and basic forms:

$$HA + H_2O = H_3O^+ + A^-$$

We say that the acidic form is HA and the basic form A^-. The law of mass action after Guldberg and Waage predicts that the equilibrium constant for this process can be expressed as—

$$\frac{[H_3O^+][A^-]}{[HA][H_2O]} = K_{eq}$$

The designation of molar concentrations is an approximation to the species activity, a value which diverges from the molarity as the concentration increases. Since our concentrations in such calculations rarely approach even 1 molar, we shall follow this approximation. Equally important, the concentration of pure water is 55.6 M and varies little when solutes at concentrations in the vicinity of 0.01 to 1.0 M are introduced. That is, the concentration (and activity) of water is nearly constant where dilute solutions are involved, so the equation above can be rewritten as

$$\frac{[H_3O^+][A^-]}{[HA]} = [H_2O] \times K_{eq} = K_a$$

and K_a is defined to be the acid dissociation constant.

If $x = y$, then $\log_{10} x = \log_{10} y$ and we can effect the following modification of the equation above

$$\log_{10} \frac{[H_3O^+][A^-]}{[HA]} = \log_{10} K_a$$

The left side can be expanded to be

$$\log_{10} [H_3O^+] + \log_{10} \left(\frac{[A^-]}{[HA]} \right) = \log_{10} K_a$$

Multiplying both sides by -1 gives us

$$-\log_{10} [H_3O^+] - \log_{10} \left(\frac{[A^-]}{[HA]} \right) = \log_{10} K_a$$

The p function of X, that is, pX is defined as $-\log_{10}X$, so the equation above can be rewritten as

$$pH - \log_{10} \left(\frac{[A^-]}{[HA]} \right) = p\, K_a$$

And this can be rearranged to give us the standard form of the Henderson Hasselbalch Equation:

$$pH = pK_a + \log_{10}\left(\frac{[A^-]}{[HA]}\right)$$

The Henderson Hasselbalch Equation often evokes strong negative responses from many teachers of chemistry. "The Henderson Hasselbalch Equation is a terrible crutch and I don't teach it," is a comment not infrequently heard. Yet the equation contains no approximations other than the fundamental one that molar concentrations approximately equal species activities. There is of course the matter of hydrolysis. In addition to the dissociation of a weak acid shown above, the conjugate base of that weak acid is often conveniently available as the sodium or potassium salt, NaA or KA, and the dissociation of NaA, for example, in an aqueous environment, follows the path:

$$NaA\ (aq) \rightarrow Na^+\ (aq) + A^-\ (aq)$$

But A^- (*aq*) hydrolyzes to some extent:

$$A^- + H_2O = HA + OH^-$$

The hydrolysis of *HA* and of A^- cause the original analytical molarities to change slightly to their species equilibrium molarities. The primary objection to the Henderson Hasselbalch Equation seems to focus on the approximation that the species equilibrium molarities equal the analytical molarities. In general, this is a valid objection. On the other hand, for weak acids with values of K_a on the order of 10^{-5} and less, the analytical molarities give a very close approximation to the values which can be used in the Henderson Hasselbalch Equation.

4.6 Errors in Chemical Analysis

Any measurement is limited by the precision of the measuring instruments and the technique and the skill of the observer. Where a measurement consists of a single reading on a simple piece of laboratory equipment, for example a burette or a thermometer, one would expect the number of variables contributing to uncertainties in that measurement to be fewer than a measurement which is the result of a multi-step process consisting of two or more weight measurements, a titration and the use of a variety of reagents.

It is important to be able to estimate the uncertainty in any measurement because not doing so leaves the investigator as ignorant as though there were no measurement at all. The phrase "not doing so" perpetuates the myth that somehow a person can make a measurement and not know anything about the variability of the measurement. That doesn't happen very often. A needle swings back and forth or a digital output shows a slight instability, so the investigator can estimate the uncertainty, but what if a gross error is made in judgment, leading one to estimate an unrealistic "safe" envelope of uncertainty in the measurement? Consider the anecdote offered by Richard Feynman about one of his experiences while working on the Manhattan Project during World War II. Although this example doesn't address the uncertainty of a particular measurement it touches on problems which can arise when there is complete ignorance of parameter boundaries:

> Some of the special problems I had at Los Alamos were rather interesting. One thing had to do with the safety of the plant at Oak Ridge, Tennessee. Los Alamos was going to make the [atomic] bomb, but at Oak Ridge they were trying to separate the isotopes of uranium—uranium 238 and uranium 235, the explosive one. They were *just* beginning to get infinitesimal amounts from an experimental thing [isotope separation] of 235, and at the same time they were practicing the chemistry. There was going to be a big plant, they were going to have vats of the stuff, and then they were going to take the purified stuff and repurify and get it ready for the next stage. (You have to purify it in several stages.) So they were practicing on the one hand, and they were just getting a little bit of U235 from one of the pieces of apparatus experimentally on the other hand. And they were trying to learn how to assay it, to determine how much uranium 235 there is in it. Though we would send them instructions, they never got it right.

So finally Emil Segrè said that the only possible way to get it right was for him to go down there and see what they were doing. The army people said, "No, it is our policy to keep all the information of Los Alamos at one place."

The people in Oak Ridge didn't know any thing about what it was to be used for (*that is, they didn't have knowledge of its range of safety—O.S.)*; they just knew what they were trying to do. I mean the higher people knew they were separating uranium, but they didn't know how powerful the bomb was, or exactly how it worked or anything. The people underneath didn't know at *all* what they were doing. And the army wanted to keep it that way. There was no information going back and forth. But Segrè insisted they'd never get the assays right, and the whole thing would go up in smoke. So he finally went down to see what they were doing, and as he was walking through he saw them wheeling a tank carboy of water, green water—which is uranium nitrate solution.

He said, "Uh, you're going to handle it like that when it's purified too? Is that what you're going to do?"

They said, "sure—why not?"

"Won't it explode?" he said.

Huh! *Explode?*

Then the army said, "You see! We shouldn't have let any information get to them! Now they are all upset."

It turned out that the army had realized how much stuff we needed to make a bomb—twenty kilograms or whatever it was—and they realized that this much material, purified, would never be in the plant, so there was no danger. But they did *not* know that the neutrons were enormously more effective when they are slowed down in water. In water it takes less than a tenth—no, a hundredth—as much material to make a reaction that makes radioactivity. It kills people around and so on. It was *very* dangerous, and they had not paid any attention to the safety at all.

Feynman's example illustrates that although there were individuals who knew something about the boundary of parameters in which one could safely work with these materials in the solid state, both they and the people who were doing the work had no knowledge of the parameters or boundaries of safety for the conditions under which the work was being done (in aqueous solution). That ignorance rendered their knowledge useless. That the nuclear accident in Japan in 1999 came about because of the ignorance of *this same characteristic of neutrons* after having been documented in many lay accounts of the history of nuclear energy testifies to the continuing illusion over the "justification of secrecy" even in the face of imminent danger to workers, equipment and the population at large.

A second recent case of knowledge rendered useless without recognizing the presence of a fatal uncertainty involved the loss of the Mars Climate Orbiter on September 23, 1999. As reported in the L.A. Times, "the $125 million spacecraft was lost because NASA navigators mistakenly thought a contractor used metric measurements. The contractor had used English units, and the probe burned up in the Martian atmosphere Sept. 23."

Information is useless if there is no knowledge of the precision of that information. Such uselessness may be the result of:

1. The presentation of a single number as a statement of the information, but lacking an estimated error.
2. Gross error in calculation or data collection.

A gross error is not necessarily one in which the investigator fails to report a precision if it is known which equipment was used, say an analytical balance with a precision of ±0.0001 g or a 50 mL buret with a precision of ±0.01 mL. Data presented to a number of significant figures less than that justifiable by the equipment certainly demonstrates carelessness but doesn't, in this writer's opinion, rise to the level demonstrated by a student doing a titration using the wrong solution, NASA engineers effecting a mid-course maneuver of a deep space probe based on the wrong system of units or Oak Ridge technicians carrying out purification procedures of ^{235}U thinking that their margin of safety is 100 times greater than it is. The latter examples illustrate the very dangerous situation of investigators not knowing what they think they know, that is, some window of confidence in their data. That's when the data become useless.

Another and shorter way of saying the same thing is that without some knowledge of the uncertainty of a measurement, the reported value could be anything. A piece of jewelry could have a weight % gold of 0% or 100%. It wouldn't matter what one were to report if there is no knowledge of the experimental uncertainty.

Once experimental uncertainty is revealed, it is a forewarning of the boundaries beyond which there may be no experimental confidence. Without that knowledge all bets are off. The engineers of the Mars Climate Orbiter didn't have *any* boundaries beyond which lay potential disaster. The disaster was everywhere and nowhere. They were reduced to crossing their fingers as a Plan A for saving the mission. Science ought not to work that way. With the knowledge of experimental uncertainty intelligent decisions which were impossible before can be made.

In this section the important concepts of *precision* and *accuracy* will be introduced. Moreover, we will be concerned with the *spread* or *range* of a series of readings, and of decisions connected with removing *outliers* from a data set. The *median* and *arithmetic mean* will be discussed in the context of reporting a best value from a data set exhibiting *random errors*. Our discussion of *accuracy* as regards the closeness of a reported result to some *true value* and how precision and accuracy may differ due to *systematic errors* will be discussed.

Often an analytical chemist is faced with a choice between time constraints and accuracy; hence the question of maximum error tolerance must be asked. The determination of quantities of DNA sequences in a forensic lab may rest heavily on the "quick and dirty" comparison of intensities of "blips" of unknown concentrations on an electrophoresis gel with those of known concentrations. On the other hand, the establishment of conspiracy in a murder case might rest on careful analysis of trace elements in a series of bullets fired from different guns to show if they came from the same batch.

The Shroud of Turin is a celebrated case of time being available to develop adequate means of analysis to establish the necessary values beyond reasonable doubt. There was no deadline to be met before some decision had to be made. Some people claimed that the Shroud had been used to wrap the body of Jesus, the prophet of Christianity, after his crucifixion, though no one disputed that its history was not known before the 12th century, when it had become the property of the cathedral at Turin, Italy. It was not an official Relic of the Church, but its reputation over the centuries had grown and it probably was responsible for many pilgrimages to the cathedral among the faithful. When radiocarbon dating was proposed as a way to determine its age, the proposal was at first rejected because such a sizeable amount of material would have to be used to carry out the determination (perhaps as much as 10 cm^2 for each sample, and at least 3 samples must be taken to assure reproducibility). The fear was that if its age could be traced to the beginning of the first millennium, then it might well be named a Church Relic—but one that would have to be mutilated to gain that stature. So identification of the image had to rest on non-destructive methods of analysis, such as microscopic observations of the surface. Some said the "blood stains" were composed of a common pigment used by 13th century artists, others claimed the discoloration to be human blood. Meanwhile, back at the lab, techniques continued to improve, until reliable radiocarbon dating could finally be done with considerably smaller samples (in the case of the shroud, just a few short strands were needed for each sample). Such small sample sizes were judged by Church authorities not to constitute mutilation and the analysis went forward. Samples were taken from the shroud and sent to several laboratories along with other samples of fabrics of known ages. The laboratories were not told which was which. The reported values showed close agreement between shroud samples and none suggested an age of the fabric having been harvested from plants before the 12th century A.D. The committee which had taken on the task of judging the validity of the analysis was sufficiently satisfied to convince local Church authorities to retire the claim that it was a Holy Shroud.

Definitions

The *arithmetic mean*, or *average*, is defined as:

$$\bar{x} = \frac{\sum_{i=1}^{N} X_t}{N}$$

The arithmetic mean is used to report a best value among a series of N replicate measurements.

The *median* is the value which divides a set of replicate measurements when the set is arranged in order from the smallest to the largest. In the set of titration volumes

23.45, 23.45, 23.47, 23.49, 23.50, 23.51, 23.55,

the arithmetic mean is found by

$$(23.45 + 23.45 + 23.47 + 23.49 + 23.50 + 23.51 + 23.55) \div 7 = 23.489$$

Since there is uncertainty in the measurements at the hundredths place, it would be best to report this value no further. The value 23.49 would suffice except possibly in the rare case where the set showed an average deviation or standard deviation somewhat less than ±0.01

The median of this set is 23.49. If there is an even number of readings in the set, the median is the mean of the middle pair.

Range or Scatter

Any group of readings may be expected to extend over a range or to show some scatter. Members of our class are routinely asked to measure the volume reading of water contained in a burette. The instructor establishes the "true" value in advance by positioning the upper black boundary of a burette card just under the silhouette of the meniscus. Correct use of a buret is mandatory if the student is to do well in this class. For some images and instructions on the use of a buret, go to http://www.csudh.edu/oliver/demos/buretuse/buretuse.htm

Over the course of a semester, it is not unusual to observe four types of scatter in student readings of a burette volume. Scatter is assumed to be the result of *random* error, influences caused by limitations in the equipment used and the limited skill of the observer. There is also the possibility of prejudice on the part of the observer and the counterpart to prejudice on the part of the instrument used: miscalibration. This second error is referred to as *systematic error.* The various types of scatter one might expect to find among a group of Quantitative Analysis students reading a burette are illustrated in the chart at the right.

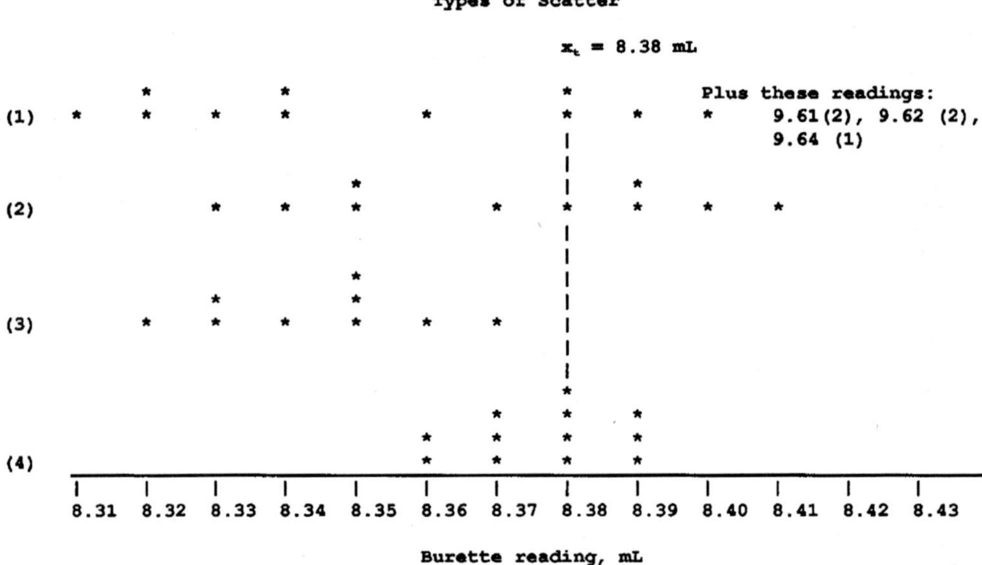

Case: (1) Low accuracy and low precision. Note that not only are the readings scattered widely about the true value of 8.38 mL, but several students have misconstrued the direction of the scale, reading upward from the 9.00 mL rather than downward from the 8.00 mL mark. This kind of scatter is often observed of student readings at the beginning of the semester. Case (2) which illustrates low precision and high accuracy. Although this case shows that the mean value of all the readings is close to the true value, it could be argued that it is by virtue of luck more than anything else that the students arrived at such a good mean value. Case (3) Low accuracy and high precision. Note in the example here that although the readings have a narrower range of scatter than in (1) and (2), there seems to be a systematic error in the low direction. If a group of readings are made without the use of a burette card, the bottom of the meniscus will appear to be higher in the burette (lower volume) and thus produce such an error. Although burette readings are corrected by subtracting the beginning volume from the ending volume, and such systematic errors would tend to cancel each other out, a burette card is necessary to produce a common, repeatable background. Finally, around the 14th week of the semester, students have had enough experience reading volumetric scales to present a set of readings like that shown in Case (4)

high precision and high accuracy, where one sees a narrow range of scatter on both sides of the true value.

Precision and Accuracy

Precision is a measure of the extent to which the values in a series of readings vary from the mean. One speaks of deviations: *average deviation* and *standard deviation* are two expressions commonly used. The term precision ought not to be used in the context of the agreement of one's average value with some "true" value. The term to be used in that case is *accuracy*, or the extent to which the mean of a series of readings varies from the "true" value.

The *absolute error* is the difference between any particular reading x_i and the true value x_t: absolute error = $x_i - x_t$.

Note that the formula is set up so that a low value produces a negative error and a high value produces a positive one. Sometimes one speaks of the absolute error of a mean:

$$\text{absolute error of the mean} = \bar{x} - x_t$$

It is often more useful to speak in terms of the relative error which relates the absolute error to the value of measurement:

$$\text{relative error} = \frac{x_i - x_e}{x_e}$$

The per cent relative error would then be given by

$$\text{per cent relative error} = \frac{x_i - x_e}{x_e} \times 100$$

Percent is of course "parts per hundred." It is useful to be able to make conversions between percent and parts per thousand (multiply by 10) or even to be able to determine the precision to parts in some other number when the other number may be close to some integral power of 10 (100, 1000 or 10000).

Where a number is expressed as 4.372 ± 0.006, the value 0.006 is the "deviation", "uncertainty" or "precision" expressed in the same units as the measured value, that is, in the same context as the absolute error above, the 0.006 would represent an "absolute deviation." or "absolute precision". So, for 4.372±0.006 g one would estimate an envelope of uncertainty limited to between 4.366 g and 4.378 g but not greater. One ought to make clear if the

0.006 is an average deviation or a standard deviation. Relative deviations are then calculated in the same manner as relative errors above except that the numerator is the absolute deviation by itself. So the relative deviation or relative precision in parts per thousand of this measured value would be $(0.006/4.372) \times 1000 = 1.4$ ppt.

The use of significant figures is ubiquitous in the language of science by virtue of its convenience. Its use will be encouraged if for no other reason than that such use provides an easily conveyed message, a verbal and written shorthand actually, the alternative for which is a mite more cumbersome. Yet the use of significant figures holds one glaring fault which we must state up front. Consider the two absorbances 0.109 and 0.901 taken from a visible absorption spectrophotometer. They both convey three significant figures because the rule says that the last digit shall be the one for which there is some uncertainty in the reading, usually the interpolated digit. Yet, at the most optimum, both would have an uncertainty of ±0.001 so that the relative uncertainty of the first would be almost nine times larger than that of the second:

$$(1)\ \ 0.001 \div 0.109 \approx 1.0 \times 10^{-2}$$

and

(2) 0.001 ¸ 0.901 » 1.1 × 10^{-2}

That having been said, let's take a look at some examples of typical reported values and our responsibility as scientists in offering reported values.

If you as a scientist report that a soluble sulfate unknown contains 21 per cent sulfate, that report conveys to the recipient the understanding that the determination is in error by at least 1 per cent, that is 21 ± 1 per cent. It might be off by 2 per cent or 3 per cent, but the last digit in the reported value is an indication of where the uncertainty lies. The fact that we *DON'T KNOW* what the uncertainty actually is, unless it is explicitly stated, represents a second defect of the use of significant figures. On the other hand, a student in Quantitative Analysis ought not to report 21.0 per cent or 21.000 per cent if the value was known only to ±1 per cent. This would convey some serious confusion to the recipient of the report, not to mention a perceptual error on the part of the student. If the student knows the percent sulfate to a hundredth of a percent, that is, if the calculations with uncertainty taken into account yielded the value of 21.37 ± 0.04 per cent sulfate, the student certainly ought not to report 21 per cent because the value is known much more precisely than 21 per cent and the report ought to reflect that.

That's fine for the investigator making the report. How about the recipient? As stated above, the recipient is to assume that any value presented will be offered according to the same rules, that it will be reported to the first uncertain digit.

With that as a jumping off point, the fundamental rule of significant figures is to report any value to the first digit for which there is some uncertainty *and that uncertainty must be reported.* If it is not, then one is left rather helplessly to assume some unknown uncertainty in the last digit.

The numbers 0.237, 4.38, 8.70 and 1.47×10^{23} all have 3 significant figures.

The expressions of the number 2.67×10^{-3}, 0.267×10^{-2}, 0.0267×10^{-1}, or 0.00267 all have 3 significant figures because, without actually saying it, the use of significant figures is a way of reducing the statement of precision to the level of *relative error* but without rigor. That is to say, if we assume for the sake of argument that the uncertainty in any of these values is ±1 for the digit "7" then 2.67±0.01, 0.267±0.001, 0.0267±0.0001 and 0.00267±0.00001 all exhibit the same value for the *ratio* of the error to the value:

$$\frac{0.001}{2.67} = \frac{0.001}{0.267} = \frac{0.0001}{0.0267} = \frac{0.00001}{0.00267} = 0.0037$$

By way of review and emphasis, 3 significant figures means that there is a *minimum* uncertainty of 1 in a number that extends from 100 to 999. So the uncertainty could represent anywhere from 0.1 per cent (1/999 x 100) to 1 per cent (1/100 x 100) of the value. But it gets worse. Saying, "My value is good to three significant figures" *doesn't state* the level of uncertainty in the last figure. In principle it could be anywhere from 1 to 9. That being the case, three significant figures could show a range of uncertainty from 0.1 per cent (1/999 × 100) to 9 per cent (9/100 x 100). That is the primary reason always to state your values with the added qualifier of the uncertainty itself, as 547 ± 6. That, then, nails down the extent to which a reported value can be trusted.

There is one other rule regarding significant figures which must be mentioned here. That deals with integers ending in one or more zeros. The *rule of thumb* is that these numbers are precise only to the last non-zero integer. Saying "The distance from the earth to the moon is 239,000 miles can be assumed to mean that the reported distance of 239,000 miles may have an uncertainty of 1000 to 9000 miles. If stated as 239,200 miles, one would take that to mean an uncertainty between 100 and 900 miles. If a writer (for example, a newspaper journalist) is forced to use integer notation to express a large whole number, then the trailing zeros must be there to establish the magnitude of the number, not necessarily its significant figures. It is for that reason that large integers ought always to be reported in scientific notation where there is little room for doubt: 2.39×10^5 miles leaves no room for doubt that the uncertainty of this figure starts at the position of the digit "9". The best way to report a number would be in any case to include the uncertainty, for example, $2.39 \pm 0.02 \times 10^5$ miles. The *rule of thumb* means "for most practical purposes." Rules of thumb always have exceptions. First of all, there are the definitions of sizes of units. That there are 1000 mL in a liter is a definition. The relationship is exact. There is no uncertainty. Secondly, there is always the case where some experimental value in the form of a large integer will come out with trailing zeros—a vote count, for example. If the total number of voters turning out in one precinct is determined to be 23000 simply by the luck of the draw, one would not be justified in saying it might have been 24,000 or then again perhaps 22,000. If the total was the result of three counts one could assume that it is either a valid exact number, or at the very most unreliable to ±1 or ±2.

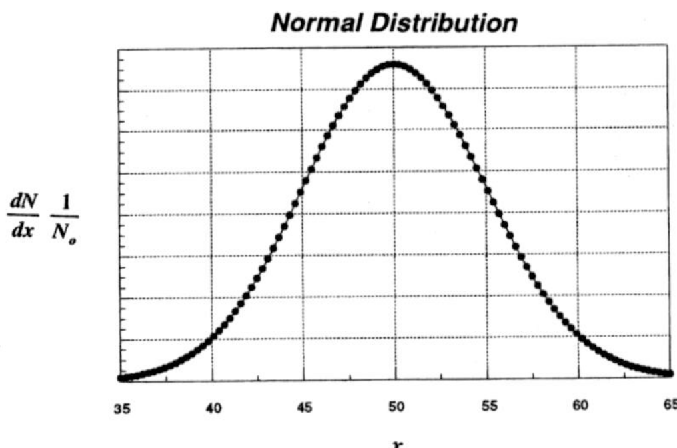

When you participate in some of our Web exercises, make sure that you follow the *rule of thumb* above to determine your answers, but store the exceptions somewhere in the back of your mind.

The Distribution of Experimental Data

The model we use to explain the tendency for values in a set of data to regress toward the mean is that in any set of readings there are multiple influences which may lead to error, some within the instrument itself, some on the part of the skill of the observer. The observed regression toward the mean, or the amassing of results somewhere toward the center of extreme readings is said to be due to the partial cancellation of some error effects against others. A simplified model to describe why there is this regression effect is that of flipping a coin. The coin flip example is not exactly the same as errors which can go either way in a scientific reading, but it does lead to a result which is self-consistent with the model, *and* the example leads to our understanding of the origin of the *normal* or *Gaussian Distribution.*

If you flip a coin once, it can be heads or tails. What happens if you flip a coin two times and you call this double flip the *reading* or the *event* or the *outcome*? There are four possibilities for the outcome: HH, HT, TH and TT. There is a probability of 25 per cent that both flips will end up as heads, 25 per cent that the two will be tails, but 50 per cent that one will be a head and one will be a tail. If two heads are considered to be one extreme and two tails the other, then an even combination of heads and tails falls in the middle. When graphed accordingly, one gets:

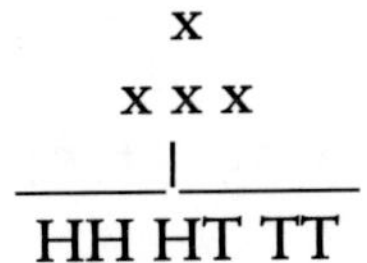

This gives you an opening which leads to the study of statistics and the normal distribution, because it is the influences, sometimes perceptible but often imperceptible, which lead to variations in any reading or taking of a measurement which leads to scatter around some mean—a mean often populated more frequently than either extreme because of the greater probability that the imperceptible influences will counteract each other.

Remember that if you actually flip a coin twice, that double flip is defined as the *event,* and if you carry out that event 4 times, you won't be guaranteed to get all of the combinations above in the frequencies indicated. Those represent probabilities only.

This program allows you to simulate the flip of a coin up to 100 times for a given trial and up to 1000 trials. Give it a try and vary the parameters to your liking. Use your word processor to read and/or print the output from the file you define. Notice that although there is a clear regression to a 50/50 mix of heads and tails, there is random variance of the mean, back and forth.

Exercise: Answer the following questions:

1. Execute the program so that the event (called a "dataset" in the program) is defined as 100 flips of a coin and define the number of events equal to 1000.
2. Draw a histogram containing the data produced by the computer

program. The x axis should be the number of heads per event and the y axis should be the number of events.

3. Determine the mean number of heads. (There is a shortcut which you must use to calculate the mean. Otherwise you'll be adding numbers of heads all night.)
4. Determine the standard deviation of the number of heads. (In this calculation there is a shortcut which you must use; it is similar in concept to the shortcut in 3, above.)
5. It is often said in books on applied statistics that the probable difference between x-bar (the mean of a small sample) and mu (the mean of a population) decreases rapidly as the number of measurements in a sample increases and that by the time the total number of observations in the sample reaches 20 to 30, this difference is negligible. There's nothing like an experimental approach to test this claim. First of all, we might ask, just what is meant by negligible? Here are 25 values for the number of heads per event independently generated by the program:
52, 53, 55, 48, 55, 53, 52, 54, 52, 51, 46, 52, 49, 51, 52, 46, 52, 49, 51, 50, 47, 51, 46, 48, 50

Determine the mean number of heads for this smaller sample and the standard deviation. Would you agree that the difference between the mean for the 1000 generated events and the smaller sample of 25 events is "negligible?" For the purpose of this exercise, "negligible" is defined as less than one standard deviation.

This exercise gives you data clearly exhibiting the beginnings of a normal curve which illustrates the scatter of an infinite number of readings over a finite range in which there is a significant frequency of events; within this range is a smaller segment which includes 68 per cent of all events and is bounded by the location of the inflection points of the population distribution. This defined range is called the *standard deviation of the population*, or (sigma) and its value on either side of the mean encompasses 68 per cent of all readings. A distance of two standard deviations or 2 sigma, encompasses 96 per cent of all readings.

The Population Mean, MU, and the Sample Mean, x Bar

Here we have three figures. The first, on the left, shows a plot of the normal distribution function with a population mean, mu, equal to 50. The figure on the right shows the same distribution function except with the abscissa in units of z=(x-mu)/sigma. By normalizing, or dividing by the value of, each unit along the abscissa is equivalent to one standard deviation of the population. The third figure, on the left shows the results of 10000 events, each event the flip of a coin 100 times. Note that the figure on the left above has been adjusted so that the standard deviation is roughly equivalent to that shown in the figure below it.

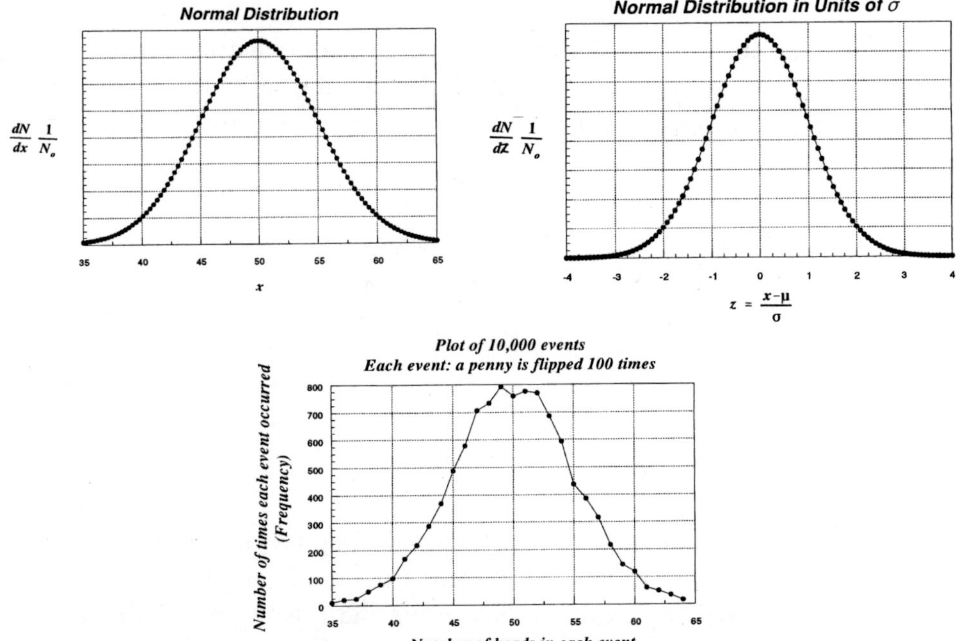

The population standard deviation, sigma and the sample standard deviation, s.

The population standard deviation which is an accepted measure of the precision of a population of data is given as:

$$\alpha = \sqrt{\frac{\sum_{i=1}^{N}(x_i - \mu)^2}{N}}$$

A small sample of data has a measure of precision given by the standard deviation, s, and uses a divisor of N-1 which is called the number of degrees of freedom. It represents the number of independent data points in the calculation of the standard deviation, *s*. If mu (the population mean) is unknown, then *x*-bar, or the mean of the sample, must be calculated. It is said that the number of degrees of freedom, originally N, is diminished by one by the calculation of x-bar because with the knowledge of x-bar there are only N-1 independent data points. Given x-bar, the Nth data point could be calculated from x-bar and the other N-1 data points.

$$s = \sqrt{\frac{\sum_{i=1}^{N}(x_i - \bar{x})^2}{N-1}}$$

There is an alternative designation for the standard deviation, s, of a small sample, for students who do not have calculators with the s function. It allows one to calculate the standard deviation without first having to calculate x-bar:

$$s = \sqrt{\frac{\sum_{i=1}^{N} x_i^2 - \frac{\left(\sum_{i=1}^{N} x_i\right)^2}{N}}{N-1}}$$

Standard ERROR of a Mean

The *standard error of a mean* is given the symbol sigma$_m$ in books on statistics and is related to the "scatter" of the means of small samples drawn from a larger collection. For the example below we shall use s_m as a substitute for that symbol. Small groups taken from a large collection will show a scatter around the mean of that collection. That scatter will diminish as the sample size increases in proportion to the function

$$\frac{1}{\sqrt{N}}$$

that is, if the standard deviation s_m of the group means is plotted vs 1/(square root N), the values ought to decrease linearly and approach zero as 1/(square root N) approaches zero. Let's see if there's any truth to the claim of linearity.

Exercise: Here is a table showing what happens if the file coin out 10k with 10000 events, each event being the flipping of a coin 100 times is read in a manner to calculate means and standard deviations of small groups of those events. Since the outcome of the events was the result of randomness and the file itself was not sorted, the groups are chosen in sequence. The following size of the groups was chosen: 3, 4, 5, 10, 50, 100, 250, 500, 1000, 2500, 5000, 10000. For groups of size 3 there are 3333 groups, for groups of size 4 there are 2500, for 5, 2000 and so on. The mean for all groups of the same size (the mean of the means), and the standard deviation produced by the individual means within each collection of groups were calculated.

For the class. On the graph paper below, label two convenient scales, putting standard deviation along the vertical axis and 1/(square root N) along the horizontal axis. Then plot the standard deviation of group means from the mean for all groups against the values of 1/(square root N). How does this agree with statistical theory?

Group size (N)	*SQRT (N)*	*1/SQRT (N)*	*Mean for all groups*	*Std. deviation of group means.*
1	1	1	50.039	4.99
2	1.414	0.707	50.039	3.53
3	1.732	0.577	50.039	2.93
4	2	0.500	50.039	2.49
5	2.236	0.447	50.039	2.25
10	3.162	0.316	50.039	1.58
50	7.071	0.141	50.039	0.68
100	10	0.10	50.039	0.48
250	15.81	0.063	50.039	0.31
500	22.36	0.045	50.039	0.22
1000	31.62	0.032	50.039	0.17
2500	50	0.020	50.039	0.13
5000	70.71	0.014	50.039	0.11
10000	100	0.010	50.039	—

The results of several experimental determinations and the subsequent calculation of the standard deviation says *nothing* about the *true value*. It is a measure solely of the reliability of the method being used. There is nevertheless what appears to be an irresistible tendency following some determination in analytical chemistry to link the standard deviation between individual determinations with their "trueness," but that tendency must be resisted. The standard deviation between individual determinations says nothing about the possibility of systematic error in the determination. It is the reliability of the method which is reflected by s.

Standard Deviation from Pooled Data

Rather than simply relying on some average standard deviation for several small samples of data, statistical theory tells us that we can go one even better. We can determine a *standard deviation from pooled data* even though those data may represent determinations on different unknowns. Consider the samples of carbonate analyzed by CHE230 students during the fall semester, 1999. Four unknowns were used, so there were four different per cent sodium carbonate values to be determined. But the same method was used for each. The level of precision would be expected to be the same for each unknown, even though their percent sodium carbonate values were widely separated. So the mean percent sodium carbonate is of no great concern, but the precision, in principle, ought to remain the same. The formula which allows us to determine a more characteristic standard deviation of the method, from pooled data, is

$$s_{polled} = \sqrt{\frac{\sum_{i=1}^{N_\alpha}\left(x_i - \bar{x}_\alpha\right)^2 + \sum_{j=1}^{N_\beta}\left(x_j - \bar{x}_\beta\right)^2 + \sum_{k=1}^{N_\alpha}\left(x_k - \bar{x}_\gamma\right)^2 ....}{N_\alpha + N_\beta + N_\gamma + ... - N_g}}$$

N_{alpha} is the number of elements in group alpha, N_{beta} is the number of elements in group beta, etc. and N_g is the number of groups that are pooled.

Exercise 1

During the fall of 1999, the following results were obtained from students carrying out the determination of sodium carbonate in samples of soda ash:

Student	*Sample 1*	*Sample 2*	*Sample 3*	*x-bar*	*s*	*(x_i—xbar)²*
1 (alpha)	39.46	39.50	39.32	39.43	0.09	0.0179
2 (beta)	29.60	29.60	29.49	29.56	0.06	0.0081
3 (gamma)	22.09	21.74	21.98	21.94	0.18	0.0641
4 (delta)	40.15	40.27	40.22	40.21	0.06	0.0073
5 (epsilon)	20.88	20.98	20.81	20.89	0.09	0.0146
6 (zeta)	38.91	38.94	38.90	38.92	0.02	0.0009
7 (eta)	32.38	31.84	32.88	32.37	0.52	0.5411
8 (theta)	38.19	38.18	38.18	38.18	0.01	0.0001
9 (iota)	48.88	48.83	48.27	48.66	0.34	0.2294
10 (kappa)	28.74	28.77	28.76	28.76	0.02	0.0005
11 (lambda)	49.90	49.81	49.79	49.83	0.06	0.0069
12 (μ)	50.42	50.38	50.45	50.42	0.04	0.0025

Using the equation above, determine the pooled standard deviation. Note that the mean for each set is used only to determine the square of the sum of the deviations of each result and that the *overall mean is of no importance* where a pooled standard deviation is to be calculated.

Exercise 2

Write a computer program to determine the pooled standard deviation of the data in the file coinout.10k. With group sizes set at 2, 3, 4, 5, 10, 50, 100, 250, 500, 1000, 2500 and 5000. Comment on the values you receive and comment on differences you observe with the standard deviation of the means offered in the table earlier in this chapter.

There are some alternative terms used for expressing the precision of sets or groups of data elements. They are important to know.

The Variance, s^2

$$s = \frac{\sum_{i=1}^{N}(x_i - \bar{x})^2}{N-1}$$

The Relative Standard Deviation

The RSD is

$$RSD\ (parts\ per\ thausand) = \frac{s}{x} \times 1000$$

The Coefficient of Variation, CV is Simply the RSD in Per cent

$$CV = \frac{s}{x} \times 100$$

The spread or range, w, is simply the difference between the lowest and highest values in a data set. There is a convenient table to estimate the standard deviation using the value of w. It is suprisingly in agreement with the calculated value for many applications:

$$S_{estimated} = K \times w$$

N	2	3	4	5	6	7	8	9	10	11	12
k	0.89	0.59	0.49	0.43	0.39	0.37	0.35	0.34	0.32	0.32	0.31

Exercise 3

A water well known to contain particularly high levels of iron has five samples drawn for spectrophotometric analysis. The results show the following levels of iron in parts per million:

134, 147, 125, 131, 152

Determine the mean, the standard deviation, the variance, the RSD, the CV the spread and the estimated standard deviation from the table above.

Exercise 4

A Quantitative Analysis student determines Cu in brass and obtains the following per cent copper: 87.85, 87.70, 87.95, 87.78 and 87.65. Calculate the same quantities requested in *Exercise* 3 above.

Determining the Calculated Uncertainty from Individual Values

The model used for the determination of the calculated uncertainty from individual values comes to us from vector algebra and is based on the assumption that if two values are to be added together, the factors resulting in small uncertainties are not linked. That is to say, a positive error in some mass reading would not somehow lead to a positive error in a later volume reading. Moreover, this model applies an estimate that there is a "most probable" error in which there might be some cancellation of effects which bring about the uncertainties. For example, in determining the anticipated error in four truckloads of oranges, each with an uncertainty of ±10 oranges, one would not assume the most probable error for the total load to be ±40 oranges because of the effects of probable cancellation of the total uncertainty between truckloads. In the final analysis, the two formulas used, one for addition and subtraction and the other for multiplication and division ought to be considered to be "best estimates" of anticipated calculated errors:

For addition and subtraction: Consider the operation $y = a + b - c + d$. Since uncertainties are considered to work in either direction symmetrically, the sign of the operation is unimportant and the function giving the uncertainty in the operation, v_y is (we shall use v here instead of s because the uncertainty expressed in each parameter may or may not be a standard deviation).

$$v_y = \sqrt{v_a^2 + v_b^2 + v_c^2 + v_d^2}$$

Note that this formula is the diagonal of a 4-dimensional rectangular solid. It is felt that such a function gives a more probable estimate of the uncertainty owing to some cancellation of error effects rather than that which would be achieved simply by adding all of the uncertainties together. In other words, it would be overkill on error estimation to state that $v_y = v_a + v_b + v_c + v_d$, because of the presumption of partial cancellation.

For multiplication and division, the formula comes to us from both a vector algebra approach as above and a differential calculus model (be forewarned that the calculus model is something of a finesse to get to the rationalization used above for addition and subtraction calculations):

Consider the Operation

$$y = \frac{a \times b}{p \times q}$$

Taking the natural logarithm of both sides, one gets

$$\ln y = \ln a + \ln b - \ln p - \ln q$$

The derivative of in y yields:

$$d(\ln y) = \frac{dy}{y} = \frac{da}{a} + \frac{db}{b} - \frac{dp}{p} - \frac{dq}{q}$$

This form is then estimated to be the *relative uncertainties* in each case:

$$\frac{\Delta y}{y} = \frac{\Delta a}{a} + \frac{\Delta b}{b} - \frac{\Delta p}{p} - \frac{\Delta q}{q} = \frac{v_a}{a} + \frac{v_b}{b} - \frac{v_p}{p} \frac{v_q}{q}$$

But since the relative uncertainties are assumed to be symmetric around the determined values and can work in either direction, these relative uncertainties are added together regardless of the sign in the derivative. Moreover, since the *most probable* error would result in some cancellation between the errors, following the model used in addition and subtraction operations above, the form for estimating the error in a process of multiplication and division is (substituting v for in each case)

$$\frac{v_y}{y} = \sqrt{\left(\frac{v_a}{a}\right)^2 + \left(\frac{v_b}{b}\right)^2 + \left(\frac{v_p}{p}\right)^2 + \left(\frac{v_q}{q}\right)^2}$$

When a problem involving all four operations is to be solved, the uncertainty in the final result ought to be calculated in the same order as the calculation of the result.

Some Exercises in Significant Figures

For the exercises below, consider each number presented to be precise to ±1 in the last digit.

Exercise 1

The sum of 3.4 + 0.020 + 7.31. To how many significant figures ought the result be reported and what is the calculated uncertainty?

Exercise 2

Consider the operation (consider all factors to be experimentally determined) (38.5 x 27)/252.3. To how many significant figures ought the result be reported and what is the calculated uncertainty?

Exercise 3

Consider the operation (again, all factors have been determined experimentally) (24 × 4.52)/100.0. To how many significant figures ought the result be reported?

The *rule of thumb* for multiplication and division is to report the result to the same number of significant figures as the smallest number of significant figures in any of the original factors, but problems arise if the result just moves beyond 1xxx.xxx or drops just below 9 xx.xxx.

Rounding Rules

If the digit following the digit to be rounded is 6, 7, 8 or 9, round to the next highest digit. If the digit following the digit to be rounded is 0, 1, 2, 3 or 4, do not change the rounding digit. If the digit following the digit to be rounded is 5, round to the even integer.

Exercise 1

Consider the following experimental values. An experimental value might be a direct observation or it might be a calculated value based on experimental observations. Next consider the estimated uncertainties. Report each experimental value as one ought to report it based on the uncertainties. State the number of significant figures indicated by the reported value. Calculate the relative uncertainty in percent in each case.

Experimental value	*Uncertainty*	*Reported value*	*Sig. Figures*	*Relative uncertainty*
3.827	±0.04			
0.08831	±0.02			
0.0243	±0.003			
2000	±10			
3.85	±0.02			
8.735	±0.01			

Significant Figure Rules with Logarithms

Two rules to remember here. (1) The logarithm ought to be reported to the same number of digits *after* the decimal point as there are digits in the original number, and (2) when taking an antilogarithm, the antilogarithm ought to be reported to a number of digits equal to the number of digits *after* the decimal point in the logarithm.

These rules work in most cases. When in doubt consider the relationship $y = \ln x$. The question is what uncertainty in y ought to be reported, knowing the uncertainty in x? Consider the calculus notation:

$$dy = d \ln x = \frac{dx}{x}$$

5

Radiological and Toxicological Chemistry

5.1 Radiology

Radiology is the specialty directing medical imaging technologies to diagnose and sometimes treat diseases. Originally it was the aspect of medical science dealing with the medical use of electromagnetic energy emitted by X-ray machines or other such radiation devices for the purpose of obtaining visual information as part of medical imaging. Radiology that involves use of x-ray is called roentgenology.

Wilhelm Conrad Röntgen first discovered x-radiation on 8 November 1895 at the Physical Institute of Wuerzburg University. He named the radiation he had discovered "X-radiation". This term is still in use today in the Anglo-American region. His work was first published in a meeting protocol of the Wuerzburg Physical-Medical Society in the 1895 volume; the article was submitted by W. C. Röntgen on 28 December 1895.

Today, following extensive training, radiologists direct an array of imaging technologies (such as ultrasound, computed tomography (CT) nuclear medicine, and magnetic resonance imaging) to diagnose or treat disease. Interventional radiology is the performance of (usually minimally invasive) medical procedures with the guidance of imaging technologies. The acquisition of medical imaging is usually carried out by the radiographer or radiologic technologist. Outside of the medical field, radiology also encompasses the examination of the inner structure of objects using X-rays or other penetrating radiation.

Subdivisions

As a medical specialty, radiology can be classified broadly into Diagnostic radiology and Therapeutic radiology.

- Diagnostic radiology is the interpretation of images of the human body to aid in the diagnosis or prognosis of disease. It is divided into subfields by anatomic location and in some cases method:

  - Chest radiology.
  - Abdominal & Pelvic radiology. Sometimes together termed "Body Imaging."
  - Interventional radiology uses imaging to guide therapeutic and angiographic procedures. Also known as Vascular & Interventional radiology.
  - Neuroradiology is the sub-specialty in the field of central nervous system, *i.e.* brain and spinal cord, peripheral nervous system, osseous spine and its neural contents, and head and neck imaging.

    - Interventional Neuroradiology uses imaging to guide therapeutic and diagnostic angiographic procedures in the head, neck and spine.

  - Musculoskeletal radiology is the sub-specialty in the field of bone, joint, and muscular imaging.
  - Pediatric radiology.
  - Mammography Subdivision of radiology that images the breast tissue.
  - Emergency radiology. Subdivision of radiology involved in the diagnosis and treatment of acutely ill or injured patients.
  - Nuclear Medicine is a subdivision of radiology that uses radioisotopes in the characterization of lesions and disease processes, and often yields functional information.

- A Radiologist is a subspecialty physician trained in all areas of diagnostic radiology. Board certification is earned through the American Board of Radiology (ABR).

  - Nuclear Medicine, Interventional radiology, Neuroradiology and Pediatric radiology have optional subspecialty Board qualifications under the American Board of Radiology.
  - Certification in Nuclear Medicine alone can be earned as a non-radiologist physician through the American Board of Nuclear Medicine.

- Therapeutic radiology utilizes radiation (radiation therapy) for therapy of diseases such as cancer.

- While originally encompassed within radiology, radiation oncology is now a separate field.
- Radiation Oncology specialty certification is earned through the American Board of Radiology.

Acquisition of Radiological Images

Patients have the following procedures to provide images for Radiological decisions to be made.

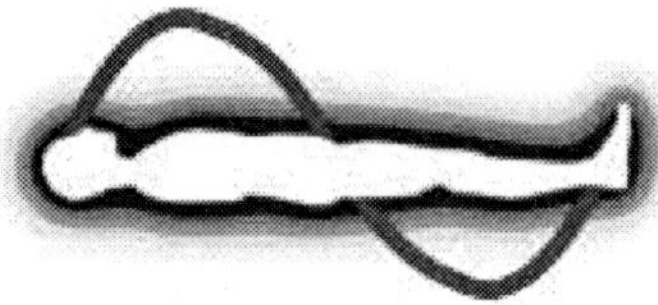

Fig. 5.1: Medical imaging.

Projection (plain) Radiography

Radiographs (or Roentgenographs, named after the discoverer of *x*-rays, Wilhelm Conrad Röntgen (1845-1923) are often used for evaluation of bony structures and soft tissues. An *x*-ray machine directs electromagnetic radiation upon a specified region in the body. This radiation tends to pass through less dense matter (air, fat, muscle, and other tissues), but is absorbed or scattered by denser materials (bones, tumors, lungs affected by severe pneumonia). In Film-Screen Radiography, radiation which has passed through a patient then strikes a cassette containing a screen of fluorescent phosphors and exposes *x*-ray film. Areas of film exposed to higher amounts of radiation will appear as black or grey on X-ray film while areas exposed to less radiation will appear lighter or white. In Computed Radiography (CR), the x-rays passing through the patient strike a sensitized plate which is then read and digitized into a computer image by a separate machine. In Digital Radiography the *x*-rays strike a plate of *x*-ray sensors producing a digital computer image directly. While all three methods are currently in use, the trend in the U.S. is away from film and toward digital imaging.

Plain radiography was the only imaging modality available during the first 50 years of Radiology. It is still the first study ordered in evaluation of the lungs, heart and skeleton because of its wide availability, speed and relative low cost.

Fluoroscopy

Fluoroscopy and angiography are special applications of X-ray imaging, in which a fluorescent screen or image intensifier tube is connected to a closed-circuit television system, which allows real-time imaging of structures in motion or augmented with a radiocontrast agent. Radiocontrast agents are administered, often swallowed or injected into the body of the patient, to delineate anatomy and functioning of the blood vessels, the genitourinary system or the gastrointestinal tract. Two radiocontrasts are presently in use. Barium (as BaSO4) may be given orally or rectally for evaluation of the GI tract. Iodine, in multiple proprietary forms, may be given by oral, rectal, intraarterial or intravenous routes. These radiocontrast agents strongly absorb or scatter X-ray radiation, and in conjunction with the real-time imaging allows demonstration of dynamic processes, such as peristalsis in the digestive tract or blood flow in arteries and veins. Iodine contrast may also be concentrated in abnormal areas more or less than in normal tissues and make abnormalities (tumors, cysts, inflammation) more conspicuous. Additionally, in specific circumstances air can be used as a contrast agent for the gastrointestinal system and carbon dioxide can be used as a contrast agent in the venous system; in these cases, the contrast agent attenuates the X-ray radiation less than the surrounding tissues.

CT Scanning

CT imaging uses X-rays in conjunction with computing algorithms to image the body. In CT, an X-ray generating tube opposite an X-ray detector (or detectors) in a ring shaped apparatus rotate around a patient producing a computer generated cross-sectional image (tomogram). CT is acquired in the axial plane, while coronal and sagittal images can be rendered by computer reconstruction. Radiocontrast agents are often used with CT for enhanced delineation of anatomy. Intravenous contrast can allow 3D reconstructions of arteries and veins. Although radiographs provide higher spatial resolution, CT can detect more subtle variations in attenuation of X-rays. CT exposes the patient to more ionizing radiation than a radiograph. Spiral Multi-detector CT utilizes 8, 16 or 64 detectors during continuous motion of the patient through the radiation beam to obtain much finer detail images in a shorter exam time. With computer manipulation these images can be reconstructed into 3D images of carotid, cerebral and coronary arteries. Faster scanning times in modern equipment has been associated with increased utilization.

The first commercially viable CT scanner was invented by Sir Godfrey

Hounsfield at EMI Central Research Labs, Great Britain in 1972. EMI owned the distribution rights to The Beatles music and it was their profits which funded the research. Sir Hounsfield and Alan McLeod McCormick shared the Nobel Prize for Medicine in 1979 for the invention of CT scanning. The first CT scanner in North America was installed at the Mayo Clinic in Rochester, MN in 1972.

Ultrasound

Medical ultrasonography uses ultrasound (high-frequency sound waves) to visualize soft tissue structures in the body in real time. No ionizing radiation is involved, but the quality of the images obtained using ultrasound is highly dependent on the skill of the person (ultrasonographer) performing the exam. Ultrasound is also limited by its inability to image through air (lungs, bowel loops) or bone. The use of ultrasound in medical imaging has developed mostly within the last 30 years. The first ultrasound images were static and two dimensional (2D), but with modern-day ultrasonography 3D reconstructions can be observed in real-time; effectively becoming 4D.

Because ultrasound does not utilize ionizing radiation, unlike radiography, CT scans, and nuclear medicine imaging techniques, it is generally considered safer. For this reason, this modality plays a vital role in obstetrical imaging. Fetal anatomic development can be thoroughly evaluated allowing early diagnosis of many fetal anomalies. Growth can be assessed over time, important in patients with chronic disease or gestation-induced disease, and in multiple gestations (twins, triplets etc.). Color-Flow Doppler Ultrasound measures the severity of peripheral vascular disease and is used by Cardiology for dynamic evaluation of the heart, heart valves and major vessels. Stenosis of the carotid arteries can presage cerebral infarcts (strokes). DVT in the legs can be found via ultrasound before it dislodges and travels to the lungs (pulmonary embolism), which can be fatal if left untreated. Ultrasound is useful for image-guided interventions like biopsies and drainages such as thoracentesis). It is also used in the treatment of kidney stones (renal lithiasis) via lithotripsy. Small portable ultrasound devices now replace peritoneal lavage in the triage of trauma victims by directly assessing for the presence of hemorrhage in the peritoneum and the integrity of the major viscera including the liver, spleen and kidneys. Extensive hemoperitoneum (bleeding inside the body cavity) or injury to the major organs may require emergent surgical exploration and repair.

MRI (Magnetic Resonance Imaging)

MRI uses strong magnetic fields to align spinning atomic nuclei (usually hydrogen protons) within body tissues, then uses a radio signal to disturb the axis of rotation of these nuclei and observes the radio frequency signal generated as the nuclei return to their baseline states plus all surounding areas. The radio signals are collected by small antennae, called coils, placed near the area of interest. An advantage of MRI is its ability to produce images in axial, coronal, sagittal and multiple oblique planes with equal ease. MRI scans give the best soft tissue contrast of all the imaging modalities. With advances in scanning speed and spatial resolution, and improvements in computer 3D algorithms and hardware, MRI has become an essential tool in musculoskeltal radiology and neuroradiology.

One disadvantage is that the patient has to hold still for long periods of time in a noisy, cramped space while the imaging is performed. Claustrophobia severe enough to terminate the MRI exam is reported in up to 5 per cent of patients. Recent improvements in magnet design including stronger magnetic fields (3 teslas), shortening exam times, wider, shorter magnet bores and more open magnet designs, have brought some relief for claustrophobic patients. However, in magnets of equal field strength there is often a trade-off between image quality and open design. MRI has great benefit in imaging the brain, spine, and musculoskeletal system. The modality is currently contraindicated for patients with pacemakers, cochlear implants, some indwelling medication pumps, certain types of cerebral aneurysm clips, metal fragments in the eyes and some metallic hardware due to the powerful magnetic fields and strong fluctuating radio signals the body is exposed to. Areas of potential advancement include functional imaging, cardiovascular MRI, as well as MR image guided therapy.

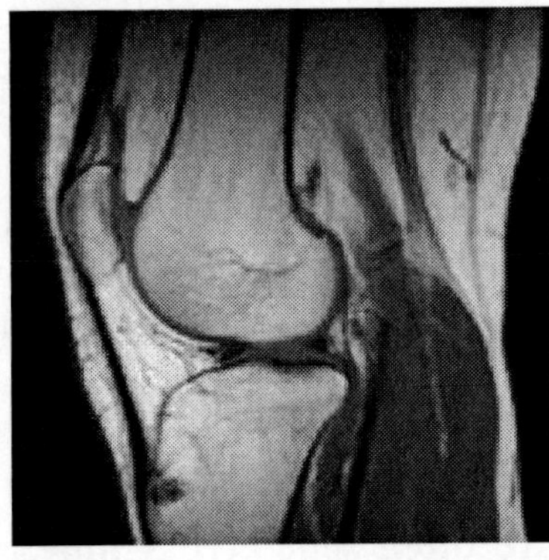

Fig. 5.2: MR image of human knee.

Nuclear Medicine

Nuclear medicine imaging involves the administration into the patient of radiopharmaceuticals consisting of substances with affinity for certain body tissues labeled with radioactive tracer. The most commonly used tracers are Technetium-99m, Iodine-123, Iodine-131 and Thallium-201. The heart, lungs, thyroid, liver, gallbladder, and bones are commonly evaluated for particular conditions using these techniques. While anatomical detail is limited in these studies, nuclear medicine is useful in displaying physiological function. The excretory function of the kidneys, iodine concentrating ability of the thyroid, blood flow to heart muscle, etc. can be measured. The principal imaging device is the gamma camera which detects the radiation emitted by the tracer in the body and displays it as an image. With computer processing, the information can be displayed as axial, coronal and sagittal images (SPECT images). In the most modern devices Nuclear Medicine images can be fused with a CT scan taken quasi-simultaneously so that the physiological information can be overlaid or co-registered with the anatomical structures to improve diagnostic accuracy.

PET scanning also falls under "nuclear medicine." In PET scanning, a radioactive biologically-active substance, most often Fluorine-18 Fluorodeoxyglucose, is injected into a patient and the radiation emitted by the patient is detected to produce multi-planar images of the body. Metabolically more active tissues, such as cancer, concentrate the active substance more than normal tissues. PET images can be combined with CT images to improve diagnostic accuracy.

The applications of nuclear medicine can include bone scanning which traditionally has had a strong role in the work-up/staging of cancers. Myocardial perfusion imaging is a sensitive and specific screening exam for reversible myocardial ischemia, which when present requires angiographic confirmation and potentially life-saving balloon angioplasty, stenting or cardiac bypass grafting. Molecular Imaging is the new and exciting frontier in this field.

Teleradiology

Teleradiology is the transmission of radiographic images from one location to another for interpretaion by a radiologist. It is most often used to allow rapid interpretation of emergency room, ICU and other emergent examinations after hours of usual operation, at night and on weekends. In these cases the images are often sent across time zones,(Spain, Australia,

India) with the receiving radiologist working his normal daylight hours. Teleradiology can also be utilized to obtain consultation with an expert or sub-specialist about a complicated or puzzling case.

Teleradiology requires a sending station, high speed Internet connection and high quality receiving station. At the sending station, plain radiographs are passed through a digitizing machine before transmission, while CT scans, MRIs, Ultrasounds and Nuclear Medicine scans can be sent directly as they are already a stream of digital data. The computer at the receiving end will need to have a high-quality display screen that has been tested and cleared for clinical purposes. Sometimes the receiving computer will have a printer so that images can be printed for convenience. The interpreting radiologist will then fax or e-mail the radiology report to the requesting physician.

The advantages of after-hours Teleradiology is the more expert and rapid examination of radiologic studies for the benefit of emergency room and hospitalized patients and their physicians. The disadvantages included limited contact between the ordering physician and the radiologist, uncertainty about the radilogists's abilities and the cost. Laws and regulations concerning the use of teleradiology vary among the states, with some states requiring a license to practise medicine in the state sending the radiologic exam while others just require an American license. Many states require the teleradiology report to be a only preliminary report and require a final report by a hospital staff radiologist.

Radiologist Training

United States

Diagnostic radiologists must complete prerequisite undergraduate training, four years of medical school, and five years of post-graduate training. The first postgraduate year is usually a transitional year of various rotations, but is sometimes a preliminary internship in medicine or surgery. A four-year diagnostic radiology residency follows. During this residency, the radiology resident must pass a medical physics board exam covering the science and technology of ultrasounds, CTs, x-rays, nuclear medicine, and MRI. Core knowledge of the radiologist includes radiobiology which is the effects of ionizing radiation on the living organisms and specifically humans. Near the completion of their residency, the radiologist in training is eligible to take board examinations (written and oral) given by the American Board of Radiology.

Following completion of residency training, radiologists either begin their practice or enter into sub-speciality training programs known as fellowships. Examples of sub-speciality training in radiology include abdominal imaging, thoracic imaging, CT/Ultrasound, MRI, musculoskeletal imaging, interventional radiology, neuroradiology, interventional neuroradiology, pediatric radiology, mammography and women's imaging. Fellowship training programs in radiology are usually 1 or 2 years in length.

Radiologists generally achieve a higher level of compensation than many medical specialties as well as a highly desirable *regular* work schedule that often does not involve many weekend or night hours. The introduction of teleradiology has significantly improved the working environment and schedules of radiologists, essentially distributing the increasing workflow into shifts. Those seeking residency positions find that entry into this field of medicine is highly competitive. The field is rapidly expanding due to advances in computer technology which is closely linked to modern imaging.

The exams (radiography) are usually performed by radiologic technologists, (also known as diagnostic radiographers) who in the United States have a 2-year Associates Degree and the UK a 3 year Honours Degree.

Veterinary radiologists are veterinarians that specialize in the use of X-rays, ultrasound, MRI and nuclear medicine for diagnostic imaging or treatment of disease in animals. Veterinary radiologists are certified in either diagnostic radiology or radiation oncology by the American College of Veterinary Radiology.

Canada

As in the United States, diagnostic radiologists usually have four years of undergraduate university, followed by three or four years of medical school and five years of residency training, one year of which can be in a field unrelated to medical imaging. Specialist examinations at the end of training are administered by the Royal College of Physicians and Surgeons of Canada. Graduates of Canadian residency programs are considered equivalent to their American counterparts, with many graduates opting to sit for American Board examinations as well. A majority of graduates from Canadian residency training programs go on to do an additional one or two years of post graduate fellowship training to allow for further subspecialization. Residency training in the United States is considered equivalent to training in Canada and US graduates are eligible to sit for the Canadian boards examinations with certain provisions related to basic medical training such as completing medical school in Canada or a year of general rotating internship in an accredited Canadian teaching hospital.

5.2 Radiation

Radiation, as used in physics, is energy in the form of waves or moving subatomic particles emitted by an atom or other body as it changes from a higher energy state to a lower energy state. Radiation can be classified as *ionizing* or *non-ionizing radiation*, depending on its effect on atomic matter. The most common use of the word "radiation" refers to ionizing radiation. Ionizing radiation has enough energy to ionize atoms or molecules while non-ionizing radiation does not. *Radioactive material* is a physical material that emits ionizing radiation.

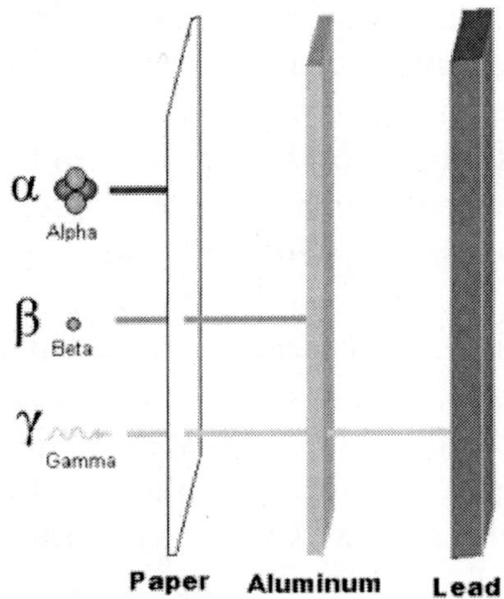

Fig. 5.3: This shows three different types of radiation and their penetration levels.

Types of Radiation

There are three principal types of ionizing radiation: alpha, beta and gamma radiation. They are all emitted from the nucleus of an unstable atom. Less commonly encountered are spontaneous nuclear fission; positron emission, which is utilized in positron emission tomography; and neutron emission. Electron capture results in the spontaneous emission of an X-ray. Certain isotopes of radium have a decay mode where they emit an entire $^{12}C_6$ nucleus.

Discovery

Wilhelm Röntgen is credited with the discovery of X-Rays. Henri Becquerel found that uranium salts caused fogging of an unexposed photographic plate, and Marie Curie discovered that only certain elements gave off these rays of energy. He named this behaviour radioactivity.

5.3 Ionizing Radiation

Ionizing radiation consists of highly-energetic particles or waves that can detach (ionize) at least one electron from an atom or molecule. Ionizing ability depends on the energy of individual particles or waves, and not on their number. A large flood of particles or waves will not, in the most common situations, cause ionization if the individual particles or waves are not by themselves ionizing.

Examples of ionizing radiation are energetic beta particles, neutrons, and alpha particles. The ability of light waves (photons) to ionize an atom or molecule varies across the electromagnetic spectrum. X-rays and gamma rays can ionize almost any molecule or atom; far ultraviolet light can ionize many atoms and molecules; near ultraviolet and visible light are ionizing to very few molecules; microwaves and radio waves are non-ionizing radiation. Visible light is so ubiquitous that molecules that are ionized by it often react nearly spontaneously unless protected by materials that block the visible spectrum. Examples include photographic film and some molecules involved in photosynthesis.

Ionizing radiation has many practical uses in medicine, research, construction, and other areas, but presents a health hazard if used improperly. If enough ionizations occur in a biological system, they can be destructive, by such means as causing DNA damage in individual cells. Extensive doses of ionizing radiation have been shown to have a mutating effect on the victim's descendants. Both helpful and harmful aspects of ionizing radiation are discussed below.

Types of Radiation

Alpha (α) radiation consists of Helium-4 (^4He) nuclei and is stopped by a sheet of paper. Beta (β) radiation, consisting of electrons, is halted by an

Fig. 5.4: Radiation hazard symbol.

Fig. 5.5: Ionizing radiation hazard symbol. (recently introduced)

aluminium plate. Gamma (**γ**) radiation, consisting of energetic photons, is eventually absorbed as it penetrates a dense material.

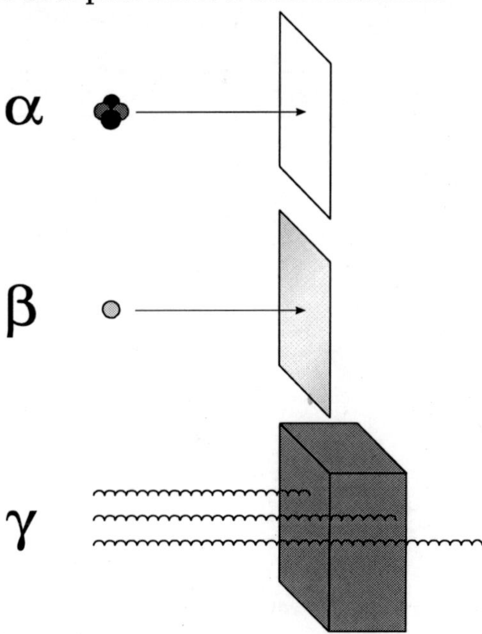

Fig. 5.6: Types of radiation.

Ionizing radiation is produced by radioactive decay, nuclear fission and nuclear fusion, by extremely hot objects (*e.g.*, the Sun produces far-ultraviolet light), and by particle accelerators that may produce, for example, fast electrons or protons or bremsstrahlung or synchrotron radiation.

In order for a particle to be ionizing, it must both have a high enough energy and interact with the atoms of a target. Photons interact strongly with charged particles, so photons of sufficiently high energy also are ionizing. The energy at which this begins to happen with photons (light) is in the ultraviolet region of the electromagnetic spectrum; sunburn is one of the effects of ionization. Charged particles such as electrons, positrons, and alpha particles also interact strongly with electrons of an atom or molecule. Neutrons, on the other hand, do not interact strongly with electrons, and so they cannot directly cause ionization by this mechanism. However, fast neutrons will interact with the protons in hydrogen (in the manner of a billiard ball hitting another, sending it away with all of the first ball's energy of motion), and this mechanism produces proton radiation (fast protons). These protons are ionizing because of their strong interaction with electrons in matter. A neutron can also interact with an atomic nucleus, depending on the nucleus and the neutron's velocity; these reactions happen with fast neutrons and slow neutrons, depending on the situation. Neutron interactions in this manner often produce radioactive nuclei, which produce ionizing radiation when they decay.

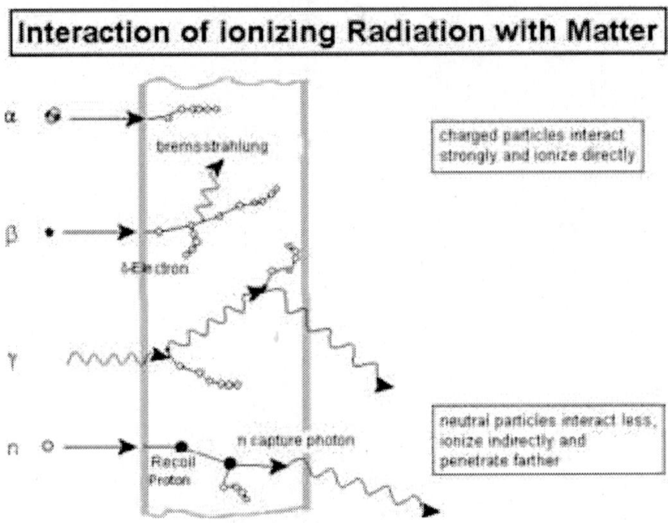

Fig. 5.7: Interaction of ionizing Radiation with Matter.

In the picture at left, gamma rays are represented by wavy lines, charged particles and neutrons by straight lines. The little circles show where ionization processes occur.

An ionization event normally produces a positive atomic ion and an electron. High-energy beta particles may produce bremsstrahlung when passing through matter, or secondary electrons (ä-electrons); both can ionize in turn.

Unlike alpha or beta particles (see particle radiation), gamma rays do not ionize all along their path, but rather interact with matter in one of three ways: the photoelectric effect, the Compton effect, and pair production. By way of example, the figure shows Compton effect: two Compton scatterings that happen sequentially. In every scattering event, the gamma ray transfers energy to an electron, and it continues on its path in a different direction and with reduced energy.

In the same figure, the neutron collides with a proton of the target material, and then becomes a fast recoil proton that ionizes in turn. At the end of its path, the neutron is captured by a nucleus in an (n, ã)-reaction that leads to a neutron capture photon.

The negatively-charged electrons and positively charged ions created by ionizing radiation may cause damage in living tissue. If the dose is sufficient, the effect may be seen almost immediately, in the form of radiation poisoning. Lower doses may cause cancer or other long-term problems. The effect of the very low doses encountered in normal circumstances (from both natural and artificial sources, like cosmic rays, medical X-rays and nuclear power plants) is a subject of current debate. A 2005 report released by the National Research Council (the BEIR VII report, summarized in) indicated that the overall cancer risk associated with background sources of radiation was relatively low.

Radioactive materials usually release alpha particles, which are the nuclei of helium, beta particles, which are quickly moving electrons or positrons, or gamma rays. Alpha and beta particles can often be stopped by a piece of paper or a sheet of aluminium, respectively. They cause most damage when they are emitted inside the human body. Gamma rays are less ionizing than either alpha or beta particles, and protection against gammas requires thicker shielding. The damage they produce is similar to that caused by X-rays, and include burns and also cancer, through mutations. Human biology resists germline mutation by either correcting the changes in the DNA or inducing apoptosis in the mutated cell.

Non-ionizing radiation is thought to be essentially harmless below the levels that cause heating. Ionizing radiation is dangerous in direct exposure, although the degree of danger is a subject of debate. Humans and animals can also be exposed to ionizing radiation internally: if radioactive isotopes are present in the environment, they may be taken into the body. For example, radioactive iodine is treated as normal iodine by the body and used by the thyroid; its accumulation there often leads to thyroid cancer. Some radioactive elements also bioaccumulate.

Uses of Ionizing Radiation

Ionizing radiation has many uses, such as to kill cancerous cells. However, although ionizing radiation has many applications, overuse can be hazardous to human health. For example, at one time, assistants in shoe shops used X-rays to check a child's shoe size, but this practice was halted when it was discovered that ionizing radiation was dangerous.

Technical uses of Ionizing Radiation

Since ionizing radiations can penetrate matter, they are used for a variety of measuring methods.

Radiography by means of gamma or X rays: This is a method used in industrial production. The piece to be radiographed is placed between the source and a photographic film in a cassette. After a certain exposure time, the film is developed and it shows internal defects of the material if there are any.

Gauges

Gauges use the exponential absorption law of gamma rays:

- Level indicators: Source and detector are placed at opposite sides of a container, indicating the presence or absence of material in the horizontal radiation path. Beta or gamma sources are used, depending on the thickness and the density of the material to be measured. The method is used for containers of liquids or of grainy substances
- Thickness gauges: if the material is of constant density, the signal measured by the radiation detector depends on the thickness of the material. This is useful for continuous production, like of paper, rubber, etc.

Applications using ionization of gases by radiation:

- To avoid the build-up of static electricity in production of paper, plastics, synthetic textiles, etc., a ribbon-shaped source of the alpha emitter ^{241}Am can be placed close to the material at the end of the production line. The source ionizes the air to remove electric charges on the material.
- *Smoke detector*: Two ionisation chambers are placed next to each other. Both contain a small source of ^{241}Am that gives rise to a small constant current. One is closed and serves for comparison, the other is open to ambient air; it has a gridded electrode. When smoke enters the open chamber, the current is disrupted as the smoke particles attach to the charged ions and restore them to a neutral electrical state. This reduces the current in the open chamber. When the current drops below a certain threshold, the alarm is triggered.
- Radioactive tracers for industry: Since radioactive isotopes behave, chemically, mostly like the inactive element, the behavior of a certain chemical substance can be followed by *tracing* the radioactivity. Examples:
    - Adding a gamma tracer to a gas or liquid in a closed system makes it possible to find a hole in a tube.
    - Adding a tracer to the surface of the component of a motor makes

it possible to measure wear by measuring the activity of the lubricating oil.

Biological and Medical Applications of Ionizing Radiation

In biology, radiation is mainly used for sterilization, and enhancing mutations. For example, mutations may be induced by radiation to produce new or improved species. A very promising field is the sterile insect technique, where male insects are sterilized and liberated in the chosen field, so that they have no descendants, and the population is reduced.

Radiation is also useful in sterilizing medical hardware or food. The advantage for medical hardware is that the object may be sealed in plastic before sterilization. For food, there are strict regulations to prevent the occurrence of induced radioactivity. The growth of a seedling may be enhanced by radiation, but excessive radiation will hinder growth.

Electrons, x rays, gamma rays or atomic ions may be used in radiation therapy to treat malignant tumors (cancer).

Tracer methods are used in nuclear medicine in a way analogous to the technical uses mentioned above.

Natural Background Radiation

Natural background radiation comes from four primary sources: cosmic radiation, solar radiation, external terrestrial sources, and radon.

Cosmic Radiation

The Earth, and all living things on it, are constantly bombarded by radiation from outside our solar system. This cosmic radiation consists of positively-charged ions from protons to iron nuclei. The energy of this radiation can far exceed that which humans can create even in the largest particle accelerators. This radiation interacts in the atmosphere to create secondary radiation that rains down, including *x*-rays, muons, protons, alpha particles, pions, electrons, and neutrons.

The dose from cosmic radiation is largely from muons, neutrons, and electrons, with a dose rate that varies in different parts of the world and based largely on the geomagnetic field, altitude, and solar cycle. The cosmic-radiation dose rate on airplanes is so high that, according to the United Nations UNSCEAR 2000 Report (see links at bottom), airline workers receive more dose on average than any other worker, including those in nuclear power plants.

Solar Radiation

While most of the Sun's output consists of light (solar radiation), particle radiation is also produced and varies with the solar cycle. These particles are mostly protons with relatively low energies (10-100 keV). Their average composition is similar to that of the Sun itself. This represents significantly lower energy particles than come from cosmic rays. Solar particles vary widely in their intensity and spectrum, increasing in strength after some solar events such as solar flares. Further, an increase in the intensity of solar cosmic rays is often followed by a *decrease* in the galactic cosmic rays, called a Forbush decrease after their discoverer, the physicist Scott Forbush. These decreases are due to the solar wind which carries the Sun's magnetic field out further to shield the earth more thoroughly from cosmic radiation.

The ionizing component of solar radiation is negligible relative to other forms of radiation on Earth's surface.

External Terrestrial Sources

Most materials on Earth contain some radioactive atoms, even if in small quantities. Most of the terrestrial non-radon-dose one receives from these sources is from gamma-ray emitters in the walls and floors when inside a house, or rocks and soil when outside. The major radionuclides of concern for *terrestrial radiation* are potassium, uranium, and thorium. Each of these sources has been decreasing in activity since the birth of the Earth so that our present dose from potassium-40 is about ½ what it would have been at the dawn of life on Earth.

Radon

Radon-222 is produced by the decay of radium-226 which is present wherever uranium is found. Since radon is a gas, it seeps out of uranium-containing soils found across most of the world and may accumulate in well-sealed homes. It is often the single largest contributor to an individual's background radiation dose and is certainly the most variable from location to location. Radon gas could be the second largest cause of lung cancer in America, after smoking.

Human-Made Radiation Sources

Natural and artificial radiation sources are similar in their effects on matter. Above the background level of radiation exposure, the U.S. Nuclear

Regulatory Commission (NRC) requires that its licensees limit human-made radiation exposure for individual members of the public to 100 mrem (1 mSv) per year, and limit occupational radiation exposure to adults working with radioactive material to 5,000 mrem (50 mSv) per year.

The average exposure for Americans is about 360 mrem (3.6 mSv) per year, 81 per cent of which comes from natural sources of radiation. The remaining 19 per cent results from exposure to human-made radiation sources such as medical X-rays, most of which is deposited in people who have CAT scans. This compares with the average dose received by people in the UK of about 2.2 mSv. As already mentioned, an important source of natural radiation is radon gas, which seeps continuously from bedrock but can, because of its high density, accumulate in poorly ventilated houses.

The background rate for radiation varies considerably with location, being as low as 1.5 mSv/a (1.5 mSv per year) in some areas and over 100 mSv/a in others. People in some parts of Ramsar, a city in northern Iran, receive an annual absorbed dose from background radiation that is up to 260 mSv/a. Despite having lived for many generations in these high background areas, inhabitants of Ramsar show no significant cytogenetic differences compared to people in normal background areas. This has led to the suggestion that high but steady levels of radiation are easier for humans to sustain than sudden radiation bursts.

Some human-made radiation sources affect the body through direct radiation, while others take the form of radioactive contamination and irradiate the body from within.

Medical procedures, such as diagnostic X-rays, nuclear medicine, and radiation therapy are by far the most significant source of human-made radiation exposure to the general public. Some of the major radionuclides used are I-131, Tc-99, Co-60, Ir-192, and Cs-137. These are rarely released into the environment. The public also is exposed to radiation from consumer products, such as tobacco (polonium-210), building materials, combustible fuels (gas, coal, etc.), ophthalmic glass, televisions, luminous watches and dials (tritium), airport X-ray systems, smoke detectors (americium), road construction materials, electron tubes, fluorescent lamp starters, and lantern mantles (thorium).

Of lesser magnitude, members of the public are exposed to radiation from the nuclear fuel cycle, which includes the entire sequence from mining and milling of uranium to the disposal of the spent fuel. The effects of such exposure have not been reliably measured due to the extremely low doses involved. Estimates of exposure are low enough that proponents of nuclear power liken them to the mutagenic power of wearing trousers for two extra minutes per year (because heat causes mutation). Opponents use a cancer per dose model to assert that such activities cause several hundred cases of cancer per year.

In a nuclear war, gamma rays from fallout of nuclear weapons would probably cause the largest number of casualties. Immediately downwind of targets, doses would exceed 300 Gy per hour. As a reference, 4.5 Gy (around 15,000 times the average annual background rate) is fatal to half of a normal population, without medical treatment.

Occupationally exposed individuals are exposed according to the sources with which they work. The radiation exposure of these individuals is carefully monitored with the use of pocket-pen-sized instruments called dosimeters.

Some of the radionuclides of concern include cobalt-60, caesium-137, americium-241, and iodine-131. Examples of industries where occupational exposure is a concern include:

- Airline crew (the most exposed population)
- Fuel cycle
- Industrial radiography
- Nuclear medicine and medical radiology departments (including nuclear oncology)
- Nuclear power plants (during maintenance work near the reactor core)
- Research laboratories (government, university and private)

Biological Effects of Ionizing Radiation

The biological effects of radiation are thought of in terms of their effects on living cells. For low levels of radiation, the biological effects are so small they may not be detected in epidemiological studies. The body repairs many types of radiation and chemical damage. Biological effects of radiation on living cells may result in a variety of outcomes, including:

1. Cells experience DNA damage and are able to detect and repair the damage.
2. Cells experience DNA damage and are unable to repair the damage. These cells may go through the process of programmed cell death, or apoptosis, thus eliminating the potential genetic damage from the larger tissue.
3. Cells experience a nonlethal DNA mutation that is passed on to subsequent cell divisions. This mutation may contribute to the formation of a cancer.

Other observations at the tissue level are more complicated. These include:

> In some cases, a small radiation dose reduces the impact of a subsequent, larger radiation dose. This has been termed an 'adaptive response' and is related to hypothetical mechanisms of hormesis.

Hormesis

Radiation hormesis is the unproven theory that a low level of ionizing radiation (*i.e.* near the level of Earth's natural background radiation) helps "immunize" cells against DNA damage from other causes (such as free radicals or larger doses of ionizing radiation), and decreases the risk of cancer. The theory proposes that such low levels activate the body's DNA repair mechanisms, causing higher levels of cellular DNA-repair proteins to be present in the body, improving the body's ability to repair DNA damage. This assertion is very difficult to prove (using, for example, statistical cancer studies) because the effects of very low ionizing radiation levels are too small to be statistically measured amid the "noise" of normal cancer rates.

Therefore, the idea of radiation hormesis is considered unproven by regulatory bodies, which generally use the standard "linear, no threshold" (LNT) model, which states that risk of cancer is directly proportional to the dose level of ionizing radiation. The LNT model is safer for regulatory purposes because it assumes worst-case damage due to ionizing radiation; therefore, if regulations are based on it, workers might be over-protected, but they will never be under-protected.

At high ionizing radiation levels, such as the acute doses received near the Hiroshima and Nagasaki bomb blasts, the risk of cancer does increase roughly linearly with dose, which is the origin of the LNT model. Thus, there is a consensus that the LNT method should continue to be used because it is safer from a regulatory perspective and because the effects of very low radiation doses are too small to be measured statistically. See the National Academies Press book.

Chronic Radiation Exposure

Exposure to ionizing radiation over an extended period of time is called chronic exposure. The natural background radiation is chronic exposure, but a normal level is difficult to determine due to variations. Geographic location and occupation often affect chronic exposure.

Acute Radiation Exposure

Acute radiation exposure is an exposure to ionizing radiation which occurs during a short period of time. There are routine brief exposures, and the boundary at which it becomes significant is difficult to identify. Extreme examples include:

- Instantaneous flashes from nuclear explosions.
- Exposures of minutes to hours during handling of highly radioactive sources.
- Laboratory and manufacturing accidents.
- Intentional and accidental high medical doses.

The effects of acute events are more easily studied than those of chronic exposure. Chronic exposure is reactant.

Radiation Levels

The associations between ionizing radiation exposure and the development of cancer are mostly based on populations exposed to relatively high levels of ionizing radiation, such as Japanese atomic bomb survivors, and recipients of selected diagnostic or therapeutic medical procedures.

Cancers associated with high dose exposure include leukemia, thyroid, breast, bladder, colon, liver, lung, esophagus, ovarian, multiple myeloma, and stomach cancers. United States Department of Health and Human Services literature also suggests a possible association between ionizing radiation exposure and prostate, nasal cavity/sinuses, pharyngeal and laryngeal, and pancreatic cancer.

The period of time between radiation exposure and the detection of cancer is known as the latent period. Those cancers that may develop as a result of radiation exposure are indistinguishable from those that occur naturally or as a result of exposure to other chemical carcinogens. Furthermore, National Cancer Institute literature indicates that other chemical and physical hazards and lifestyle factors, such as smoking, alcohol consumption, and diet, significantly contribute to many of these same diseases.

Although radiation may cause cancer at high doses and high dose rates, public health data regarding lower levels of exposure, below about 1,000 mrem (10 mSv), are harder to interpret. To assess the health impacts of lower radiation doses, researchers rely on models of the process by which radiation causes cancer; several models have emerged which predict differing levels of risk.

Studies of occupational workers exposed to chronic low levels of radiation, above normal background, have provided mixed evidence regarding cancer and transgenerational effects. Cancer results, although uncertain, are consistent with estimates of risk based on atomic bomb survivors and suggest that these workers do face a small increase in the probability of developing leukemia and other cancers. One of the most recent and extensive studies of workers was published by Cardis *et al.* in 2005.

The linear dose-response model suggests that any increase in dose, no matter how small, results in an incremental increase in risk. The linear no-threshold model (LNT) hypothesis is accepted by the Nuclear Regulatory Commission (NRC) and the EPA and its validity has been reaffirmed by a National Academy of Sciences Committee. (See the BEIR VII report, summarized in) Under this model, about 1 per cent of a population would develop cancer in their lifetime as a result of ionizing radiation from background levels of natural and man-made sources.

Ionizing radiation damages tissue by causing ionization, which disrupts molecules directly and also produces highly reactive free radicals, which attack nearby cells. The net effect is that biological molecules suffer local disruption; this may exceed the body's capacity to repair the damage and may also cause mutations in cells currently undergoing replication.

Two widely studied instances of large-scale exposure to high doses of ionizing radiation are: atomic bomb survivors in 1945; and emergency workers responding to the 1986 Chernobyl accident.

Approximately 134 plant workers and fire fighters engaged at the Chernobyl power plant received high radiation doses (70,000 to 1,340,000 mrem or 700 to 13,400 mSv) and suffered from acute radiation sickness. Of these, 28 died from their radiation injuries.

Longer term effects of the Chernobyl accident have also been studied. There is a clear link (see the UNSCEAR 2000 Report, Volume 2: Effects) between the Chernobyl accident and the unusually large number, approximately 1,800, of thyroid cancers reported in contaminated areas, mostly in children. These were fatal in some cases. Other health effects of the Chernobyl accident are subject to current debate.

Ionizing Radiation Level Examples

Recognized effects of acute radiation exposure are described in the article on radiation poisoning. The exact units of measurement vary, but light radiation sickness begins at about 50-100 rad (0.5-1 gray (Gy), 0.5-1 Sv, 50-100 rem, 50,000-100,000 mrem).

Although the SI unit of radiation dose equivalent is the sievert, chronic radiation levels and standards are still often given in millirems, 1/1000th of a rem (1 mrem = 0.01 mSv).

The following table includes some short-term dosages for comparison purposes.

Level (mSv)	*Duration*	*Description*
0.001-0.01	Hourly	Cosmic ray dose on high-altitude flight, depends on position and solar sunspot phase.
0.01	Annual	USA dose from nuclear fuel and nuclear power plants
0.01	Daily	Natural background radiation, including radon
0.1	Annual	Average USA dose from consumer products
0.15	Annual	USA EPA cleanup standard
0.25	Annual	USA NRC cleanup standard for individual sites/sources
0.27	Annual	USA dose from natural cosmic radiation (0.16 coastal plain, 0.63 eastern Rocky Mountains)
0.28	Annual	USA dose from natural terrestrial sources
0.39	Annual	Global level of human internal radiation due to radioactive potassium
0.46	Acute	Estimated largest off-site dose possible from March 28, 1979 Three Mile Island accident
0.48	Day	USA NRC public area exposure limit
0.66	Annual	Average USA dose from human-made sources
1	Annual	Limit of dose from all DOE facilities to a member of the public who is not a radiation worker
1.1	Annual	1980 average USA radiation worker occupational dose
2	Annual	USA average medical and natural background
		Human internal radiation due to radon, varies with radon levels
2.2	Acute	Average dose from upper gastrointestinal diagnostic X-ray series
3	Annual	USA average dose from all natural sources
3.66	Annual	USA average from all sources, including medical diagnostic radiation doses
few	Annual	Estimate of cobalt-60 contamination within about 0.5 mile of dirty bomb
5	Annual	USA NRC occupational limit for minors (10% of adult limit)
		USA NRC limit for visitors
		Orvieto town, Italy, natural
5	Pregnancy	USA NRC occupational limit for pregnant women
6.4	Annual	High Background Radiation Area (HBRA) of Yangjiang, China
7.6	Annual	Fountainhead Rock Place, Santa Fe, NM natural
10-50	Acute	USA EPA nuclear accident emergency action level
50	Annual	USA NRC occupational limit (10 CFR 20)
100	Acute	USA EPA acute dose level estimated to increase cancer risk 0.8%
120	30 years	Exposure, long duration, Ural mountains, lower limit, lower cancer mortality rate
150	Annual	USA NRC occupational eye lens exposure limit
175	Annual	Guarapari, Brazil natural radiation sources
250	Acute	USA EPA voluntary maximum dose for emergency non-life-saving work
260	Annual	Ramsar, Iran, natural background peak dose
500	Annual	USA NRC occupational whole skin, limb skin, or single organ exposure limit
500	30 years	Exposure, long duration, Ural mountains, upper limit (exposed population lower cancer mortality rate)
750	Acute	USA EPA voluntary maximum dose for emergency life-saving work
500-1000	Acute	Low-level radiation sickness due to short-term exposure
500-1000	Detonation	World War II nuclear bomb victims
4500-5000	Acute	LD_{50} in humans (from radiation poisoning), with medical treatment.

Monitoring and Controlling Exposure to Ionizing Radiation

Radiation has always been present in the environment and in our bodies. The human body cannot sense ionizing radiation, but a range of instruments exists which are capable of detecting even very low levels of radiation from natural and man-made sources.

Dosimeters measure an absolute dose received over a period of time. Ion-chamber dosimeters resemble pens, and can be clipped to one's clothing. Film-badge dosimeters enclose a piece of photographic film, which will become exposed as radiation passes through it. Ion-chamber dosimeters must be periodically recharged, and the result logged. Film-badge dosimeters must be developed as photographic emulsion so the exposures can be counted and logged; once developed, they are discarded.

Geiger counters and scintillation counters measure the dose rate of ionizing radiation directly.

There are four standard ways to limit exposure:

1. *Time*: For people who are exposed to radiation in addition to natural background radiation, limiting or minimizing the exposure time will reduce the dose from the radiation source.
2. *Distance*: Radiation intensity decreases sharply with distance, according to an inverse square law.
3. *Shielding*: Barriers of lead, concrete, or water give effective protection from radiation formed of energetic particles such as gamma rays and neutrons. Some radioactive materials are stored or handled underwater or by remote control in rooms constructed of thick concrete or lined with lead. There are special plastic shields which stop beta particles and air will stop alpha particles. Shielding can be designed using halving thicknesses, the thickness of material that reduces the radiation by half. Halving thicknesses for gamma rays are discussed in the article gamma rays.
4. *Containment*: Radioactive materials are confined in the smallest possible space and kept out of the environment. Radioactive isotopes for medical use, for example, are dispensed in closed handling facilities, while nuclear reactors operate within closed systems with multiple barriers which keep the radioactive materials contained. Rooms have a reduced air pressure so that any leaks occur into the room and not out of it.

In a nuclear war, an effective fallout shelter reduces human exposure at least 1,000 times. Most people can accept doses as high as 1 Gy, distributed over several months, although with increased risk of cancer later in life. Other civil defense measures can help reduce exposure of populations by reducing ingestion of isotopes and occupational exposure during war time. One of these available measures could be the use of potassium iodide (KI) tablets which effectively block the uptake of dangerous radioactive iodine into the human thyroid gland.

5.4 Nuclear Forensics

Nuclear forensics and attribution are becoming increasingly important tools in the fight against illegal smuggling and trafficking of radiological and nuclear materials. These include materials intended for industrial and medical use (radiological), nuclear materials such as those produced in the nuclear fuel cycle, and much more dangerous weapons-usable nuclear materials—plutonium and highly enriched uranium.

Livermore scientists are among the leaders in nuclear forensics—the chemical, isotopic, and morphological analysis of interdicted illicit nuclear or radioactive materials and any associated materials. (See the box below.) They are also supporting the national effort in nuclear attribution, which is the challenging discipline of combining input from nuclear and conventional forensics to identify the source of nuclear and radiological materials and determine their points of origin and routes of transit.

Nuclear forensics and attribution go beyond determining the physical, chemical, and isotopic characteristics of intercepted nuclear or radiological materials. Simply determining that an interdicted material is, for example, enriched nuclear power plant fuel often tells law-enforcement authorities little about the origin of the material, where it has been, or who is responsible for the theft. "We want to know the exact nature of the interdicted material, who the perpetrators were, and where legitimate control was lost," says David Smith, a geochemist in the Nonproliferation, Homeland and International Security (NHI) Directorate and a leader of the Livermore nuclear forensics team. Knowledge of the pathways involved in transporting the material is also vital to nonproliferation and counterterrorism efforts.

Nuclear forensics and attribution can be difficult because of the large amount of radioactive materials in the industrialized and developing world. For example, the nuclear fuel cycle produces a variety of materials at different processing steps, ranging from processed ore (commonly known as "yellow cake") to uranium oxide fuel enriched in uranium-235. Opportunities exist at many points during the fuel cycle for materials to be diverted outside the channels authorized for their legitimate manufacture, handling, and protection.

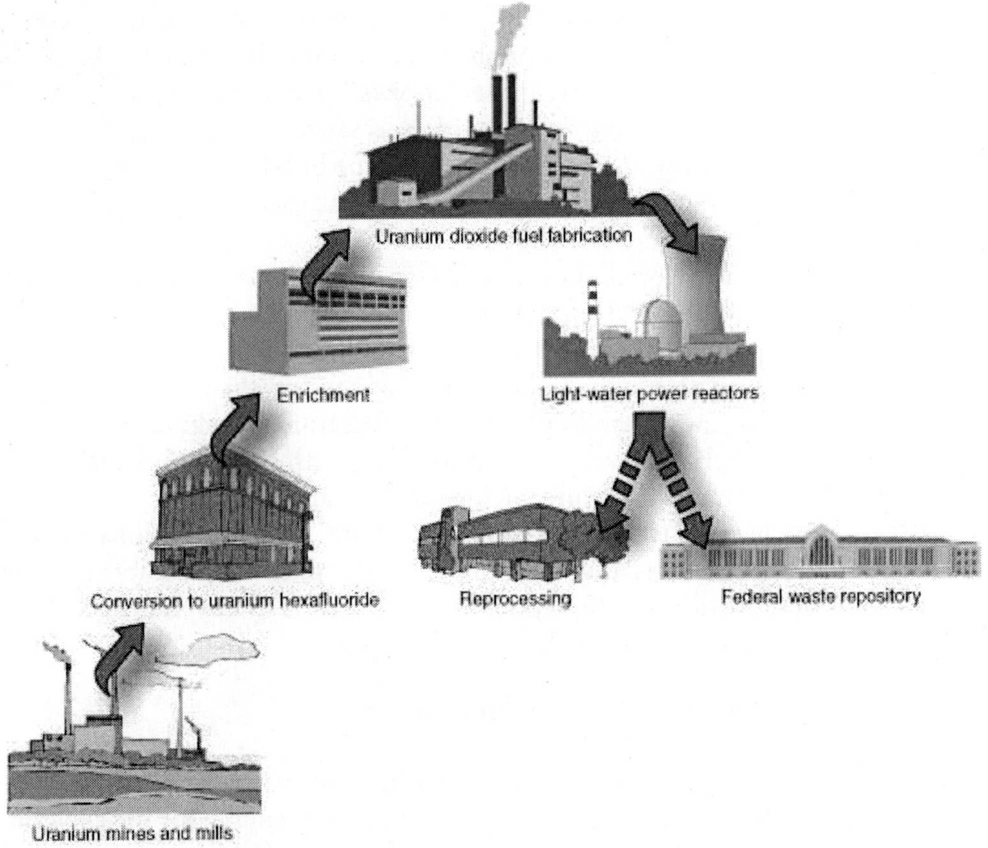

Fig. 5.8: Nuclear Forensics.

Successful nuclear forensics and attribution contribute to the prosecution of those responsible and strengthen national efforts in nuclear nonproliferation and counterterrorism. According to former Livermore physicist Jay Davis, the first director of the federal Defense Threat Reduction Agency (DTRA), "Nuclear attribution, with the accompanying possibility of prosecution and retribution, may be one of our greatest deterrent tools and hence a vital and compelling component of our defense against terrorism."

Livermore and seven other Department of Energy (DOE) national laboratories have been tasked by the Federal Bureau of Investigation (FBI) and the Department of Homeland Security (DHS) with further developing the nation's technical forensics capability for nuclear and radiological materials. Major U.S. government partners in addressing domestic and foreign obligations include DOE, the National Nuclear Security Administration (NNSA), the Department of State, and DTRA. Technical nuclear forensics includes the analysis of conventional evidence that is radiologically contaminated as well as the analysis of the radiological materials themselves.

Lawrence Livermore and Savannah River national laboratories serve as a "hub" for nuclear forensics, with the "spokes" of the hub represented by six other national laboratories that provide specialized analyses and serve in supporting roles. A sample identified by one of the hub laboratories could be sent to experts at one or more of the spoke laboratories for complementary analyses and further characterization.

"We're part of a national nuclear forensic and attribution capability that ties nuclear and radiological materials to people, places, and events," says Smith. The effort is also part of a national strategy to counter nuclear terrorism. Other activities at the Laboratory that support the national strategy include developing more capable radiation detectors, designing detection systems for ports and other venues, and working with Russia and nations of the former Soviet Union to secure their nuclear materials.

Because nuclear material goes through several transformations during the uranium concentration and enrichment cycle, specific signatures may be unique to different stages of the cycle. Scientists want to develop signatures across the entire nuclear fuel cycle.

A Textbook for Nuclear Forensic Scientists

Nuclear Forensic Analysis, written by Livermore scientists Ken Moody, Ian Hutcheon, and Pat Grant, is the first primary reference source for the growing specialty of nuclear forensics. In addition to being a resource for nuclear forensic scientists, the book also serves as a textbook for traditional radiochemistry science courses that increasingly include material on nuclear forensics. The book covers the principles of the chemical, physical, and nuclear characteristics associated with the production and interrogation of a radioactive sample; protocols and procedures; and attribution. The authors discuss principles and techniques used in numerous case studies of nuclear investigations conducted at Livermore.

Hutcheon is deputy director of the Glenn T. Seaborg Institute. Moody, who studied with Nobel Laureate Glenn Seaborg (the discoverer of plutonium), is a radiochemist and co-discoverer of four heavy elements. Grant serves as the deputy director of the Forensic Science Center. "Pat is trained in classical forensics, Ken is a nuclear chemist, and I specialize in instrumentation," says Hutcheon. "Together, we give a balanced viewpoint."

According to Hutcheon, "The book gives one an idea of the analytic techniques we use and the types of material we are asked to investigate. It also describes case studies and the ways we assign attribution." Hutcheon says the book was written to be understandable by nonspecialists.

Focus on Forensics

Nuclear forensics began at the Laboratory under the leadership of scientist Sid Niemeyer in the mid-1990s, and Livermore has maintained a leading role because of a collective group effort. The Laboratory has been involved in nuclear forensics and attribution for more than 15 years, since the collapse of the Soviet Union sparked concerns about the diversion of nuclear materials from former Soviet nuclear laboratories and other sites. Livermore's capabilities in radiochemistry and nuclear physics, originally developed for the nation's underground nuclear testing program, were adapted for use in nuclear forensics and attribution.

The strength of the national capability rests on the ability of Livermore to work in concert with the other laboratories, drawing on a core group of 30 to 50 scientists from diverse backgrounds to work collaboratively on nuclear forensics and attribution cases. This depth of knowledge and breadth of experience are essential to tackling current issues in nuclear forensics.

Technical nuclear forensics can be performed on a broad spectrum of possible substances. Some interdicted samples are stolen containers of uranium diverted during one of the mining, milling, conversion, enrichment, or fuel fabrication steps used to convert uranium ore to enriched fuel for nuclear power plants. Alternatively, the interdicted sample could well be a commercial radioactive material such as cesium-137, strontium-90, cobalt-60, or americium-241. These isotopes and others are used in applications such as medical diagnostics, nondestructive analysis, food sterilization, and thermoelectric generators.

"We support investigations with the extraction of evidence that can be used by law-enforcement agencies," says Smith. "Our casework validates and verifies our technical approaches." Livermore and the other national laboratories subject an interdicted sample to a host of extremely sensitive and accurate measurement techniques. Researchers analyze the material's chemical and isotopic composition, which includes measuring the amounts of trace elements as well as the ratio of parent isotopes to daughter isotopes. These measurements help to determine the source location and sample's age. They also examine the material's morphological characteristics such as shape, size, and texture. Together, these characteristics are indicative of the specific processes used to produce the material.

Analytical methods include electron microscopy, x-ray diffraction, and mass spectrometry. "As we obtain new tools, we open new frontiers of technical nuclear forensics," says Ian Hutcheon, deputy director of the Glenn T. Seaborg Institute and senior scientist in the Forensic Science Center. For example, Livermore's NanoSIMS, a secondary-ion mass spectrometer, provides a 50-nanometer (a billionth of a meter) spatial resolution. This capability enables researchers to analyze sub-samples of less than 1 microgram and perform particle-by-particle characterization to elucidate additional signatures beyond those gained through dissolving the bulk sample. "We are advancing from the micro to the nano level in analyzing samples to obtain forensic clues," says Hutcheon.

Livermore researchers use techniques such as electron microscopy, x-ray diffraction, and mass spectrometry to analyze interdicted radiological and nuclear materials. In this photo, Ian Hutcheon (left), a senior scientist working with Livermore's Forensic Science Center, and analytical chemist Peter Weber use NanoSIMS, a secondary-ion mass spectrometer with nanometer-scale resolution.

Contaminated Evidence

An important element of nuclear forensics is the analysis of nonnuclear materials found with the radioactive material. Livermore scientists have been involved in developing forensics on radiologically contaminated evidence, which enables conventional law- enforcement forensics to be performed on materials that are radioactive. In addition, as a sample is moved from place to place, it picks up clues such as pollen, cloth fibers, and organic compounds. These so-called route materials provide information about who has handled a sample and the path it has traveled.

Nuclear forensics takes advantage of capabilities in Livermore's Forensic Science Center to tease out fingerprints, paper, hair, fibers, pollens, dust, plant DNA, and chemical explosives from contaminated substrates for the FBI and other government agencies. Established in 1991, the center supports DOE in verifying compliance with international treaties. It has also assisted federal, state, and local law enforcement on a wide range of criminal cases, including the 1993 World Trade Center bombing and the Unabomber investigations. The Forensic Science Center incorporates subject-matter experts from the Laboratory's NHI and Chemistry, Materials, Earth, and Life Sciences (CMELS) directorates to build a complete profile of a sample.

Livermore's nuclear forensics program uses laboratories equipped to handle large objects that are contaminated with radioactive materials. These laboratories make it possible, for example, to analyze a contaminated truck axle while still maintaining the evidentiary chain of custody required in a court of law. In addition, the program operates a mobile van outfitted to ferry samples around the site or, if necessary, transport material between the Laboratory and offsite locations.

Developing Signatures

A major focus of nuclear forensics is identifying signatures, which are the physical, chemical, and isotopic characteristics that distinguish one nuclear or radiological material from another. Signatures enable researchers to identify the processes used to initially create a material.

Livermore scientists are developing signatures for a variety of nuclear materials. Enriched uranium is of particular concern. The goal is to develop validated signatures across the entire nuclear fuel cycle by experimental measurement or by simulation. Because nuclear material is transformed at different points during uranium concentration and enrichment processing, clues may be unique to different stages.

For example, unirradiated uranium reactor fuel pellets have inherent elemental oxygen content. Because the ratio of naturally occurring isotopes of oxygen-18 to oxygen-16 varies worldwide, these ratios could correlate with the locations of production sites. Similarly, the variations of lead isotopes could provide clues about where a uranium compound was produced. Age, color, density, trace elements, and surface characteristics of the uranium compound are other important characteristics. A related investigation is studying the effects of different uranium manufacturing processes on the grain size and microstructure of the finished product.

When comparing a sample's signature against known signatures from uranium mines and fabrication plants, researchers can benefit by assembling a library of nuclear materials of known origin from around the world. Livermore scientists, with the help of the Office of Laboratory Counsel, have developed relationships with domestic suppliers of nuclear materials (uranium hexafluoride and uranium oxide reactor fuel) to assemble such a library. Contracts with major U.S. uranium fuel suppliers have provided researchers with samples and manufacturing data. Livermore scientists are using statistics to identify distinguishing signatures in nuclear materials using samples and data obtained from manufacturers.

Livermore forensic scientists are also seeking to obtain samples of uranium products worldwide to analyze the products' isotopic and trace-element content, grain size, and microstructure. Nations with nuclear capabilities are beginning to share information about their nuclear fuel processes and materials. (See the box below.) For example, Livermore representatives signed a five-year agreement in January 2006 with KazAtomProm, the national atomic energy enterprise in Kazakhstan. KazAtomProm has long provided uranium ores to Russia and other nations of the former Soviet Union. With legal and technical support from NNSA, DHS funded the Laboratory to contract with KazAtomProm to provide Livermore with uranium ore samples and data. In addition, KazAtomProm officials have identified experts available for consultation on nuclear forensic issues.

A similar agreement was signed in 2006 with nuclear power officials in Tajikistan. This agreement provides samples to enable comparison between known and questioned sources. "We want to expand our agreements to all the Central Asian nations that process uranium," Smith says.

The importance of international cooperation was underscored in July 2006, when U.S. President George W. Bush and Russian President Vladimir Putin announced the creation of a Global Initiative to Combat Nuclear Terrorism. The initiative's goal is to strengthen worldwide cooperation in making nuclear materials more secure and preventing terrorist acts that involve nuclear or radioactive substances.

The Cosmic Connection to Nuclear Forensics

Scientific research in cosmochemistry and isotope geochemistry continues to technically validate Livermore's nuclear forensics capabilities. Cosmochemistry is the study of the origin and development of the elements and their isotopes in the universe and the formation of our solar system. Livermore cosmochemistry projects, funded by the Laboratory Directed Research and Development (LDRD) Program and the National Aeronautics and Space Administration, allow researchers to use the same instruments and procedures that nuclear forensic scientists use when analyzing interdicted nuclear and radiological samples.

One LDRD effort is studying inclusions in Australian zircons more than 4 billion years old to determine when conditions suitable for life first emerged on Earth. The work, led by scientist Ian Hutcheon, involves dating and isotopically and chemically analyzing the zircons and the mineral inclusions in them. The effort focuses on measuring the abundance of trace elements, notably uranium and other actinides in these ancient zircons. This analysis will help researchers to understand the evolution of the atmosphere and hydrosphere during the earliest epoch of Earth's history and to evaluate the evidence for volcanic activity as far back as 4.4 billion years.

Hutcheon says this LDRD effort develops and enhances microanalytical capabilities needed for nuclear forensics. The investigation uses tools that also support Livermore's nuclear forensics work, such as the Laboratory's nanometer-scale secondary-ion mass spectrometer (NanoSIMS) and a new, ultrahigh-resolution scanning electron microscope. "Whether we're measuring oxygen isotopes in Australian zircons or in interdicted uranium yellow cake, the techniques are similar," he says. "Our ancient zircon research is published in peer-reviewed journals, and the suggestions and criticism made in response to these published papers strengthen our nuclear forensics work."

Hutcheon was involved in another LDRD cosmochemistry effort headed by John Bradley of Livermore's Institute of Geophysics and Planetary Physics. Using a transmission electron microscope, Laboratory researchers detected a 2,175-angstrom extinction feature (or bump) in interstellar grains embedded within interplanetary dust particles. The Livermore team identified organic carbon and amorphous silica-rich material as responsible for the bump. (See *S&TR*, September 2005, Dust That's Worth Keeping.)

These measurements may help explain how interstellar organic matter was incorporated into the solar system. Interplanetary dust particles are roughly the same size scales of interdicted radiological and nuclear materials, such as powdered nuclear enriched uranium. By studying these dust particles, scientists gain confidence in using advanced tools to characterize other materials of interest to nuclear forensics work.

Under an agreement signed in 2006, KazAtomProm is providing Livermore scientists with uranium ore samples and data and has identified experts available for consultation on nuclear forensic issues. Front row, left to right: Dave Herr from Livermore's Procurement Office and Sergei Yashin, vice president of KazAtomProm. Back row, left to right: Mike Kristo and David Smith from Livermore and representatives from KazAtomProm. (background) KazAtomProm is the leading uranium mining and fuel production facility in Kazakhstan.

Building a Knowledge Base

The data obtained from U.S. and Central Asian uranium producers are part of a growing technical nuclear forensics knowledge base involving Livermore and seven other DOE national laboratories. "We are establishing a knowledge management system for nuclear signatures, processes, origins, and pathways," says Livermore nuclear engineer Frank Wong, who is leading the knowledge base effort, which now includes about 30 experts nationwide.

Rapid and credible interpretation of nuclear forensics data is predicated on a knowledge base that accesses and analyzes information derived from measurements or simulation of the full spectrum of the nuclear fuel cycle. Knowledge management allows for the ready comparison of analytical signatures obtained from suspect nuclear samples against known signatures from nuclear production, reprocessing, manufacturing, and storage. It also allows for the analysis of existing knowledge and data with the goal of identifying the most diagnostic nuclear forensic signatures.

The knowledge base is being designed so that it can be accessed by a simple query on a computer. "As an example, when key words are entered, the system will find the relevant information," Wong says. He notes that the knowledge base will use a distributed architecture. That is, the information will not be centralized in one location. Instead, information will be distributed across several government sites, such as DHS, FBI, and DOE, with several levels of controlled access. The knowledge base will include a list of subject-matter experts worldwide who could be called upon to assist in specific nuclear forensics casework. Another proposed element is a U.S. evidence archive, where nuclear materials from past cases would be stored as references.

The Lawrence Livermore technical team that addresses the technical nuclear forensics of interdicted materials includes (left to right) in the front row: Ian Hutcheon, Erick Ramon, and Michael Kristo; second row: Brett Isselhardt, Giles Graham, and Lars Borg; third row: Ross Williams and Michael Singleton; and fourth row: David Smith and Lee Davisson. Not pictured: Glenn Fox, Pat Grant, Leonard Gray, Steven Kreek, Ken Moody, Sid Niemeyer, Martin Robel, Steve Steward, Louann Tung, Alan Volpe, Philip Wilk, Nathan Wimer, and Frank Wong.

Growing Capabilities

Nuclear forensics and attribution are two important elements among several DHS and NNSA efforts in border protection, radiation monitoring, emergency response, and consequence management to provide the nation with the best protection against nuclear and radiological threats. Nuclear chemist Michael Kristo says that the nuclear forensics and attribution effort has gained strength with the launching of the Global Nuclear Energy Partnership (GNEP) in February 2006.

As part of President Bush's Advanced Energy Initiative, GNEP is a strategy to expand nuclear energy worldwide by demonstrating and deploying new technologies to recycle nuclear fuel, minimize waste, and improve the ability to keep nuclear technologies and materials out of the hands of terrorists. GNEP would build recycling technologies that enhance energy security in a safe and environmentally responsible manner. A basic goal of GNEP is to make it impossible to divert nuclear materials or modify systems without immediate detection. One option is to incorporate tags into nuclear materials at different stages of the fuel cycle so that the materials could be tracked and traced.

Looking to the future, Smith notes that existing Laboratory technical efforts define an emerging Nuclear Forensics Analysis Center funded primarily by DHS. This center would build on Livermore's experience in nuclear assessment, nuclear weapons and materials, and isotope and trace-element science as well as on the capabilities and expertise of the Forensic Science Center and the NHI and CMELS directorates. "Livermore is uniquely positioned," he says. "We have science-based signatures, expertise, and facilities. All the pieces are here for such a center and for making an even greater contribution to the prevention of nuclear materials trafficking and nuclear terrorism."

Strengthening the Worldwide Effort

The Nuclear Smuggling International Technical Working Group (ITWG) was chartered in 1996 to foster international cooperation in combating illicit trafficking of nuclear materials. "The ITWG was formed with the recognition that nations must work together," says geochemist David Smith of the Nonproliferation, Homeland and International Security Directorate. The ITWG was cofounded by Livermore scientist Sid Niemeyer and has been co chaired by Lawrence Livermore since its inception.

The ITWG works closely with the International Atomic Energy Agency (IAEA) to provide member countries with support for forensic analyses. Priorities include the development of common protocols for the collection of evidence and laboratory investigations, organization of forensic exercises, and technical assistance to requesting nations. Experts from participating nations and organizations meet annually to work on issues concerning illicit trafficking of nuclear materials. The 2006 meeting was sponsored by the European Commission's Institute for Transuranium Elements in Karlsruhe, Germany.

To promote the science of nuclear forensics within the ITWG, the Nuclear Forensics Laboratory Group was organized in 2004. In that year, Livermore scientists wrote a comprehensive description of a model action plan to guide member states in their own nuclear forensic investigations. The plan provides recommendations governing incident response, sampling and distribution of materials, radioactive materials analysis, traditional forensic analysis, and nuclear forensic interpretation of signatures. In 2006, the IAEA published the model action plan as a Nuclear Security Series Technical Document. Participating countries have adopted the plan and used it in their own nuclear forensics investigations.

Exercises are critical to operational readiness within ITWG participants. An international exercise is planned for later this year, when participants will receive identical radiological samples supplied by Pacific Northwest National Laboratory. "It's essential for laboratories to discuss the results from their

5.5 Non-ionizing Radiation

Non-ionizing radiation (or, esp. in British English, *non-ionising radiation*) refers to any type of electromagnetic radiation that does not carry enough energy per quantum to ionize atoms or molecules—that is, to completely remove an electron from an atom or molecule. Instead of producing charged ions when passing through matter, the electromagnetic radiation has sufficient energy only for excitation, the movement of an electron to a higher energy state. Nevertheless, different biological effects are observed for different types of non-ionizing radiation.

Near ultraviolet, visible light, infrared, microwave, radio waves, low frequency RF and static fields are all examples of non-ionizing radiation. Visible and near ultraviolet may induce photochemical reactions, ionize some

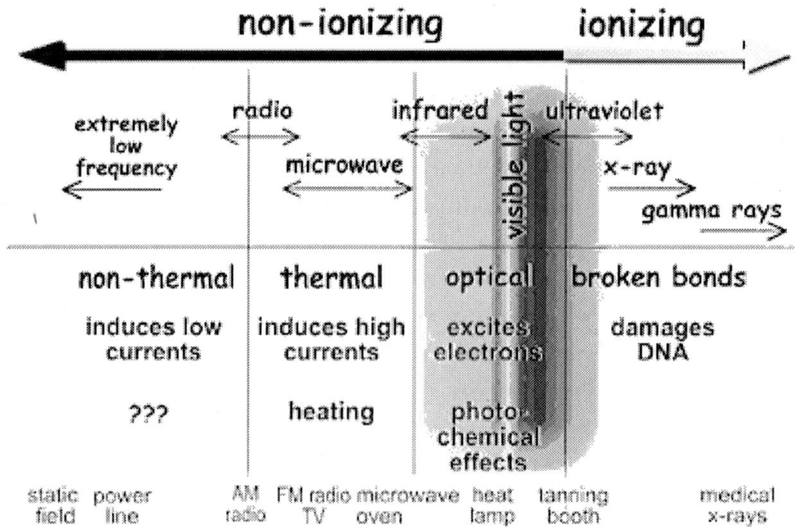

Fig. 5.9: Different types of electromagnetic radiation.

molecules or accelerate radical reactions, such as photochemical aging of varnishes or the breakdown of flavoring compounds in beer to produce the 'lightstruck flavor'. The light from the Sun that reaches the earth is largely composed of non-ionizing radiation, with the notable exception of some ultraviolet rays. However, most ionizing radiation is filtered out by the atmosphere. Static fields do not radiate.

	Source	*Wavelength*	*Frequency*	*Biological Effects*
UVC	Germicidal light	100 nm-280 nm	1075 THz-3000 THz	Skin-Erythematic, inc. pigmentation; Eye-Photokeratitis (inflammation of cornea)
UVB	Tanning booth	280 nm-315 nm	950 THz-1075 THz	Eye-Photokeratitis (inflammation of cornea) Skin-Erythema, inc. pigmentation Skin cancer, Photosensitive skin reactions, Production of vitamin D
UVA	Black light, sunlight	315 nm-400 nm	750 THz-950 THz	Eye-Photochemical cataract Skin-Erythema, inc. pigmentation
Visible Light	lasers, sunlight, fire, LEDs, Light Bulbs	400 nm-780 nm	385 THz-750 THz	Skin photo-ageing, Skin cancer; Eye—Photochemical & thermal retinal injury
IR-A	lasers, remote controls	780 nm-1.4 μm	215 THz-385 THz	Eye—Thermal retinal injury, thermal cataract; Skin burn
IR-B	lasers, long-distance telecommunications	1.4 μm-3 μm	100 THz-215 THz	Eye—Corneal burn, cataract; Skin burn
IR-C	Far-infrared laser	3 μm-1 mm	300 GHz-100 THz	Eye—Corneal burn, cataract; Heating of body surface
Microwave	PCS phones, some mobile/cell phones, microwave ovens, cordless phones, motion detectors, radar, Wi-Fi	1 mm-33 cm	1 GHz-300 GHz	Heating of body surface
Radio Frequency Radiation	Mobile/Cell phones, television, FM, AM, Shortwave, CB, cordless phones	33 cm-3 km	100 kHz-1 GHz	Heating with 'penetration depth' of 10 mm, Raised body temperature
Low frequency RF	power lines	> 3 km	< 100 kHz	Accumulation of charge on body surface Disturbance of nerve & muscle responses
Static Field	strong magnets, MRI	infinite	0 Hz	Magnetic-vertigo/nausea, Electric-charge on body surface

Ultraviolet Radiation

Ultraviolet light can cause burns to skin and cataracts to the eyes. Ultraviolet is classified into near, medium and far UV according to energy, where near ultraviolet is non-ionizing. Ultraviolet light produces free radicals that induce cellular damage, which can be carcinogenic. Ultraviolet light also induces melanin production from melanocyte cells to cause sun tanning of skin. Vitamin D is produced on the skin by a radical reaction initiated by UV radiation.

Plastic sunglasses (polycarbonate) generally absorb UV radiation. UV overexposure to the eyes causes snow blindness, which is a risk particularly on the sea or when there is snow on the ground.

Visible and Infrared, Lasers

Visible light causes few effects to the human body. Bright visible light irritates the eyes. Visible-light lasers have much more powerful effects and may damage the eyes even at small powers. Very strong visible light is used for cauterizing hair follicles.

Microwave and Radio Frequency Radiation

- ❖ Biological effects:
  - Effects on the Skin
  - Effects on the Eyes
  - Other Hazards
  - Occupational Exposure Standards

Low Frequency ELF

- ❖ Biological effects:
  - Effects on the Skin
  - Effects on the Eyes
  - Other Hazards
  - Occupational Exposure Standards

Static Fields

- ❖ Biological effects:
  - Effects on the Skin
  - Effects on the Eyes

- Other Hazards

c.f Electric Power Transmission.

- Occupational Exposure Standards

5.6 Radiological and Environmental Sciences Laboratory

The *Radiological and Environmental Sciences Laboratory* (RESL) is a federally-owned and operated laboratory by the United States Department of Energy (DOE) located in the Central Facilities Area of the Idaho National Laboratory near Idaho Falls, Idaho. The laboratory's focus is primarily in analytical chemistry, radiation protection, and as a reference laboratory for numerous Performance Evaluation Programs. RESL provides technical support and quality assurance metrology, which is directly traceable to the National Institute of Standards and Technology (NIST).

RESL staff consists of professional chemists, physicists (including health physicists), engineers, computer programmers, and related technicians. The staff has academic training to the doctorate level. Individual staff members participate in professional society activities, in working groups and committees, such as American Society for Testing and Materials (ASTM), International Organization for Standardization (ISO) and American National Standards Institute (ANSI) committees, the Health Physics Society, and on editorial and accreditation boards.

5.7 Radioactive Decay

Radioactive decay is the process in which an unstable *atomic nucleus* loses energy by emitting radiation in the form of particles or electromagnetic waves. This decay, or loss of energy, results in an atom of one type, called the *parent nuclide* transforming to an atom of a different type, called the *daughter nuclide*. For example: a carbon-14 atom (the "parent") emits radiation and transforms to a nitrogen-14 atom (the "daughter"). This is a random process on the atomic level, in that it is impossible to predict when a given atom will decay, but given a large number of similar atoms, the decay rate, on average, is predictable.

The SI unit of radioactive decay (the phenomenon of natural and artificial radioactivity) is the becquerel (Bq). One Bq is defined as one transformation (or decay) per second. Since any reasonably-sized sample of radioactive material contains many atoms, a Bq is a tiny measure of activity; amounts on the order of TBq (terabecquerel) or GBq (gigabecquerel) are commonly used. Another unit of (radio) activity is the curie, Ci, which was originally defined as the

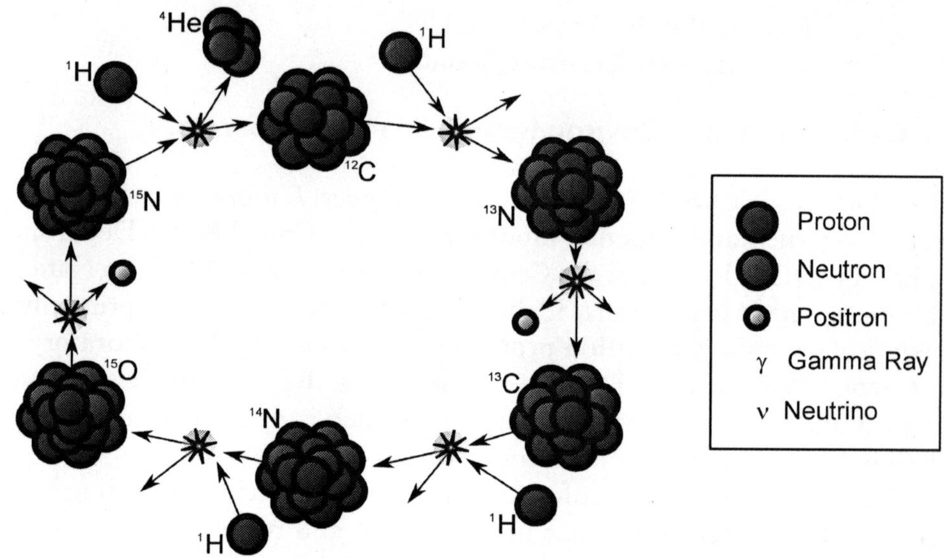

activity of one gram of pure radium, isotope Ra-226. At present it is equal (by definition) to the activity of any radionuclide decaying with a disintegration rate of 3.7×10^{10} Bq. The use of Ci is presently discouraged by SI.

Explanation

The neutrons and protons that constitute nuclei, as well as other particles that may approach them, are governed by several interactions. The strong nuclear force, not observed at the familiar macroscopic scale, is the most powerful force over subatomic distances. The electrostatic force is also significant, while the weak nuclear force is responsible for beta decay.

The interplay of these forces is simple. Some configurations of the particles in a nucleus have the property that, should they shift ever so slightly, the particles could fall into a lower-energy arrangement (with the extra energy moving elsewhere). One might draw an analogy with a snowfield on a mountain: while friction between the snow crystals can support the snow's weight, the system is inherently unstable with regard to a lower-potential-energy state, and a disturbance may facilitate the path to a greater entropy state (*i.e.*, towards the ground state where heat will be produced, and thus total energy is distributed over a larger number of quantum states). Thus, an avalanche results. The total energy does not change in this process, but because of entropy effects, avalanches only happen in one direction, and the end of this direction, which is dictated by the largest number of chance-mediated ways to distribute available energy, is what we commonly refer to as the "ground state".

Such a collapse (a *decay event*) requires a specific activation energy. In the case of a snow avalanche, this energy classically comes as a disturbance from outside the system, although such disturbances can be arbitrarily small. In the case of an excited atomic nucleus, the arbitrarily small disturbance comes from quantum vacuum fluctuations. A nucleus (or any excited system in quantum mechanics) is unstable, and can thus *spontaneously stabilize* to a less-excited system. This process is driven by entropy considerations: the energy does not change, but at the end of the process, the total energy is more diffused in spacial volume. The resulting transformation alters the structure of the nucleus. Such a reaction is thus a nuclear reaction, in contrast to chemical reactions, which also are driven by entropy, but which involve changes in the arrangement of the outer electrons of atoms, rather than their nuclei.

Some nuclear reactions do involve external sources of energy, in the form of collisions with outside particles. However, these are not considered *decay*. Rather, they are examples of induced nuclear reactions. Nuclear fission and fusion are common types of induced nuclear reactions.

Discovery

Radioactivity was first discovered in 1896 by the French scientist Henri Becquerel, while working on phosphorescent materials. These materials glow in the dark after exposure to light, and he thought that the glow produced in cathode ray tubes by X-rays might be connected with phosphorescence. He wrapped a photographic plate in black paper and placed various phosphorescent minerals on it. All results were negative until he used uranium salts. The result with these compounds was a deep blackening of the plate.

It soon became clear that the blackening of the plate had nothing to do with phosphorescence, because the plate blackened when the mineral was in the dark. Non-phosphorescent salts of uranium and metallic uranium also blackened the plate. Clearly there was a form of radiation that could pass through paper that was causing the plate to blacken.

At first it seemed that the new radiation was similar to the then recently discovered X-rays. Further research by Becquerel, Marie Curie, Pierre Curie, Ernest Rutherford and others discovered that radioactivity was significantly more complicated. Different types of decay can occur, but Rutherford was the first to realize that they all occur with the same mathematical approximately exponential formula.

As for types of radioactive radiation, it was found that an electric or magnetic field could split such emissions into three types of beams. For lack of better terms, the rays were given the alphabetic names alpha, beta and gamma, still in use today. It was obvious from the direction of electromagnetic forces that alpha rays carried a positive charge, beta rays carried a negative charge, and gamma rays were neutral. From the magnitude of deflection, it was clear that alpha particles were much more massive than beta particles. Passing alpha particles through a very thin glass window and trapping them in a discharge tube allowed researchers to study the emission spectrum of the resulting gas, and ultimately prove that alpha particles are helium nuclei. Other experiments showed the similarity between beta radiation and cathode rays; they are both streams of electrons, and between gamma radiation and X-rays, which are both high energy electromagnetic radiation.

Although alpha, beta, and gamma are most common, other types of decay were eventually discovered. Shortly after discovery of the neutron in 1932, it was discovered by Enrico Fermi that certain rare decay reactions yield neutrons as a decay particle. Isolated proton emission was eventually observed in some elements. Shortly after the discovery of the positron in cosmic ray products, it was realized that the same process that operates in classical beta decay can also produce positrons (positron emission), analogously to negative electrons. Each of the two types of beta decay acts to move a nucleus toward a ratio of neutrons and protons which has the least energy for the combination. Finally, in a phenomenon called cluster decay, specific combinations of neutrons and protons other than alpha particles were spontaneously emitted from atoms on occasion.

Still other types of radioactive decay were found which emit previously seen particles, but by different mechanisms. An example is internal conversion, which results in electron and sometimes high energy photon emission, even though it involves neither beta nor gamma decay.

The early researchers also discovered that many other chemical elements besides uranium have radioactive isotopes. A systematic search for the total radioactivity in uranium ores also guided Marie Curie to isolate a new element polonium and to separate a new element radium from barium. The two elements' chemical similarity would otherwise have made them difficult to distinguish.

The dangers of radioactivity and of radiation were not immediately recognized. Acute effects of radiation were first observed in the use of X-rays when the Serbo-Croatian-American electric engineer Nikola Tesla intentionally subjected his fingers to X-rays in 1896. He published his observations concerning the burns that developed, though he attributed them to ozone rather than to X-rays. His injuries healed later.

Fig. 5.10: The danger classification sign of radioactive materials.

The genetic effects of radiation, including the effects on cancer risk, were recognized much later. In 1927 Hermann Joseph Muller published research showing genetic effects, and in 1946 was awarded the Nobel prize for his findings.

Before the biological effects of radiation were known, many physicians and corporations had begun marketing radioactive substances as patent medicine and radioactive quackery. Examples were radium enema treatments, and radium-containing waters to be drunk as tonics. Marie Curie spoke out against this sort of treatment, warning that the effects of radiation on the human body were not well understood (Curie later died from aplastic anemia assumed due to her work with radium, but later examination of her bones showed that she had been a careful laboratory worker and had a low burden of radium. A more likely cause was her exposure to unshielded X-ray tubes while a volunteer medical worker in WWI). By the 1930s, after a number of cases of bone necrosis and death in enthusiasts, radium-containing medical products had nearly vanished from the market.

Modes of Decay

Radionuclides can undergo a number of different reactions. These are summarized in the following table. A nucleus with mass number A and atomic number Z is represented as (A, Z). The column "Daughter nucleus" indicates the difference between the new nucleus and the original nucleus. Thus, $(A\text{-}1, Z)$ means that the mass number is one less than before, but the atomic number is the same as before.

Radioactive decay results in a reduction of summed rest mass, which is converted to energy (the *disintegration energy*) according to the formula $E = mc^2$. This energy is released as kinetic energy of the emitted particles. The energy remains associated with a measure of mass of the decay system invariant mass, inasmuch the kinetic energy of emitted particles contributes also to the total invariant mass of systems. Thus, the sum of rest masses of particles is not conserved in decay, but the *system* mass or system invariant mass (as also system total energy) is conserved.

Mode of decay	*Participating particles*	*Daughter nucleus*
Decays with emission of nucleons:		
Alpha decay	An alpha particle (A=4, Z=2) emitted from nucleus	(A-4, Z-2)
Proton emission	A proton ejected from nucleus	(A-1, Z1)
Neutron emission	A neutron ejected from nucleus	(A1, Z)
Double proton emission	Two protons ejected from nucleus simultaneously	(A-2, Z-2)
Spontaneous fission	Nucleus disintegrates into two or more smaller nuclei and other particles	-
Cluster decay	Nucleus emits a specific type of smaller nucleus (A_1, Z_1) smaller than, or larger than, an alpha particle	($A-A_1$, $Z-Z_1$) + (A_1, Z_1)
Different modes of beta decay:		
Beta-Negative decay	A nucleus emits an electron and an antineutrino	(A, Z+1)
Positron emission, also Beta-Positive decay	A nucleus emits a positron and a neutrino	(A, Z-1)
Electron capture	A nucleus captures an orbiting electron and emits a neutrino—The daughter nucleus is left in an excited and unstable state	(A, Z-1)
Double beta decay	A nucleus emits two electrons and two antineutrinos	(A, Z+2)
Double electron capture	A nucleus absorbs two orbital electrons and emits two neutrinos—The daughter nucleus is left in an excited and unstable state	(A, Z-2)
Electron capture with positron emission	A nucleus absorbs one orbital electron, emits one positron and two neutrinos	(A, Z-2)
Double positron emission	A nucleus emits two positrons and two neutrinos	(A, Z-2)
Transitions between states of the same nucleus:		
Gamma decay	Excited nucleus releases a high-energy photon (gamma ray)	(A, Z)
Internal conversion	Excited nucleus transfers energy to an orbital electron and it is ejected from the atom	(A, Z)

Decay Chains and Multiple Modes

The daughter nuclide of a decay event may also be unstable (radioactive). In this case, it will also decay, producing radiation. The resulting second daughter nuclide may also be radioactive. This can lead to a sequence of several decay events. Eventually a stable nuclide is produced. This is called a *decay chain*.

An example is the natural decay chain of uranium-238 which is as follows:

- decays, through alpha-emission, with a half-life of 4.5 billion years to thorium-234,
- which decays, through beta-emission, with a half-life of 24 days to protactinium-234,

- which decays, through beta-emission, with a half-life of 1.2 minutes to uranium-234,
- which decays, through alpha-emission, with a half-life of 240 thousand years to thorium-230,
- which decays, through alpha-emission, with a half-life of 77 thousand years to radium-226,
- which decays, through alpha-emission, with a half-life of 1.6 thousand years to radon-222,
- which decays, through alpha-emission, with a half-life of 3.8 days to polonium-218,
- which decays, through alpha-emission, with a half-life of 3.1 minutes to lead-214,
- which decays, through beta-emission, with a half-life of 27 minutes to bismuth-214,
- which decays, through beta-emission, with a half-life of 20 minutes to polonium-214,
- which decays, through alpha-emission, with a half-life of 160 microseconds to lead-210,
- which decays, through beta-emission, with a half-life of 22 years to bismuth-210,
- which decays, through beta-emission, with a half-life of 5 days to polonium-210, and
- which decays, through alpha-emission, with a half-life of 140 days to lead-206, which is a stable nuclide.

Some radionuclides may have several different paths of decay. For example, approximately 36 per cent of bismuth-212, decays, through alpha-emission, to thallium-208 while approximately 64 per cent of bismuth-212 decays, through beta-emission, to polonium-212. Both the thallium-208 and the polonium-212 are radioactive daughter products of bismuth-212, and both decay directly to stable lead-208.

Occurrence and Applications

According to the Big Bang theory, stable isotopes of the lightest five elements (H, He, and traces of Li, Be, and B) were produced very shortly after the emergence of the universe, in a process called Big Bang nucleosynthesis. These lightest stable nuclides (including deuterium) survive to today, but any radioactive isotopes of the light elements produced in the Big Bang (such as tritium) have long since decayed. Isotopes of elements heavier than boron were not produced at all in the Big Bang, and these first five elements do not have any long-lived radioisotopes. Thus, all radioactive nuclei are therefore relatively young with respect to the birth of the universe, having formed later in various other types of nucleosynthesis in stars (particularly supernovae), and also during ongoing interactions between stable isotopes and energetic particles. For example, carbon-14, a radioactive nuclide with a half-life of only 5730 years, is constantly produced in Earth's upper atmosphere due to interactions between cosmic rays and nitrogen.

Radioactive decay has been put to use in the technique of radioisotopic labeling, used to track the passage of a chemical substance through a complex system (such as a living organism). A sample of the substance is synthesized with a high concentration of unstable atoms. The presence of the substance in one or another part of the system is determined by detecting the locations of decay events.

On the premise that radioactive decay is truly random (rather than merely chaotic), it has been used in hardware random-number generators. Because the process is not thought to vary significantly in mechanism over time, it is also a valuable tool in estimating the absolute ages of certain materials. For geological materials, the radioisotopes and some of their decay products become trapped when a rock solidifies, and can then later be used (subject to many well-known qualifications) to estimate the date of the solidification. These include checking the results of several simultaneous processes and their products against each other, within the same sample. In a similar fashion, and also subject to qualification, the rate of formation of carbon-14 in various eras, the date of formation of organic matter within a certain period related to the isotope's half-live may be estimated, because the carbon-14 becomes trapped when the organic matter grows and incorporates the new carbon-14 from the air. Thereafter, the amount of carbon-14 in organic matter decreases according to decay processes which may also be independently cross-checked by other means (such as checking the carbon-14 in individual tree rings, for example).

Radioactive Decay Rates

The *decay rate*, or *activity*, of a radioactive substance are characterized by:

Constant Quantities

- *Half life*—symbol $t_{1/2}$—the time taken for the activity of a given amount of a radioactive substance to decay to half of its initial value.

- *Mean lifetime* (symbol τ)—the average lifetime of a radioactive particle.
- *Decay constant* (symbol λ)—the inverse of the mean lifetime. (*Note* that although these are constants, they are associated with statistically random behavior of substances, and predictions using these constants are less accurate for small number of atoms.)

Time-variable Quantities

- *Total activity* (symbol *A)*—number of decays an object undergoes per second.
- *Number of particles* (symbol *N)*—the total number of particles in the sample.
- *Specific activity* (symbol S_A)—number of decays per second per amount of substance. (The "*amount of substance*" can be the unit of either mass or volume.)

These are related as follows:

$$t_{1/2} = \frac{\ln(2)}{\lambda} = \tau \ln(2)$$

$$A = \frac{dN}{dt} = \lambda N$$

$$S_A a_0 = -\left.\frac{dN}{dt}\right|_{t=0} = \lambda N_0$$

where

a_0 is the initial amount of active substance—substance that has the same percentage of unstable particles as when the substance was formed.

Activity Measurements

The units in which activities are measured are: becquerel (symbol *Bq*) = number of disintegrations per second; curie (Ci) = 3.7×10^{10} disintegrations per second. Low activities are also measured in *disintegrations per minute* (dpm).

Decay Timing

As discussed above, the decay of an unstable nucleus is entirely random and it is impossible to predict when a particular atom will decay. However, it is equally likely to decay at any time. Therefore, given a sample of a particular radioisotope, the number of decay events—*dN* expected to occur in a small interval of time *dt* is proportional to the number of atoms present. If *N* is the number of atoms, then the probability of decay (- *dN*/*N*) is proportional to *dt*:

$$\left(-\frac{dN}{N}\right) = -\lambda.\ dt$$

Particular radionuclides decay at different rates, each having its own decay constant (λ). The negative sign indicates that N decreases with each decay event. The solution to this first-order differential equation is the following function:

$$\mathrm{N}(t) = N_0 e^{-\lambda t} = N_0 e^{-t/\tau}.$$

Where N_0 is the amount of N at time zero ($t = 0$). The second equation recognizes that the differential decay constant λ has units of 1/time, and can thus also be represented as 1/τ, where τ is a characteristic time for the process. This characteristic time is called the time constant of the process. In radioactive decay, this process time constant it also the mean lifetime for decaying atoms. Each atom "lives" for a finite amount of time before it decays, and it may be shown that this mean lifetime is the arithmetic mean of all the atoms' lifetimes, and that it is τ, which again is related to the decay constant as follows:

$$\tau = \frac{1}{\lambda}.$$

The previous exponential function generally represents the result of exponential decay. It is only an approximate solution, for two reasons. Firstly, the exponential function is continuous, but the physical quantity *N* can only take non-negative integer values. Secondly, because it describes a random process, it is only statistically true. However, in most common cases, *N* is a very large number and the function is a good approximation.

A more commonly used parameter is the half-life. Given a sample of a particular radionuclide, the half-life is the time taken for half the radionuclide's atoms to decay. The half life is related to the decay constant as follows:

$$t½ = \frac{\ln 2}{\lambda} = \tau \ln 2.$$

This relationship between the half-life and the decay constant shows that highly radioactive substances are quickly spent, while those that radiate weakly endure longer. Half-lives of known radionuclides vary widely, from more than 10^{19} years (such as for very nearly stable nuclides, *e.g.* ^{209}Bi), to 10^{-23} seconds for highly unstable ones.

5.8 Toxicological Chemistry

Toxicology → the science of poisons poison, toxicant—a substance that is harmful to living organisms Poisonous effects depends on:

1. type of organism exposed
2. the amount of the substance
3. the rate of the exposure:
   - skin
   - inhalation
   - injection

Physical forms of toxicants:

- gases
- vapors
- dusts
- fumes
- mists

Exposure:

- dose
- toxicant concentration
- duration
- frequency
- rate

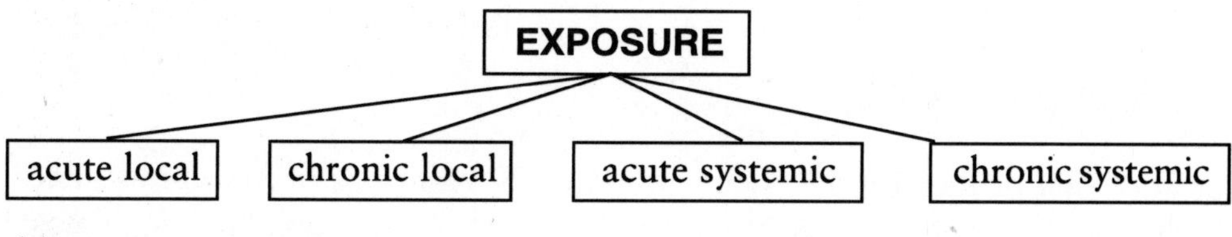

dose <——> response

Dose

Amount (per unit mass of body) of a toxicant to which an organism is exposed

Response

Effect upon an organism resulting from exposure to a toxicant dose that would kill 50 per cent of subjects.

Toxicity Rating

based on LD_{50}

10^5 (nontoxic) —-> 10^{-2} or smaller (supertoxic)

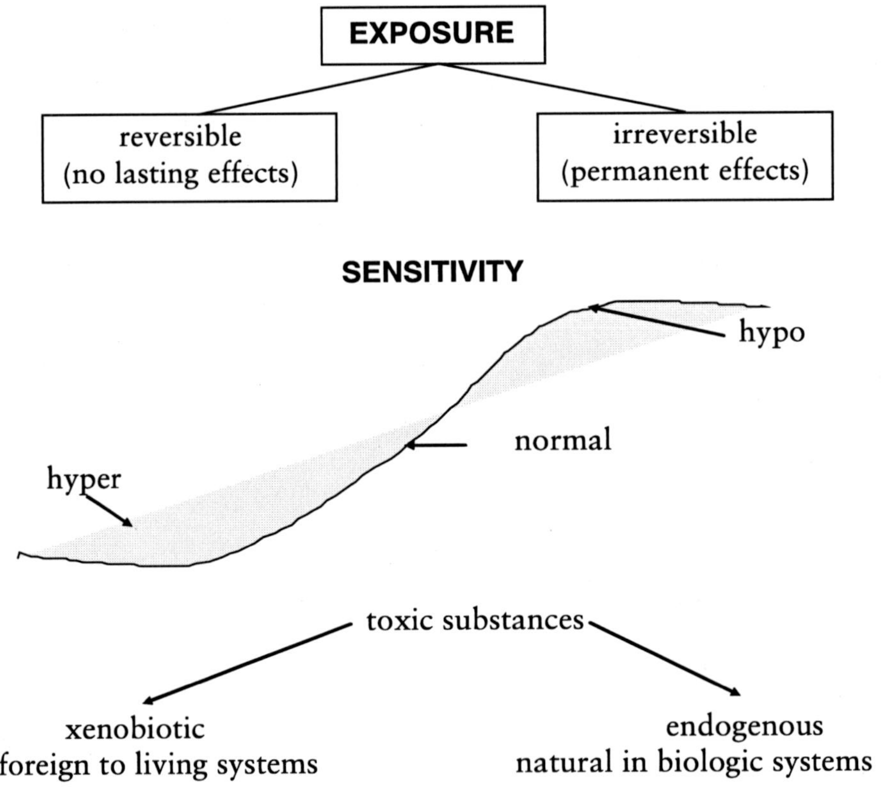

Toxicological Chemistry

Science that deals with the chemical nature & reactions of toxic substances, including their origins, uses, & chemical aspects of exposure, fates & disposal

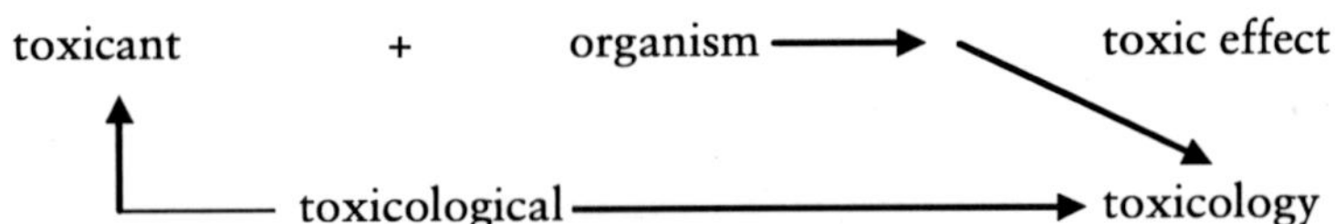

Metabolism of Xenobiotic Species

1. *Phase I reactions*: Attachment of functional groups ===> more water soluble species.
2. *Phase II reaction*: Polar functional groups are sites for those reactions (conjugation) enzymes attach conjugating agent to xenobiotics ===> conjugation product (less toxic).

Kinetic Phase:

- ❖ adsorption
- ❖ metabolism
- ❖ temporary storage
- ❖ distribution
- ❖ excretion

Dynamic Phase:

1. primary reaction
2. biochemical response
3. observable effects (behavioral or physiological response)

irreversible | reversible | $O_2Hb + CO \Longleftrightarrow COHb + O_2$

- ❖ impairment of enzyme function
- ❖ alteration of cell membrane
- ❖ interference with metabolism
- ❖ interference with respiration
- ❖ interference with regulatory process

alterations in vital signs:

- ❖ temperature
- ❖ pulse rate
- ❖ respiratory rate
- ❖ blood pressure
- ❖ odors
- ❖ changes in eyes
- ❖ convulsions
- ❖ paralysis

- ❖ hallucinations
- ❖ coma

Teratogens

Chemical species that cause birth defects:

- ❖ enzyme inhibition
- ❖ interference with energy supply
- ❖ depravation of the fetus of essential substances

Mutagenes

Alter DNA to produce inheritable traits:

- ❖ replacement of amino groups in DNA bases by hydroxy groups — dimethylnitroso amine
- ❖ alkylation — methylemetane sulfonate

Carcinogenic Agents

Cancer ===> a condition characterized by the uncontrolled replication & growth of the body's own cells.

Classification of Carcinogens:

- ❖ chemical agents (PAH)
- ❖ biological agents (retroviruses)
- ❖ ionizing radiation (X-ray)
- ❖ genetic factors (selective breeding)

Biochemistry of Carcinogens

1. initiation stage:
   - ❖ epigenetic carcinogens
   - ❖ genotoxic carcinogens
   - ❖ paracarcinogens (require metabolic activation)
   - ❖ ultimate carcinogens (metabolic species)
2. promotional stage promoters:
   - ❖ increase in the number of tumor cells
   - ❖ decrease in the length of time for tumor to develop

Chemical carcinogens—-> have the ability to form covalent bonds with macromolecular life molecules

|

alkylating agents—attach alkyl group to DNA

Immune System Response

acts as the body's natural defense system to protect it from xenobiotic chemicals. infectious agents, neoplastic cells:

toxicants → immunosupression impairment of the body's natural defense mechanisms

↓

allergy (hypersensitivity) immune system overacts to the presence of a foreign agent or its metabolites in a self-destructive manner

Risk Assessment

- identification of hazard
- dose-response assessment
- exposure assessment
- risk characterization

Toxic Elements (& elemental forms)

- ozone
- white phosphorus
- elemental halogens
- heavy metals—Be, Cd, Pb, Hg

Toxic Inorganic

- cyanide (HCN)
- carbon monoxide (CO)
- nitrogen oxides (lipid peroxidation)
- hydrogen halides, HX
- inorganic compounds of Si
- silica (SiO_2)

- silanes (H-Si)
- silicon tetrahalides
- phosphines (PH_3)
- tetraphosphorus dexoide (P_4O_{10})
- phosphorus halides (PX_3)
- phosphorus oxyhalides ($POCl_3$)
- hydrogen sulfide (H_2S)
- sulfur dioxide.

Organometallics

- organolead—tetratethyl lead
- organotin—TBT (tibutyltin)
- carbonyl—[$Ni(CO_4)$]

Toxic Organic Compounds

- alkane hydrocarbons (no-hexane),
- alkene & alkyne hydrocarbons,
- benzene & aromatic compounds:
    - toluene,
    - naphtalene,
    - PAH.
- oxygen containing organics:
    - oxides.

5.9 Radioactive Waste

Radioactive wastes are waste types containing radioactive chemical elements that do not have a practical purpose. They are sometimes the products of nuclear processes, such as nuclear fission. However, industries not directly connected to the nuclear industry can produce large quantities of radioactive waste. It has been estimated, for instance, that the past 20 years the oil-producing endeavors of the United States have accumulated eight million tons of radioactive wastes. The majority of radioactive waste is "low-level waste", meaning it contains low levels of radioactivity per mass or volume. This type of waste often consists of used protective clothing, which is only slightly contaminated but still dangerous in case of radioactive contamination of a human body through ingestion, inhalation, absorption, or injection. In the United States alone, the Department of Energy states that there are "millions of gallons of radioactive waste" as well as "thousands of tons of spent nuclear fuel and material" and also "huge quantities of contaminated soil and water". Despite these copious quantities of waste, the DOE has a goal of cleaning all presently contaminated sites successfully by 2025. The Fernald, Ohio site for example had "31 million pounds of uranium product", "2.5 billion pounds of waste", "2.75 million cubic yards of contaminated soil and debris", and a "223 acre portion of the underlying Great Miami Aquifer had uranium levels above drinking standards". The United States currently has at least 108 sites it currently designates as areas that are contaminated and unusable, sometimes many thousands of acres. The DOE wishes to try and clean or mitigate many or all by 2025, however the task can be difficult and it acknowledges that some will never be completely remediated, and just in one of these 108 larger designations, Oak Ridge National Laboratory, there were for example at least "167 known contaminant release sites" in one of the three subdivisions of the 37,000-acre (150 km^2) site. Some of the U.S. sites were smaller in nature, however, and cleanup issues were simpler to address, and the DOE has successfully completed cleanup, or at least closure, of several sites.

The issue of disposal methods for nuclear waste was one of the most pressing current problems the international nuclear industry faced when trying to establish a long term energy production plan, yet there was hope it could be safely solved. A recent research report on the Nuclear Industry perspective of the current state of scientific knowledge in predicting the extent that waste would find its way from the deep burial facility—back to soil and drinking water (such that it presents a direct threat to the health of human beings—as well as to other forms of life) is presented in a document from the IAEA (The International Atomic Energy Agency)—which was published in October 2007 This document states "The capacity to model all the effects involved in the dissolution of the waste form, in conditions similar to the disposal site, is the final goal of all the research undertaken by many research groups over many years. As we will see in this report, this kind of investigation is far from being finished". In the United States, the DOE acknowledges much progress in addressing the waste problems of the industry, and successful remediation of some contaminated sites, yet also major uncertainties and sometimes complications and setbacks in handling the issue properly, cost effectively, and in the projected time frame. In other countries with lower ability or will to maintain environmental integrity the issue would be even more problematic.

The Nature and Significance of Radioactive Waste

Radioactive waste typically comprises a number of radioisotopes: unstable configurations of elements that decay, emitting ionizing radiation which can be harmful to human health and to the environment. Those isotopes emit different types and levels of radiation, which last for different periods of time.

Physics

Medium-lived fission products				
Property: $t^{½}$	Unit: (a)	Yield (%)	Q * (KeV)	βγ*
^{155}Eu	4.76	.0803	252	βγ
^{85}Kr	10.76	.2180	687	βγ
^{113m}Cd	14.1	.0008	316	β
^{90}Sr	28.9	4.505	2826	β
^{137}Cs	30.23	6.337	1176	βγ
^{121m}Sn	43.9	.00005	390	βγ
^{151}Sm	90	.5314	77	β

Long-Lived Fission Products				
Property: $t^{½}$	Unit: (Ma)	Yield (%)	Q * (KeV)	βγ*
^{99}Tc	.211	6.1385	294	β
^{126}Sn	.230	.1084	4050	βγ
^{79}Se	.295	.0447	151	β
^{93}Zr	1.53	5.4575	91	βγ
^{135}Cs	2.3	6.9110	269	β
^{107}Pd	6.5	1.2499	33	β
^{129}I	15.7	.8410	194	βγ

The radioactivity of all nuclear waste diminishes with time. All radioisotopes contained in the waste have a half-life—the time it takes for any radionuclide to lose half of its radioactivity and eventually all radioactive waste decays into non-radioactive elements. Certain radioactive elements (such as plutonium-239) in "spent" fuel will remain hazardous to humans and other living beings for hundreds of thousands of years. Other radioisotopes will remain hazardous for millions of years. Thus, these wastes must be shielded for centuries and isolated from the living environment for hundreds of millennia. Some elements, such as Iodine-131, have a short half-life (around 8 days in this case) and thus they will cease to be a problem much more quickly than other, longer-lived, decay products but their activity is much greater initially. The two tables show some of the major radioisotopes, their half-lives, and their radiation yield as a proportion of the yield of fission of Uranium-235.

The faster a radioisotope decays, the more radioactive it will be. The energy and the type of the ionizing radiation emitted by a pure radioactive substance are important factors in deciding how dangerous it will be. The chemical properties of the radioactive element will determine how mobile the substance is and how likely it is to spread into the environment and contaminate human bodies. This is further complicated by the fact that many radioisotopes do not decay immediately to a stable state but rather to a radioactive decay product leading to decay chains.

Chemistry

The chemical properties of the radioactive substance and the other substances found within (and near) the waste store has a great effect upon the ability of the waste to cause harm to humans or other organisms. For instance TcO_4^- tends to adsorb on the surfaces of steel objects which reduces its ability to move out of the waste store in water.

Pharmacokinetics

Exposure to high levels of radioactive waste may cause serious harm or death. Treatment of an adult animal with radiation or some other mutation-causing effect, such as a cytotoxic anti-cancer drug, may cause cancer in the animal. In humans it has been calculated that a 1 sievert dose has a 5 per cent chance of causing cancer and a 1 per cent chance of causing a mutation in a gamete which can be passed to the next generation. If a developing organism such as an unborn child is irradiated, then it is possible to induce a birth defect but it is unlikely that this defect will be in a gamete or a gamete forming cell.

Depending on the decay mode and the pharmacokinetics of an element (how the body processes it and how quickly), the threat due to exposure to a given activity of a radioisotope will differ. For instance Iodine-131 is a short-lived beta and gamma emitter but because it concentrates in the thyroid gland, it is more able to cause injury than cesium-137 which, being water soluble, is rapidly excreted in urine. In a similar way, the alpha emitting actinides and radium are considered very harmful as they tend to have long biological half-lives and their radiation has a high linear energy transfer value. Because of such differences, the rules determining biological injury differ widely according to the radioisotope, and sometimes also the nature of the chemical compound which contains the radioisotope.

Philosophy

The main objective in managing and disposing of radioactive (or other) waste is to protect people and the environment. This means isolating or diluting the waste so that the rate or concentration of any radionuclide returned to the biosphere is harmless. To achieve this the preferred technology to date has been deep and secure burial for the more dangerous wastes; transmutation, long-term retrievable storage, and removal to space have also been suggested. Management options for waste are discussed below.

Radioactivity by definition reduces over time, so in principle the waste needs to be isolated for a particular period of time until its components have decayed such that it no longer poses a threat. In practice this can mean periods of hundreds of thousands of years, depending on the nature of the waste involved.

Though an affirmative answer is often taken for granted, the question as to whether or not we should endeavor to avoid causing harm to remote future generations, perhaps thousands upon thousands of years hence, is essentially one which must be dealt with by philosophy.

Sources of Waste

Radioactive waste comes from a number of sources. The majority originates from the nuclear fuel cycle and nuclear weapon reprocessing. However, other sources include medical and industrial wastes, as well as naturally occurring radioactive materials (NORM) that can be concentrated as a result of the processing or consumption of coal, oil and gas, and some minerals.

Nuclear Fuel Cycle

Front End

Waste from the front end of the nuclear fuel cycle is usually alpha emitting waste from the extraction of uranium. It often contains radium and its decay products.

Uranium dioxide (UO_2) concentrate from mining is not very radioactive—only a thousand or so times as radioactive as the granite used in buildings. It is refined from yellowcake (U_3O_8), then converted to uranium hexafluoride gas (UF_6). As a gas, it undergoes enrichment to increase the U-235 content from 0.7 per cent to about 4.4 per cent (LEU). It is then turned into a hard ceramic oxide (UO_2) for assembly as reactor fuel elements.

The main by-product of enrichment is depleted uranium (DU), principally the U-238 isotope, with a U-235 content of ~0.3 per cent. It is stored, either as UF_6 or as U_3O_8. Some is used in applications where its extremely high density makes it valuable, such as the keels of yachts, and anti-tank shells. It is also used (with recycled plutonium) for making mixed oxide fuel (MOX) and to dilute highly enriched uranium from weapons stockpiles which is now being redirected to become reactor fuel. This dilution, also called downblending, means that any nation or group that acquired the finished fuel would have to repeat the (very expensive and complex) enrichment process before assembling a weapon.

Back End

The back end of the nuclear fuel cycle, mostly spent fuel rods, contains fission products that emit beta and gamma radiation, and actinides that emit alpha particles, such as uranium-234, neptunium-237, plutonium-238 and americium-241, and even sometimes some neutron emitters such as californium (Cf). These isotopes are formed in nuclear reactors.

It is important to distinguish the processing of uranium to make fuel from the reprocessing of used fuel. Used fuel contains the highly radioactive products of fission (see high level waste below). Many of these are neutron absorbers, called neutron poisons in this context. These eventually build up to a level where they absorb so many neutrons that the chain reaction stops, even with the control rods completely removed. At that point the fuel has to be replaced in the reactor with fresh fuel, even though there is still a substantial quantity of uranium-235 and plutonium present. In the United States, this used fuel is stored, while in countries such as the United Kingdom, France, and Japan, the fuel is reprocessed to remove the fission products, and the fuel can then be re-used. This reprocessing involves handling highly radioactive materials, and the fission products removed from the fuel are a concentrated form of high-level waste as are the chemicals used in the process.

Proliferation Concerns

When dealing with uranium and plutonium, the possibility that they may be used to build nuclear weapons is often a concern. Active nuclear reactors and nuclear weapons stockpiles are very carefully safeguarded and controlled. However, high-level waste from nuclear reactors may contain plutonium. Ordinarily, this plutonium is reactor-grade plutonium, containing a mixture of plutonium-239 (highly suitable for building nuclear weapons), plutonium-240 (an undesirable contaminant and highly radioactive), plutonium-241 and plutonium-238; these isotopes are difficult to separate. Moreover, high-level waste is full of highly radioactive fission products. However, most fission products are relatively short-lived. This is a concern since if the waste is stored, perhaps in deep geological storage, over many years the fission products decay, decreasing the radioactivity of the waste and making the plutonium easier to access. Moreover, the undesirable contaminant Pu-240 decays faster than the Pu-239, and thus the quality of the bomb material increases with time (although its quantity decreases during that time as well). Thus, some have argued, as time passes, these deep storage areas have the potential to become "plutonium mines", from which material for nuclear weapons can be acquired with relatively little difficulty. Critics of the latter idea point out that the half-life of Pu-240 is 6,560 years and Pu-239 is 24,110 years, and thus the relative enrichment of one isotope to the other with time occurs with a half-life of 9,000 years (that is, it takes 9000 years for the *fraction* of Pu-240 in a sample of mixed plutonium isotopes, to spontaneously decrease by half—a typical enrichment needed to turn reactor-grade into weapons-grade Pu). Thus "weapons grade plutonium mines" would be a problem for the very far future (>9,000 years from now), so that there remains a great deal of time for technology to advance to solve this problem, before it becomes acute.

Pu-239 decays to U-235 which is suitable for weapons and which has a very long half life (roughly 10^9 years). Thus plutonium may decay and leave uranium-235. However, modern reactors are only moderately enriched with U-235 relative to U-238, so the U-238 continues to serve as denaturation agent for any U-235 produced by plutonium decay.

One solution to this problem is to recycle the plutonium and use it as a fuel *e.g.* in fast reactors. But in the minds of some, the very existence of the nuclear fuel reprocessing plant needed to separate the plutonium from the other elements represents a proliferation concern. In pyrometallurgical fast reactors, the waste generated is an actinide compound that cannot be used for nuclear weapons.

Nuclear Weapons Reprocessing

Waste from nuclear weapons reprocessing (as opposed to production, which requires primary processing from reactor fuel) is unlikely to contain much beta or gamma activity other than tritium and americium. It is more likely to contain alpha emitting actinides such as Pu-239 which is a fissile material used in bombs, plus some material with much higher specific activities, such as Pu-238 or Po.

In the past the neutron trigger for a bomb tended to be beryllium and a high activity alpha emitter such as polonium; an alternative to polonium is Pu-238. For reasons of national security, details of the design of modern bombs are normally not released to the open literature. It is likely however that a D-T fusion reaction in either an electrically driven device or a D-T fusion reaction driven by the chemical explosives would be used to start up a modern device.

Some designs might well contain a radioisotope thermoelectric generator using Pu-238 to provide a longlasting source of electrical power for the electronics in the device.

It is likely that the fissile material of an old bomb which is due for refitting will contain decay products of the plutonium isotopes used in it, these are likely to include alpha-emitting Np-236 from Pu-240 impurities, plus some U-235 from decay of the Pu-239; however, due to the relatively long half-life of these Pu isotopes, these wastes from radioactive decay of bomb core material would be very small, and in any case, far less dangerous (even in terms of simple radioactivity) than the Pu-239 itself.

The beta decay of Pu-241 forms Am-241; the in-growth of americium is likely to be a greater problem than the decay of Pu-239 and Pu-240 as the americium is a gamma emitter (increasing external-exposure to workers) and is an alpha emitter which can cause the generation of heat. The plutonium could be separated from the americium by several different processes; these would include pyrochemical processes and aqueous/organic solvent extraction. A truncated PUREX type extraction process would be one possible method of making the separation.

Medical

Radioactive medical waste tends to contain beta particle and gamma ray emitters. It can be divided into two main classes. In diagnostic nuclear medicine a number of short-lived gamma emitters such as technetium-99m are used. Many of these can be disposed of by leaving it to decay for a short time before disposal as normal trash. Other isotopes used in medicine, with half-lives in parentheses:

- Y-90, used for treating lymphoma (2.7 days).
- I-131, used for thyroid function tests and for treating thyroid cancer (8.0 days).
- Sr-89, used for treating bone cancer, intravenous injection (52 days).
- Ir-192, used for brachytherapy (74 days).
- Co-60, used for brachytherapy and external radiotherapy (5.3 years).
- Cs-137, used for brachytherapy, external radiotherapy (30 years).

Industrial

Industrial source waste can contain alpha, beta, neutron or gamma emitters. Gamma emitters are used in radiography while neutron emitting sources are used in a range of applications, such as oil well logging.

Naturally Occurring Radioactive Material (NORM)

Processing of substances containing natural radioactivity; this is often known as NORM. A lot of this waste is alpha particle-emitting matter from the decay chains of uranium and thorium. The main source of radiation in the human body is potassium-40 (^{40}K).

Coal

Coal contains a small amount of radioactive uranium, barium, thorium and potassium, but, in the case of pure coal, this is significantly less than the average concentration of those elements in the Earth's crust. However, the surrounding strata, if shale or mudstone, often contains slightly more than average and this may also be reflected in the ash content of 'dirty' coals. The more active ash minerals become concentrated in the fly ash precisely because they do not burn well. However, the radioactivity of fly ash is still very low. It is about the same as black shale and is less than phosphate rocks, but is more of a concern because a small amount of the fly ash ends up in the atmosphere where it can be inhaled.

Oil and Gas

Residues from the oil and gas industry often contain radium and its daughters. The sulphate scale from an oil well can be very radium rich, while the water, oil and gas from a well often contains radon. The radon decays to form solid radioisotopes which form coatings on the inside of pipework. In an oil processing plant the area of the plant where propane is processed is often one of the more contaminated areas of the plant as radon has a similar boiling point as propane.

Types of Radioactive Waste

Although not significantly radioactive, uranium mill tailings are waste. They are byproduct material from the rough processing of uranium-bearing ore. They are sometimes referred to as 11(e)2 wastes, from the section of the U.S.

Atomic Energy Act that defines them. Uranium mill tailings typically also contain chemically-hazardous heavy metals such as lead and arsenic. Vast mounds of uranium mill tailings are left at many old mining sites, especially in Colorado, New Mexico, and Utah.

Low level waste (LLW) is generated from hospitals and industry, as well as the nuclear fuel cycle. It comprises paper, rags, tools, clothing, filters, etc., which contain small amounts of mostly short-lived radioactivity. Commonly, LLW is designated as such as a precautionary measure if it originated from any region of an 'Active Area', which frequently includes offices with only a remote possibility of being contaminated with radioactive materials. Such LLW typically exhibits no higher radioactivity than one would expect from the same material disposed of in a non-active area, such as a normal office block. Some high activity LLW requires shielding during handling and transport but most LLW is suitable for shallow land burial. To reduce its volume, it is often compacted or incinerated before disposal. Low level waste is divided into four classes, class A, B, C and GTCC, which means "Greater Than Class C".

Intermediate level waste (ILW) contains higher amounts of radioactivity and in some cases requires shielding. ILW includes resins, chemical sludge and metal reactor fuel cladding, as well as contaminated materials from reactor decommissioning. It may be solidified in concrete or bitumen for disposal. As a general rule, short-lived waste (mainly non-fuel materials from reactors) is buried in shallow repositories, while long-lived waste (from fuel and fuel-reprocessing) is deposited in deep underground facilities. U.S. regulations do not define this category of waste; the term is used in Europe and elsewhere.

High level waste (HLW) is produced by nuclear reactors. It contains fission products and transuranic elements generated in the reactor core. It is highly radioactive and often thermally hot. LLW and ILW accounts for over 95 per cent of the total radioactivity produced in the process of nuclear electricity generation. The amount of HLW worldwide is currently increasing by about 12,000 metric tons every year, which is the equival to about 100 double-decker busses or a two-story structure built on top of a basketball court.

Transuranic waste (TRUW) as defined by U.S. regulations is, without regard to form or origin, waste that is contaminated with alpha-emitting transuranic radionuclides with half-lives greater than 20 years, and concentrations greater than 100 nCi/g (3.7 MBq/kg), excluding High Level Waste. Elements that have an atomic number greater than uranium are called transuranic ("beyond uranium"). Because of their long half-lives, TRUW is disposed more cautiously than either low level or intermediate level waste. In the U.S. it arises mainly from weapons production, and consists of clothing, tools, rags, residues, debris and other items contaminated with small amounts of radioactive elements (mainly plutonium).

Under U.S. law, TRUW is further categorized into "contact-handled" (CH) and "remote-handled" (RH) on the basis of radiation dose measured at the surface of the waste container. CH TRUW has a surface dose rate not greater than 200 mrem per hour (2 mSv/h), whereas RH TRUW has a surface dose rate of 200 mrem per hour (2 mSv/h) or greater. CH TRUW does not have the very high radioactivity of high level waste, nor its high heat generation, but RH TRUW can be highly radioactive, with surface dose rates up to 1000000 mrem per hour (10000 mSv/h). The United States currently permanently disposes of TRUW generated from nuclear power plants and military facilities at the Waste Isolation Pilot Plant.

Management of Waste

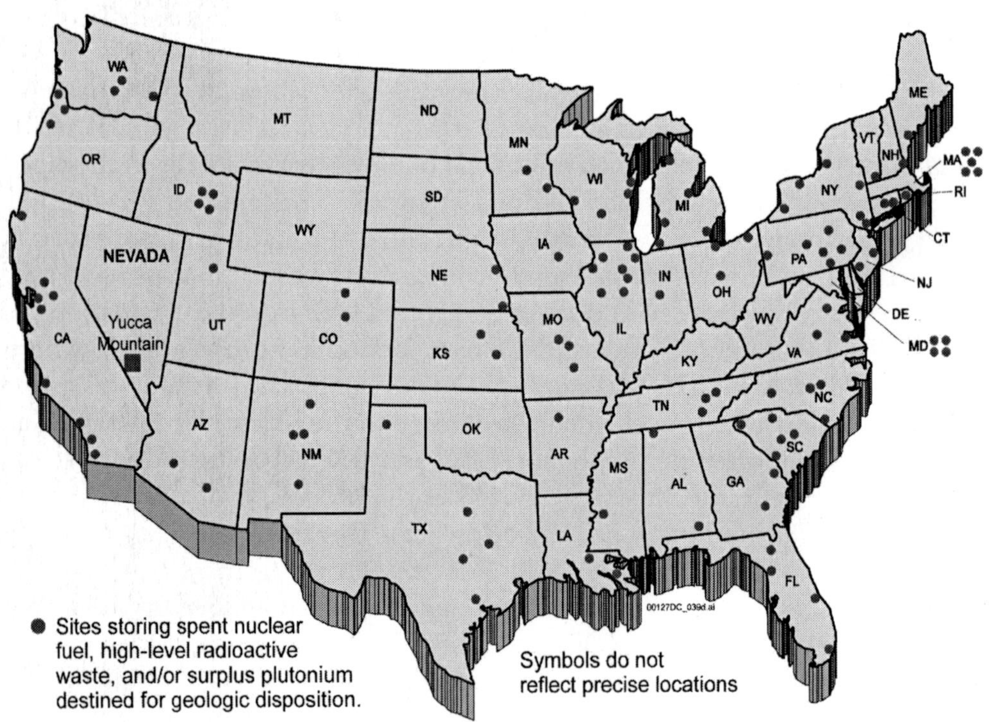

Fig. 5.11: Nuclear waste locations in USA.

Nuclear waste requires sophisticated treatment and management in order to successfully isolate it from interacting with the biosphere. This usually necessitates treatment, followed by a long-term management strategy involving storage, disposal or transformation of the waste into a non-toxic form.

Initial Treatment of Waste

Vitrification

Long-term storage of radioactive waste requires the stabilization of the waste into a form which will not react, nor degrade, for extended periods of time. One way to do this is through vitrification. Currently at Sellafield the high-level waste (PUREX first cycle raffinate) is mixed with sugar and then calcined. Calcination involves passing the waste through a heated, rotating tube. The purposes of calcination are to evaporate the water from the waste, and de-nitrate the fission products to assist the stability of the glass produced.

The 'calcine' generated is fed continuously into an induction heated furnace with fragmented glass. The resulting glass is a new substance in which the waste products are bonded into the glass matrix when it solidifies. This product, as a molten fluid, is poured into stainless steel cylindrical containers ("cylinders") in a batch process. When cooled, the fluid solidifies ("vitrifies") into the glass. Such glass, after being formed, is very highly resistant to water. According to the ITU, it will require about 1 million years for 10 per cent of such glass to dissolve in water.

After filling a cylinder, a seal is welded onto the cylinder. The cylinder is then washed. After being inspected for external contamination, the steel cylinder is stored, usually in an underground repository. In this form, the waste products are expected to be immobilized for a very long period of time (many thousands of years).

The glass inside a cylinder is usually a black glossy substance. All this work (in the United Kingdom) is done using hot cell systems. The sugar is added to control the ruthenium chemistry and to stop the formation of the volatile RuO_4 containing radio ruthenium. In the west, the glass is normally a borosilicate glass (similar to Pyrex), while in the former Soviet bloc it is normal to use a phosphate glass. The amount of fission products in the glass must be limited because some (palladium, the other Pt group metals, and tellurium) tend to form metallic phases which separate from the glass. In Germany a vitrification plant is in use; this is treating the waste from a small demonstration reprocessing plant which has since been closed down.

Ion Exchange

It is common for medium active wastes in the nuclear industry to be treated with ion exchange or other means to concentrate the radioactivity into a small volume. The much less radioactive bulk (after treatment) is often then discharged. For instance, it is possible to use a ferric hydroxide floc to remove radioactive metals from aqueous mixtures. After the radioisotopes are absorbed onto the ferric hydroxide, the resulting sludge can be placed in a metal drum before being mixed with cement to form a solid waste form. In order to get better long-term performance (mechanical stability) from such forms, they may be made from a mixture of fly ash, or blast furnace slag, and portland cement, instead of normal concrete (made with portland cement, gravel and sand).

Synroc

The Australian Synroc (synthetic rock) is a more sophisticated way to immobilize such waste, and this process may eventually come into commercial use for civil wastes (it is currently being developed for U.S. military wastes). Synroc was invented by the late Prof Ted Ringwood (a geochemist) at the Australian National University. The Synroc contains pyrochlore and cryptomelane type minerals. The original form of Synroc (Synroc C) was designed for the liquid high level waste (PUREX raffinate) from a light water reactor. The main minerals in this Synroc are hollandite ($BaAl_2Ti_6O_{16}$), zirconolite ($CaZrTi_2O_7$) and perovskite ($CaTiO_3$). The zirconolite and perovskite are hosts for the actinides. The strontium and barium will be fixed in the perovskite. The caesium will be fixed in the hollandite.

Long Term Management of Waste

We are talking here of durations ranging from 10,000 to 1,000,000 years, according to studies based on the effect of estimated radiation doses. It is worthwhile noting that state of the art only allows geological considerations for such long periods. Researchers suggest that forecasts of health detriment for such periods *should be examined critically*. Practical studies only consider up to 100 years as far as effective planning and cost evaluations are concerned.

Storage

High-level radioactive waste is stored temporarily in spent fuel pools and in dry cask storage facilities. This allows the shorter-lived isotopes to decay before further handling.

In 1997, in the 20 countries which account for most of the world's nuclear power generation, spent fuel storage capacity at the reactors was 148,000 tonnes, with 59 per cent of this utilized. However, a number of nuclear power plants in countries that do not reprocess had nearly filled their spent fuel pools, and resorted to Away-from-reactor storage (AFRS). AFRS capacity in

1997 was 78,000 tonnes, with 44 per cent utilized, and annual additions of about 12,000 tonnes. AFRS cannot be expanded forever, and the lead times for final disposal sites have proven to be unpredictable (see below).

In 1989 and 1992, France commissioned commercial plants to vitrify HLW left over from reprocessing oxide fuel, although there are adequate facilities elsewhere, notably in the United Kingdom and Belgium. The capacity of these western European plants is 2,500 canisters (1000 t) a year, and some have been operating for 18 years.

Geological Disposal

The process of selecting appropriate deep final repositories for high level waste and spent fuel is now under way in several countries with the first expected to be commissioned some time after 2010. However, many people remain uncomfortable with the immediate stewardship cessation of this management system. In Switzerland, the Grimsel Test Site is an international research facility investigating the open questions in radioactive waste disposal. Sweden is well advanced with plans for direct disposal of spent fuel, since its Parliament decided that this is acceptably safe, using the KBS-3 technology. In Germany, there is a political discussion about the search for an *Endlager* (final repository) for radioactive waste, accompanied by loud protests especially in the Gorleben village in the Wendland area, which was seen ideal for the final repository until 1990 because of its location next to the border to the former German Democratic Republic. Gorleben is presently being used to store radioactive waste non-permanently, with a decision on final disposal to be made at some future time. The U.S. has opted for a final repository at Yucca Mountain in Nevada, but this project is widely opposed and is a hotly debated topic, with some of the main concerns being the long distance transportation of the waste from across the United States to this area, and the possibility of accidents over time that could occur. The Waste Isolation Pilot Plant in the United States is the world's first underground repository for transuranic waste. There is also a proposal for an international HLW repository in optimum geology, with Australia or Russia as possible locations, although the proposal for a global repository for Australia has raised fierce domestic political objections.

The Canadian government, for example, is seriously considering this method of disposal, known as the *Deep Geological Disposal* concept. Under the current plan, a vault is to be dug 500 to 1000 meters below ground, under the Canadian Shield, one of the most stable landforms on the planet. The vaults are to be dug inside geological formations known as *batholiths*, formed about a billion years ago. The used fuel bundles will be encased in a corrosion-resistant container, and further surrounded by a layer of *buffer material*, possibly of a special kind of clay (bentonite clay). The case itself is designed to last for thousands of years, while the clay would further slow the corrosion rates of the container. The batholiths themselves are chosen for their low ground-water movement rates, geological stability, and low economic value.

The Finnish government has already started building a vault to store nuclear waste 500 to 1000 meters below ground, not far from the Olkiluoto Nuclear Power Plant.

In the EU, Covra is negotiating about a European-wide waste disposal system with single disposal sites that can be used by several EU-countries. This EU-wide storage possibility is being researched under the SAPIERR-2 program.

Storing high level nuclear waste above ground for a century or so is considered appropriate by many scientists. This allows for the material to be more easily observed and any problems detected and managed, while the decay over this time period significantly reduces the level of radioactivity and the associated harmful effects to the container material. It is also considered likely that over the next century newer materials will be developed which will not break down as quickly when exposed to a high neutron flux thus increasing the longevity of the container once it is permanently buried.

Sea-based options for disposal of radioactive waste include burial beneath a stable abyssal plain, burial in a subduction zone that would slowly carry the waste downward into the Earth's mantle, and burial beneath a remote natural or human-made island. While these approaches all have merit and would facilitate an international solution to the vexing problem of disposal of radioactive waste, they are currently not being seriously considered because of the legal barrier of the Law of the Sea and because in North America and Europe sea-based burial has become taboo from fear that such a repository could leak and cause widespread damage. Dumping of radioactive waste from ships has reinforced this concern, as has contamination of islands in the Pacific. However, sea-based approaches might come under consideration in the future by individual countries or groups of countries that cannot find other acceptable solutions.

Article 1 (Definitions), 7., of the 1996 Protocol to the Convention on the Prevention of Marine Pollution by Dumping of Wastes and Other Matter, (the London Dumping Convention) states:

"Sea" means all marine waters other than the internal waters of States, as well as the seabed and the subsoil thereof; it does no include sub-seabed repositories accessed only from land."

A subduction zone accessed from land as is the embodiment of the subductive waste disposal method http://www3.telus.net/subductionservices/ is not nor ever has been prohibited by international agreement.

This method has been described as the most viable means of disposing of radioactive and the state-of-the-art in nuclear waste disposal technology.

Another approach termed Remix & Return would blend high-level waste with uranium mine and mill tailings down to the level of the original radioactivity of the uranium ore, then replace it in empty uranium mines. This approach has the merits of totally eliminating the problem of high-level waste, of providing jobs for miners who would double as disposal staff, and of facilitating a cradle-to-grave cycle for all radioactive materials.

Transmutation

There have been proposals for reactors that consume nuclear waste and transmute it to other, less-harmful nuclear waste. In particular, the Integral Fast Reactor was a proposed nuclear reactor with a nuclear fuel cycle that produced no transuranic waste and in fact, could consume transuranic waste. It proceeded as far as large-scale tests but was then canceled by the U.S. Government. Another approach, considered safer but requiring more development, is to dedicate subcritical reactors to the transmutation of the left-over transuranic elements.

Transmutation was banned in the US on April 1977 by President Carter due to the danger of plutonium proliferation however, President Reagan rescinded the ban in 1981. Due to the economic losses and risks, construction of reprocessing plants during this time did not resume. Due to high energy demand, work on the method has continued in the EU. This has resulted in a practical nuclear research reactor called Myrrha in which transmutation is possible. Additionally, a new research program called ACTINET has been started in the EU to make transmutation possible on a large, industrial scale. According to President Bush's Global Nuclear Energy Partnership (GNEP) of 2007, the US is now actively promoting research on transmutation technologies needed to markedly reduce the problem of nuclear waste treatment.

There have also been theoretical studies involving the use of fusion reactors as so called "actinide burners" where a fusion reactor plasma such as in a tokamak, could be "doped" with a small amount of the "minor" transuranic atoms which would be transmuted (meaning fissioned in the actinide case) to lighter elements upon their successive bombardment by the very high energy neutrons produced by the fusion of deuterium and tritium in the reactor. It was recently found by a study done at MIT, that only 2 or 3 fusion reactors with parameters similar to that of the International Thermonuclear Experimental Reactor (ITER) could transmute the entire annual minor actinide production from all of the light water reactors presently operating in the United States fleet while simultaneously generating approximately 1 gigawatt of power from each reactor

Reuse of Waste

Another option is to find applications of the isotopes in nuclear waste so as to reuse them. Already, caesium-137, strontium-90 and a few other isotopes are extracted for certain industrial applications such as food irradiation and radioisotope thermoelectric generators. While re-use does not eliminate the need to manage radioisotopes, it may reduce the quantity of waste produced.

Space Disposal

Space disposal is an attractive notion because it permanently removes nuclear waste from the environment. However, it has significant disadvantages, not least of which is the potential for catastrophic failure of a launch vehicle. Furthermore, the high number of launches that would be required—due to the fact that no individual rocket would be able to carry very much of the material relative to the material needed to be disposed of—makes the proposal impractical (for both economic and risk-based reasons). To further complicate matters, international agreements on the regulation of such a program would need to be established.

It has been suggested that through the use of a stationary launch system many of the risks of catastrophic launch failure could be avoided. A promising concept is the use of high power lasers to launch "indestructible" containers from the ground into space. Such a system would require no rocket propellant, with the launch vehicle's payload making up a near entirety of the vehicle's mass. Without the use of rocket fuel on board there would be little chance of the vehicle exploding.

Another form of safe removal would possibly be the space elevator. Encasing the waste in glassified form inside a steel shell 9 inches (230 mm) thick, which in turn is tiled with shuttle tile to its exterior. If the launch vehicle fails just before reaching orbit, the waste ball will safely re-enter the earth's atmosphere. The steel shell would deform on impact, but would not rupture due to the density of the shell. Also, this would potentially allow the waste to be shot into the Sun.

Accidents Involving Radioactive Waste

A number of incidents have occurred when radioactive material was disposed of improperly, shielding during transport was defective, or when it was simply abandoned or even stolen from a waste store. In the former Soviet Union, waste stored in Lake Karachay was blown over the area during a dust storm after the lake had partly dried out. At Maxey Flat, a low-level radioactive waste facility located in Kentucky, containment trenches covered with dirt, instead of steel or cement, collapsed under heavy rainfall into the trenches and filled with water. The water that invaded the trenches became radioactive and had to be disposed of at the Maxey Flat facility itself. In other cases of radioactive waste accidents, lakes or ponds with radioactive waste accidentally overflowed into the rivers during exceptional storms.

Scavenging of abandoned radioactive material has been the cause of several other cases of radiation exposure, mostly in developing nations, which may have less regulation of dangerous substances (and sometimes less general education about radioactivity and its hazards) and a market for scavenged goods and scrap metal. The scavengers and those who buy the material are almost always unaware that the material is radioactive and it is selected for its aesthetics or scrap value. Irresponsibility on the part of the radioactive material's owners, usually a hospital, university or military, and the absence of regulation concerning radioactive waste, or a lack of enforcement of such regulations, have been significant factors in radiation exposures. For an example of an accident involving radioactive scrap originating from a hospital see the Goiânia accident.

Transportation accidents involving spent nuclear fuel from power plants are unlikely to have serious consequences due to the strength of the spent nuclear fuel shipping casks.

Radioactive Waste in Fiction and Popular Culture

In fiction, radioactive waste is often cited as the reason for gaining super-human powers and abilities. An example of this fictional scenario is the 1981 movie "Modern Problems" in which actor Chevy Chase portrays a jealous, harried air traffic controller Max Fiedler. Fiedler, recently dumped by his girlfriend, comes into contact with nuclear waste and is granted the power of telekinesis, which he uses to not only win her back, but to gain a little revenge.

In reality, of course, exposure to radioactive waste instead would lead to illness and/or death.

In the science fiction television series, "Space: 1999," a massive nuclear waste dump on the Moon explodes, hurtling the Moon, and the inhabitants of "Moonbase Alpha" out of the Solar System at interstellar speeds.

In the television comedy series Family Guy, the Griffin family all get super-human powers from toxic waste. When the local mayor Adam West tries to do the same thing, he gets lymphoma.

In The Simpsons, many mutant three-eyed fish live near the Springfield Nuclear Power Plant. The owner of the plant, Mr Burns, is also repeatedly shown disposing of his plants waste in an improper manner, either dumping it in the river or hiding it in trees at the local park.

5.10 Toxicology

Toxicology (from the Greek words *toxicos* and *logos*) is the study of the adverse effects of chemicals on living organisms. It is the study of symptoms, mechanisms, treatments and detection of poisoning, especially the poisoning of people.

History

Mathieu Orfila is considered to be the modern father of toxicology, having given the subject its first formal treatment in 1813 in his *Traité des poisons*, also called *Toxicologie générale.*

Theophrastus Phillipus Auroleus Bombastus von Hohenheim (1493—1541) (also referred to as Paracelsus, from his belief that his studies were above or beyond the work of Celsus—the Roman physician from the first century) is also considered "the father" of toxicology. He is credited with the classic toxicology maxim, "*Alle Dinge sind Gift und nichts ist ohne Gift; allein die Dosis macht, dass ein Ding kein Gift ist.*" which translates as, "All things are poison and nothing is without poison; only the dose makes a thing not a poison." This is often condensed to: "The dose makes the poison".

An even earlier writer on toxicology was Ibn Wahshiya, who wrote the *Book on Poisons* in the 9th or 10th century.

Relationship between Dose and Toxicity

Toxicology is the study of the relationship between dose and its effects on the exposed organism. The chief criterion regarding the toxicity of a chemical is the dose, *i.e.* the amount of exposure to the substance. Almost all substances are toxic under the right conditions as Paracelsus, the father of modern toxicology said, *Sola dosis facit venenum* (only dose makes the poison).

Paracelsus, who lived in the 16th century, was the first person to explain the dose-response relationship of toxic substances. The term LD_{50} refers to the dose of a toxic substance that kills 50 percent of a test population (typically rats or other surrogates when the test concerns human toxicity). LD_{50} estimations in animals are no longer required for regulatory submissions as a part of pre-clinical development package.

Toxicity of Metabolites

Many substances regarded as poisons are toxic only indirectly. An example is "wood alcohol," or methanol, which is chemically converted to formaldehyde and formic acid in the liver. It is the formaldehyde and formic acid that cause the toxic effects of methanol exposure. Many drug molecules are made toxic in the liver, a good example being acetaminophen (paracetamol), especially in the presence of alcohol. The genetic variability of certain liver enzymes makes the toxicity of many compounds differ between one individual and the next. Because demands placed on one liver enzyme can induce activity in another, many molecules become toxic only in combination with others. A family of activities that engages many toxicologists includes identifying which liver enzymes convert a molecule into a poison, what are the toxic products of the conversion and under what conditions and in which individuals this conversion takes place.

Chemical Toxicology

Chemical toxicology is a scientific discipline involving the study of structure and mechanism related to the toxic effects of chemical agents, and encompasses technology advances in research related to chemical aspects of *toxicology*. Research in this area is strongly multidisciplinary, spanning computational chemistry and synthetic chemistry, proteomics and metabolomics, drug discovery, drug metabolism and mechanisms of action, bioinformatics, bioanalytical chemistry, chemical biology, and molecular epidemiology.

5.11 Transport of Radioactive Materials

- About twenty million packages of all sizes containing radioactive materials are routinely transported worldwide annually on public roads, railways and ships.
- These use robust and secure containers. At sea, they are generally carried in purpose-built ships.

- Since 1971 there have been more than 20 000 shipments of used fuel and high-level wastes (over 80 000 tonnes) over many million kilometres.
- There has never been any accident in which a container with highly radioactive material has been breached, or has leaked.

About 20 million transports of radioactive material (which may be either a single package or a number of packages sent from one location to another at the same time) take place around the world each year. Radioactive material is not unique to the nuclear fuel cycle and most transports of such material are not fuel cycle related. Radioactive materials are used extensively in medicine, agriculture, research, manufacturing, non-destructive testing and minerals' exploration.

The regulatory control of shipments of radioactive material is independent of its intended application and the same safety procedures are employed, whatever the intended end-use.

Nuclear fuel cycle facilities are located in various parts of the world and materials of many kinds need to be transported between them. Many of these are similar to materials used in other industrial activities. However, the nuclear industry's fuel and waste materials are radioactive, and it is these 'nuclear materials' about which there is most public concern.

Nuclear materials have been transported since before the advent of nuclear power over fifty years ago. The procedures employed are designed to ensure the protection of the public and the environment. For the generation of a given quantity of electricity, the amount of nuclear fuel required is very much smaller than the amount of any other fuels. Therefore, the conventional risks and environmental impacts associated with fuel transport are greatly reduced with nuclear power.

Materials being Transported

Transport is an integral part of the nuclear fuel cycle. There are some 430 nuclear power reactors in operation in 32 countries but uranium mining is viable in only a few areas. Furthermore, in the course of over forty years of operation by the nuclear industry, a number of specialised facilities have been developed in various locations around the world to provide fuel cycle services. It is clear that there is a need to transport nuclear fuel cycle materials to and from these facilities. Indeed, most of the material used in nuclear fuel is transported several times during its progress through the fuel cycle. Transports are frequently international, and are often over large distances. Nuclear materials are generally transported by specialised transport companies.

The term 'transport' is used in this document only to refer to the movement of material between facilities, *i.e.* through areas outside such facilities. Most transports of nuclear fuel material occur between different stages of the cycle, but occasionally a material may be transported between similar facilities. When the stages are directly linked (such as mining and milling), it is sometimes advantageous to construct facilities for the different stages on the same site and no transport is then required.

With very few exceptions, nuclear fuel cycle materials are transported in solid form. The following table shows the principal nuclear material transport activities:

From:	*To:*	*Material:*	*Notes:*
Mining	Milling	Ore	Rare: usually on the same site
Milling	Conversion	Uranium oxide concentrate ("Yellowcake")	
Conversion	Enrichment	Uranium hexafluoride (UF_6)	
Enrichment	Fuel fabrication	Enriched UF_6	
Fuel fabrication	Power generation	Fresh (unused) fuel	
Power generation	Used fuel storage	used fuel	After on-site storage
Used fuel storage	Disposal*	used fuel	
Used fuel storage	Reprocessing	used fuel	
Reprocessing	Conversion	Uranium oxide	Called reprocessed uranium
Reprocessing	Fuel fabrication	Plutonium oxide	
Reprocessing	Disposal*	Fission products	Vitrified (incorporated into glass)
All facilities	Storage/disposal	Waste materials	Sometimes on the same site

* Not yet taking place.

Although some waste disposal facilities are located adjacent to the facilities that they serve, utilising one disposal site to manage the wastes from several facilities usually reduces environmental impacts. When this is the case, transport of the wastes from the facilities to the disposal site will be required.

Packaging

The principal assurance of safety in the transport of nuclear materials is the design of the packaging, which must allow for foreseeable accidents. The consignor bears primary responsibility for this. Many different nuclear materials are transported and the degree of potential hazard from these materials varies considerably. Different packaging standards have been developed to recognise that increased potential hazard calls for increased protection.

'Type A' packages are designed to withstand minor accidents and are used for medium-activity materials such as medical or industrial radioisotopes. Ordinary industrial containers are used for low-activity material such as U_3O_8.

Packages for high-level waste (HLW) and used fuel are robust and very secure containers are known as 'Type B' packages. They also maintain shielding from gamma and neutron radiation, even under extreme conditions. There are over 150 kinds of Type B packages, and the larger ones cost some US$ 1.6 million each.

In France alone, there are some 750 shipments each year of Type B packages, among 15 million shipments classified as 'dangerous materials', 300,000 of these being radioactive materials of some kind.

Smaller amounts of high-activity materials (including plutonium) transported by aircraft will be in 'Type C' packages, which give greater protection in all respects than Type B packages in accident scenarios.

Radiation Protection

Since nuclear materials are radioactive, it is important to ensure that radiation exposure of both those involved in the transport of such materials and the general public along transport routes is limited. Packaging for nuclear materials includes, where appropriate, shielding to reduce potential radiation exposures. In the case of some materials, such as fresh uranium fuel assemblies, the radiation levels are negligible and no shielding is required. Other materials, such as used fuel and high-level waste, are highly radioactive and purpose-designed containers with integral shielding are used. To limit the risk in handling of highly radioactive materials, dual-purpose containers (casks), which are appropriate for both storage and transport of used nuclear fuel, are often used.

As with other hazardous materials being transported, packages of nuclear materials are labeled in accordance with the requirements of national and international regulations. These labels not only indicate that the material is radioactive, by including a radiation symbol, but also give an indication of the radiation field in the vicinity of the package.

Personnel directly involved in the transport of nuclear materials are trained to take appropriate precautions and to respond in case of an emergency.

Environmental Protection

Packages used for the transport of nuclear materials are designed to retain their integrity during the various conditions that may be encountered while they are being transported and to ensure that an accident will not have any major consequences. Conditions which packages are tested to withstand include: fire, impact, wetting, pressure, heat and cold. Packages of radioactive material are checked prior to shipping and, when it is found to be necessary, cleaned to remove contamination.

Although not required by transport regulations, the nuclear industry chooses to undertake some shipments of nuclear material using dedicated, purpose-built transport vehicles or vessels.

Regulation of Transport

Since 1961 the International Atomic Energy Agency (IAEA) has published advisory regulations for the safe transport of radioactive material. These regulations have come to be recognised throughout the world as the uniform basis for both national and international transport safety requirements in this area. Requirements based on the IAEA regulations have been adopted in about 60 countries, as well as by the International Civil Aviation Organisation (ICAO), the International Maritime Organisation (IMO), and regional transport organisations.

The IAEA has regularly issued revisions to the transport regulations in order to keep them up to date. The main publication on which other IAEA transport regulations are based is Safety Series No. ST-l, Regulations for the Safe Transport of Radioactive Material.

The objective of the regulations is to protect people and the environment from the effects of radiation during the transport of radioactive material.

Protection is achieved by:

- Containment of radioactive contents;
- Control of external radiation levels;
- Prevention of criticality; and
- Prevention of damage caused by heat.

The fundamental principle applied to the transport of radioactive material is that the protection comes from the design of the package, regardless of how the material is transported.

Transport of uranium oxide from mines and uranium hexafluoride

Uranium oxide concentrate, sometimes called yellowcake, is transported from the mines to conversion plants in 200-litre drums packed into normal shipping containers. No radiation protection is required beyond having the steel drums clean and within the steel container.

From the conversion plant, the uranium is in the form of uranium hexafluoride, which again is barely radioactive but has significant chemical toxicity. It is in special containers, which also function for storage.

Transport of uranium fuel assemblies

Uranium fuel assemblies are manufactured at fuel fabrication plants. The fuel assemblies are made up of ceramic pellets formed from pressed uranium oxide that has been sintered at a high temperature (over 1400Â°C). The pellets are aligned within long, hollow, metal rods, which in turn are arranged in the fuel assemblies, ready for introduction into the reactor. Different types of reactors require different types of fuel assembly, so when the fuel assemblies are transported from the fuel fabrication facility (where they are manufactured) to the nuclear power reactor, the contents of the shipment will vary with the type of reactor receiving it.

In Western Europe, Asia and the US, the most common means of transporting uranium fuel assemblies is by truck. A typical truckload supplying a light water reactor contains 6 tonnes of fuel. In the countries of the former Soviet Union, rail transport is most often used. Intercontinental transports are mostly by sea, though occasionally transport is by air.

The annual operation of a 1000 MWe light water reactor requires an average fuel load of 27 tonnes of uranium dioxide, containing 24 tonnes of enriched uranium. The assemblies containing this are normally supplied in one consignment occupying 4 to 5 trucks.

The fuel assemblies are transported in packages specially constructed to protect the precision-made fuel assemblies from damage during transport. Uranium fuel assemblies have a low radioactivity level and radiation shielding is not necessary.

Fuel assemblies contain fissile material and in some circumstances fissile material can spontaneously become critical, *i.e.* start a self-sustaining, nuclear chain reaction, releasing energy. Criticality is prevented by the design of the package, the arrangement of the fuel assemblies within the package, limitations on the amount of material contained within the package, and on the number of packages carried in one shipment.

Transport of LLW and ILW

Low-level and intermediate-level wastes (LLW and ILW) are generated throughout the nuclear fuel cycle. The transport of these wastes is commonplace and they are safely transported to waste treatment facilities and storage sites.

Low-level radioactive wastes are a variety of materials that emit low levels of radiation, slightly above normal background levels. They often consist of solid materials, such as clothing, tools, or contaminated soil. Low-level waste is transported from its origin to waste treatment sites, or to an intermediate or final storage facility.

A variety of radionuclides give low-level waste its radioactive character. However, the radiation levels from these materials are very low and the packaging used for the transport of low-level waste does not require special shielding.

Low-level wastes are moved by road, rail, and internationally, by sea. However, most low-level waste is only transported within the country where it is produced.

Low-level wastes are transported in drums, often after being compacted in order to reduce the total volume of waste. The drums commonly used contain up to 200 litres of material. Typically, 36 standard, 200 litre drums go into a 6-metre transport container.

The composition of intermediate-level wastes is broad, but they require shielding. Much ILW comes from nuclear power plants and reprocessing facilities.

Intermediate-level wastes are taken from their source to an interim storage site, a final storage site (as in Sweden), or a waste treatment facility. They are transported by road, rail and sea.

The radioactivity level of intermediate-level waste is higher than low-level waste. The classification of radioactive wastes is decided for disposal purposes, not on transport grounds. The transport aspects of intermediate-level waste take into account any specific properties of the material, and provide shielding.

Classification of Radioactive Wastes

There are several systems of nomenclature in use, but the following is generally accepted:

- *Exempt waste*—excluded from regulatory control because radiological hazards are negligible.
- *Low-level waste* (LLW)—contains enough radioactive material to require action for the protection of people, but not so much that it requires shielding in handling or storage.
- *Intermediate-level waste* (ILW)—requires shielding. If it has more than 4000 Bq/g of long-lived (over 30 year half-life) alpha emitters it is categorised as "long-lived" and requires more sophisticated handling and disposal.

- *High-level waste* (HLW)—sufficiently radioactive to require both shielding and cooling,
- Generates >2 kW/m 3 of heat and has a high level of long-lived alpha-emitting isotopes.

Transport of used Fuel

When used fuel is unloaded from a nuclear power reactor, it contains: 96 per cent uranium, 1 per cent plutonium and 3 per cent of fission products (from the nuclear reaction) and transhumances).

used fuel looks the same as fresh fuel but when the fuel assembly is removed from a reactor it will be emitting high levels of both radiation and heat. It is stored in water pools adjacent to the reactor to allow the initial heat and radiation levels to decrease. Typically, used fuel is stored for at least five months before it can be transported, although it may be stored there long-term.

From the reactor site, used fuel is transported by road, rail or sea to either an interim storage site or a reprocessing plant where it will be reprocessed.

used fuel assemblies are shipped in Type B casks. These casks are shielded with steel, or a combination of steel and lead, and can weigh up to 110 tonnes each when empty. A typical transport cask holds up to 6 tonnes of used fuel.

Since 1971 there have been some 7000 shipments of used fuel (over 80 000 tonnes) over many million kilometres with no property damage or personal injury, no breach of containment, and very low dose rate to the personnel involved (*e.g.* 0.33 mSv/yr per operator at La Hague). This includes 40,000 tonnes of used fuel shipped to Areva's La Hague reprocessing plant, at least 30,000 tonnes of mostly UK used fuel shipped to UK's Sellafield reprocessing plant, 7140 t used fuel in 160 shipments from Japan to Europe by sea and 4500 tonnes of used fuel shipped around the Swedish coast.

In the USA alone, one percent of the 300 million packages of hazardous material shipped each year contain radioactive materials. Of this, about 250,000 contain radioactive wastes from US nuclear power plants, and 25 to 100 packages contain used fuel. Most of these are in robust 125-tonne Type B casks carried by rail, each containing 20 tonnes of used fuel.

Transport of Plutonium

Plutonium is separated during the reprocessing of used fuel. It is normally then made into mixed oxide (MOX) fuel.

Plutonium is transported, following reprocessing, as an oxide as this is its most stable form. Plutonium oxide is a solid, and normally transported as a powder in sealed packages. It is insoluble in water and only harmful to humans if it enters the lungs.

Plutonium oxide is transported in several different types of packages and each can contain several kilograms of material. Criticality is prevented by the design of the package, limitations on the amount of material contained within the package, and on the number of packages carried on a transport vessel.

Plutonium is subjected to physical protection controls and special physical protection measures apply to plutonium transports.

A typical transport consists of one truck carrying one protected shipping container. The container holds a number of packages with a total weight varying from 80 to 200 kg of plutonium oxide.

A sea shipment may consist of several containers, each of them holding between 80 to 200 kg of plutonium in sealed packages.

Transport of vitrified Waste

The highly radioactive wastes (especially fission products) created in the nuclear reactor are segregated and recovered during the reprocessing operation. These wastes are incorporated in a glass matrix by a process known as 'verification', which stabilizes the radioactive material.

The molten glass is then poured into a stainless steel canister where it cools and solidifies. A lid is welded into place to seal the canister. The canisters are then placed inside a Type B cask, similar to those used for the transport of used fuel.

The quantity per shipment depends upon the capacity of the transport cask. Typically a vitrified waste transport cask contains up to 28 canisters of glass. The main characteristics of the canister are as follows:

- height (with lid) = 1.34 m
- outside diameter = 0.43 m
- weight (empty) = 90 kg

So far, France is the only country that has carried out transports of vitrified waste. Since 1995, there have been five transports (two to Germany by rail, six to Japan by sea). This figure should increase in the years to come.

In the UK, storage is on the same site as reprocessing and verification.

Sea Shipments of Wastes

Some 300 sea voyages have been made carrying used nuclear fuel or separated high-level waste over a distance of more than 8 million kilometres. The major company involved has transported over 4000 casks, each of about 100 tonnes, carrying 8000 tonnes of used fuel or separated high-level wastes. A quarter of these have been through the Panama Canal.

In Sweden alone, more than 80 large transport casks are shipped annually from nuclear power stations (all on the coast) to a central interim waste storage facility called CLAB. Each 80 tonne cask has steel walls 30 cm thick and holds 17 BWR or 7 PWR fuel assemblies. The used fuel is shipped to CLAB after it has been stored for about a year at the reactor, during which time heat and radioactivity diminish considerably. A purpose-built 2000 tonne ship is used for moving the used fuel. Some 4500 tonnes of used fuel had been shipped around the coast to CLAB by the end of 2007.

Shipments of used fuel from Japan to Europe for reprocessing use 94-tonne Type B casks, each holding a number of fuel assemblies (*e.g.* 12 PWR assemblies, total 6 tonnes, with each cask 6.1 metres long, 2.5 metres diameter, and with 25 cm thick forged steel walls). More than 160 of these shipments took place from 1969 to the 1990s, involving more than 4000 casks, and moving several thousand tonnes of highly radioactive used fuel—4200t to UK and 2940t to France.

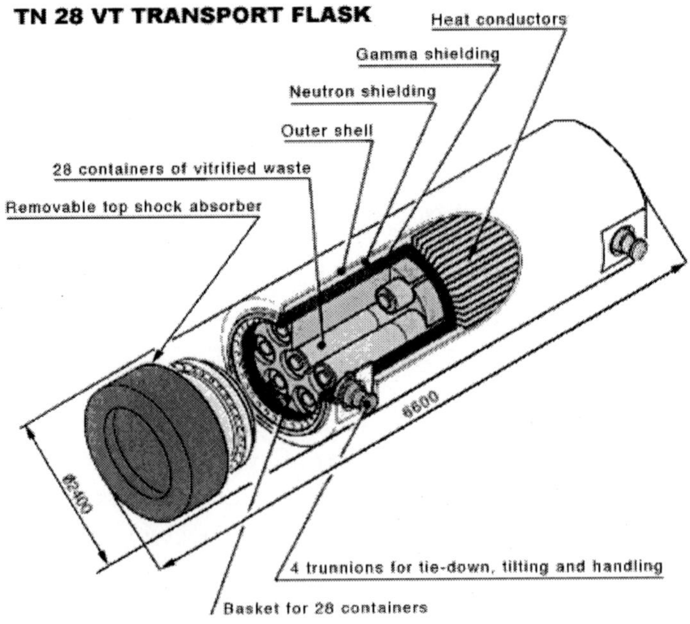

Return shipments from Europe to Japan since 1995 are of vitrified high-level wastes in stainless steel canisters. Up to 28 canisters (total 14 tonnes) are packed in each 98-tonne steel transport cask. Each is 6.6 metres long and 2.4 metres diameter, with a 25 cm thick wall. Over 1995-2007 twelve shipments were made from France of vitrified HLW comprising 1310 canisters containing almost 700 tonnes of glass. In 2008 return shipments from the UK are due to commence, and there will be about 11 shipments to 2016.

Apart from Sweden's dedicated small ship, there are five purpose-built 5100 tonne ships, with elaborate safety provisions, which carry the casks. These have double hulls with impact-resistant structures between the hulls, together with duplication and separation of all essential systems to provide high reliability and also survivability in the event of an accident. Twin engines operate independently. Each ship can carry up to 17 used fuel flasks or 14 waste transport flasks.

The ships are owned by Pacific Nuclear Transport Ltd. They conform to all relevant international safety standards, notably one known as INF-3 (Irradiated Nuclear Fuel class 3) set by the International Maritime Organisation. This allows them to carry highly radioactive materials such as high-level wastes, used nuclear fuel, mixed-oxide (MOX) fuel, and plutonium. PNTL is now owned by International Nuclear Services Ltd (INS, 62.5%), Japanese utilities (25%) and Area (12.5%). PNTL is currently renewing its fleet. INS is 51% owned by Sellafield Ltd and 49 per cent by the UK's Nuclear Decommissioning Authority and is managed by Sellafield Ltd.

Purpose-build vessel for transport of spent nuclear fuel

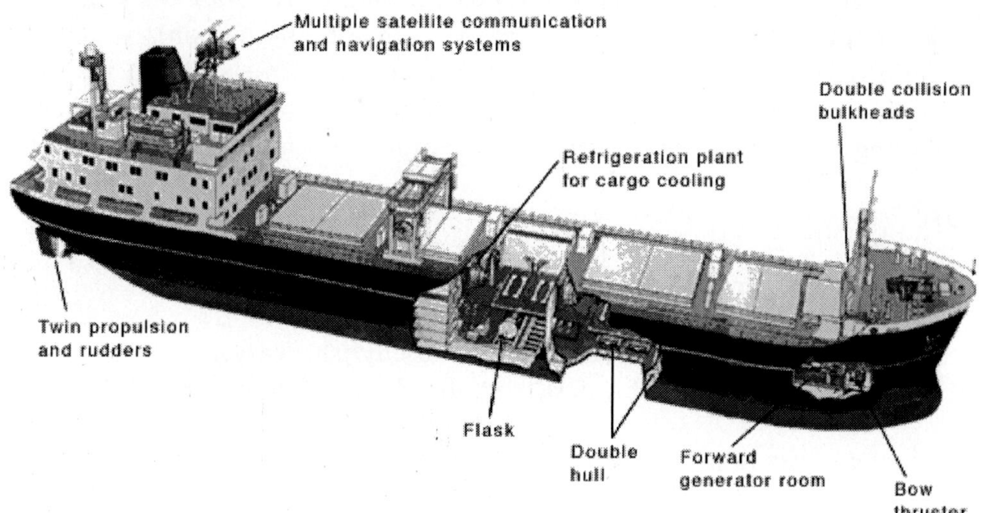

Within Europe, used fuel in casks has often been carried on normal ferries, *e.g.* across the English Channel.

Accident Scenarios

There has never been any accident in which a Type B transport cask containing radioactive materials has been breached or has leaked.

For the radioactive material in a large Type B package in sea transit to become exposed, the ship's hold (inside double hulls) would need to rupture, the 25 cm thick steel cask would need to rupture, and the stainless steel flask or the fuel rods would need to be broken open. Either borosilicate glass (for reprocessed wastes) or ceramic fuel material would then be exposed, but in either case these materials are very insoluble.

The transport ships are designed to withstand a side-on collision with a large oil tanker. If the ship did sink, the casks will remain sound for many years and would be relatively easy to recover since instrumentation including location beacons would activate and monitor the casks.

5.12 Toxin

A *toxin* (Greek: ôïîéêüí, *toxikon*, lit. (poison) for use on arrows) is a poisonous substance produced by living cells or organisms that is active at very low concentrations. Toxins can be small molecules, peptides, or proteins and are capable of causing disease on contact or absorption with body tissues by interacting with biological macromolecules such as enzymes or cellular receptors. Toxins vary greatly in their severity, ranging from usually minor and acute (as in a bee sting) to almost immediately deadly (as in botulinum toxin).

Biotoxins vary greatly in purpose and mechanism, and can be highly complex (the venom of the cone snail contains dozens of small proteins, each targeting a specific nerve channel or receptor), or relatively small protein.

Use

Biotoxins in nature have two primary functions:

- Predation (spider, snake, scorpion, jellyfish, wasp),
- Defense (bee, poison dart frog, deadly nightshade, honeybee, wasp).

Some of the more well known types of biotoxins include:

- Cyanotoxins, produced by cyanobacteria.

❖ *Hemotoxins* target and destroy red blood cells, and are transmitted through the bloodstream. Organisms that possess hemotoxins include:

- Pit Vipers, such as rattlesnakes.

❖ *Necrotoxins* cause necrosis (*i.e.*, death) in the cells they encounter and destroy all types of tissue. Necrotoxins spread through the bloodstream, but infect all tissues. In humans, skin and muscle tissues are most sensitive to necrotoxins. Organisms that possess necrotoxins include:

- The brown recluse or "fiddle back" spider.
  Necrotizing fasciitis (the "flesh eating" bacteria).
- *Neurotoxins* primarily affect the nervous systems of animals. Organisms that possess neurotoxins include:
  - The Black Widow and other widow spiders.
  - Most scorpions.
  - The box jellyfish.
  - Elapid snakes.
  - The Cone Snail.

Ricinis is a plant toxin found in the castor bean plant.

Environmental Toxins

Non-Technical Usage

When used non-technically, the term "toxin" is often applied to any toxic substances. Toxic substances not of biological origin are more properly termed poisons. Many non-technical and lifestyle journalists also follow this usage to refer to toxic substances in general, though some specialist journalists at publishers such as the BBC and *The Guardian* maintain the distinction that toxins are only those produced by living organisms.

In the context of alternative medicine the term is often used non-specifically to refer to any substance claimed to cause ill health, ranging anywhere from trace amounts of pesticides to common food items like refined sugar or additives like artificial sweeteners and MSG.

5.13 Radioactive Chemicals

Radioactivity was discovered near the turn of the twentieth century through the work of Wilhelm Röentgen (1895, discovers X-rays), Antoine Becquerel

(1896, discovers radioactivity), Marie and Pierre Curie (1898, isolates polonium and radium), and Ernest Rutherford (1899 and following years, identifies alpha, beta, and gamma radiation). Since that time, use of radioactive materials has made significant impacts in such areas as biology, chemistry, medicine, energy production, and nuclear weapons production.

While use of radioactive materials has been essential to many recent advances in these fields, use of radioactive materials has also resulted in generation of various waste materials that have the potential for impact on human health and the environment. Indeed, one of the major concerns with the use of radioactive materials is their potential release to the natural environment. *Radionuclides* may be introduced to surface water and *groundwater* resources from both natural and human (anthropogenic) sources.

Radioactivity and Types of Radiation

The term "radioactive" means that certain *isotopes* of some chemical elements have an unstable nucleus that will spontaneously decay with the concurrent emission of ionizing radiation. The ionizing radiation, which can cause cellular damage in humans and other species, includes types called alpha, beta, and gamma radiation. These different types of radiation have vastly different penetrating power through matter.

With regard to human health impacts, alpha radiation is of most concern through internal (inhalation) exposure, whereas beta and gamma radiation may cause health impacts through either internal or external exposure. In particular, gamma radiation has significant penetrating powers, and one must shield gamma-radiating materials to reduce human exposure. Different radioactive materials emit different combinations of radiation types (that is, different ratios of alpha, beta, and gamma radiation).

Quantifying Radioactivity and Radiation

The decay of radioactive materials follows a simple mathematical law. Observation has shown that the rate of decrease in the number of radiation emissions is proportional to the amount of radioactive material present. The amount of radiation of a certain type that is present in a sample of matter is called its "activity," and the standard unit for measuring activity is the "Becquerel" (often abbreviated Bq), which is 1 radiation emission per second. The frequency of radiation emission is characterized by the element's half-life, which is the time duration over which the amount of radioactive material decreases by a factor of two.

For radioactive materials, the half-life ranges from a fraction of a second to billions of years. The potential of the radiation to damage human health depends on both the amount of material present (its activity) and the type of radiation emitted.

Classifying Radioactive Materials

To help manage their potential impacts, a number of classes of radioactive materials have been defined for regulatory purposes. These classes include:

Fig. 5.12: Greenpeace activists protest in 2001 against the imminent sailing of the vessel Pacific Swan. The ship was loaded with a highly radioactive product made of waste material from nuclear reactors. Protestors feared an accident would release large quantities of radioactivity to ocean waters.

Naturally occurring radioactive material (NORM), uranium mill tailings, low-level radioactive waste (LLRW), high-level radioactive waste (HLRW), transuranic (TRU) waste, and radioactive material that has been classified as below regulatory concern (BRC).

The Connection between NORM and Radon

As its name implies, naturally occurring radioactive material (NORM) is naturally found in most soils, surface water, and groundwater throughout the world. Some NORM has existed since the origin of the Earth, while additional NORM is continually being generated within the atmosphere from cosmic radiation.

The primordial NORM category includes decay products of three primary decay chains that start with the isotopes ^{238}Uranium, ^{232}Thorium, and ^{235}Uranium. In evaluating the potential NORM impacts, particular attention has been paid to ^{226}Radium and ^{228}Radium, which are daughters of the^{238}Uranium and ^{232}Thorium decay chains. Various technological and industrial processes may result in the material concentration of radium above natural background levels.

Radon, a natural decay product of radium, is a *noble gas* that will also dissolve in water. It exists worldwide in surface water and groundwater, and can be present in concentrations that are of concern for human health (see box on next page).

Anthropogenic Radioactive Waste

High-level radioactive waste (HLRW) includes fuel rods and associated material from commercial nuclear power generation. Transuranic (TRU) waste consists of elements with atomic numbers greater than that of uranium, including plutonium, neptunium, and americium. These materials were generated as part of the nuclear weapons program.

Waste that is below regulatory concern (BRC) consists primarily of short-lived materials that are used in medicine and research. Low-level radioactive waste (LLRW) is a catch-all category and consists of everything that is not included in the other waste categories. Low-level waste does not necessarily mean low activity waste, though much of it is. Certain low-level waste streams may contain high levels of radioactivity, such as resins, filters, and certain pieces of equipment from the nuclear power industry.

Storing Radioactive Wastes Underground

Within the United States, the first low-level radioactive waste disposal site was licensed in 1962 to help manage waste generated from the widespread use of radioactive material over the previous half century. During the remainder of the 1960s, five additional commercial sites were developed and opened throughout the country. All of these sites relied on shallow land burial technology in which the waste was placed in trenches and covered with earthen material. Experience from these facilities has highlighted the significance of local geology, *hydrology*, and site engineering in the performance of disposal facilities.

Licensing requirements for a shallow land burial facility are specified in Title 10, Part 61, of the Code of Federal Regulations. Among the provisions required in the license application are documentation that the site will be geologically stable, that the site will be safe from inadvertent intrusion, and that normal releases of radioactivity will not cause an unacceptable risk to the public.

Despite a number of licensing efforts, a combination of public concern and lack of political will has resulted in no licensing of additional facilities since the 1960s, while only two facilities remain open. In an attempt to address public concerns, policy appears to be turning toward the concept of "assured isolation" wherein radioactive waste would be placed in long-term storage. During the period of storage, most of the radioactivity would be:

Radiologists measure soil radioactivity levels near Ukraine's Chernobyl Nuclear Plant, where a nuclear reactor exploded in 1986, releasing 100 times the radiation released by the atomic bombs dropped on Hiroshima and Nagasaki during World War II. Approximately 125,000 to 146,000 square kilometers (48,000 to 56,000 square miles) were affected by the contamination, including the region's water resources.

lost, or decayed. The ultimate disposal of the remaining active material would likely be shallow land burial.

Disposal and Cleanup of Radioactive Wastes

The U.S. Department of Energy is investigating Yucca Mountain, Nevada as a site for subsurface disposal of high-level radioactive waste. The department has recently received a license for disposal of transuranic waste at the Waste Isolation Pilot Plant (WIPP) site near Carlsbad, New Mexico. Waste that is below regulatory concern is not regulated, and these materials may be disposed with normal municipal solid waste. Disposal of lowlevel radioactive waste is the responsibility of the states rather than the federal government.

Prime Candidates for Nuclear Waste Cleanup

The most significant sources of concentrated radioactive materials in the environment have been facilities involved with nuclear weapons production. Cleaning up these installations in the United States and the former Soviet Union will be the most costly environmental restoration project in world history. In 1977, the U.S. Department of Energy was given stewardship of the nation's nuclear weapons arsenal, and it also inherited responsibility for cleaning up the legacy of environmental contamination associated with nuclear weapons production.

The processes that ultimately lead to soil and groundwater contamination include uranium mining and processing, isotope and chemical separations, component fabrication, and weapons testing. Like other industries, the U.S. Department of Energy and its predecessors frequently disposed of wastes in landfills, lagoons, or underground injection wells, and spills of byproduct materials were not uncommon.

The department is charged with cleanup of 113 installations in 30 states. However, the following five installations are expected to account for the majority (approximately two-thirds) of the costs for cleanup: the Rocky Flats Environmental Technology Site in Colorado; the Idaho National Engineering and Environmental Laboratory; the Savannah River Site in South Carolina; the Oak Ridge Reservation in Tennessee; and the Hanford Site in Washington.

While waste management programs have been greatly improved, cleanup of these legacy wastes remains a significant environmental and political challenge for the twenty-first century.

Radioactivity in Drinking Water

Radon is a naturally occurring noble gas which emits radiation and that, if present in sufficient concentrations, may be harmful to humans. Most of the radon found in a home comes from radioactive decay in the soil and geological materials beneath the house. Radon then enters the home, particularly via the basement or crawlspace. The greatest exposure risk to humans is through inhalation of radon via the indoor air. The U.S. Surgeon General has warned that breathing radon is the second leading cause of lung cancer.

Radon from tap water is a much smaller source (approximately 1 to 2 percent) of radon in the home than that derived from geologic materials beneath the house. Further, the risk applies only to groundwater-based water systems (wells and springs); the risk of radon in surface-water sources (streams, lakes, and reservoirs) is very low.

The U.S. Environmental Protection Agency has proposed a radon standard for drinking water that would incorporate the potential for radon contribution from indoor air as well as that from drinking water. The agency also has established maximum contaminant levels for beta particle and photon activity (4 mrem), gross alpha particle activity (15 pCi/liter), and combined ^{226}Ra and ^{228}Ra (5 pCi/liter).

6

Atmospheric Chemistry and Air Pollution

6.1 Atmospheric Chemistry

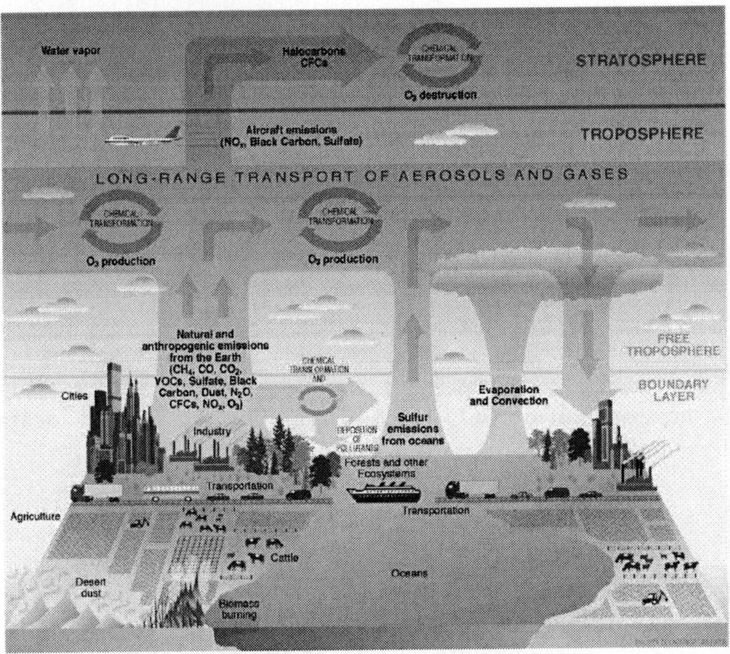

Fig. 6.1: Schematic of chemical and transport processes related to atmospheric composition.

Atmospheric chemistry is a branch of atmospheric science in which the chemistry of the Earth's atmosphere and that of other planets is studied. It is a multidisciplinary field of research and draws on environmental chemistry, physics, meteorology, computer modeling, oceanography, geology and volcanology and other disciplines. Research is increasingly connected with other areas of study such as climatology.

The composition and chemistry of the atmosphere is of importance for several reasons, but primarily because of the interactions between the atmosphere and living organisms. The composition of the Earth's atmosphere has been changed by human activity and some of these changes are harmful to human health, crops and ecosystems. Examples of problems which have been addressed by atmospheric chemistry include acid rain, photochemical smog and global warming. Atmospheric chemistry seeks to understand the causes of these problems, and by obtaining a theoretical understanding of them, allow possible solutions to be tested and the effects of changes in government policy evaluated.

Atmospheric Composition

Average composition of dry atmosphere, by volume:

Gas	*per NASA*
Nitrogen, N_2	78.084%
Oxygen, O_2	20.946%
Argon, Ar	0.934%
Water vapour	Highly variable; typically makes up about 1%
Minor constituents in ppmv.	
Carbon Dioxide, CO_2	383
Neon, Ne	18.18
Helium, He	5.24
Methane, CH_4	1.7
Krypton, Kr	1.14
Hydrogen, H_2	0.55

Notes: The concentration of CO_2 and CH_4 vary by season and location. ppmv represents parts per million by volume.

The mean molecular mass of air is 28.97 g/mol.

History

The ancient Greeks regarded air as one of the four elements, but the first scientific studies of atmospheric composition began in the 18th century. Chemists such as Joseph Priestley, Antoine Lavoisier and Henry Cavendish made the first measurements of the composition of the atmosphere.

In the late 19th and early 20th centuries interest shifted towards trace constituents with very small concentrations. One particularly important discovery for atmospheric chemistry was the discovery of ozone by Christian Friedrich Schoenbein in 1840.

In the 20th century atmospheric science moved on from studying the composition of air to a consideration of how the concentrations of trace gases in the atmosphere have changed over time and the chemical processes which create and destroy compounds in the air. Two particularly important examples of this were the explanation of how the ozone layer is created and maintained by Sydney Chapman and Gordon Dobson, and the explanation of Photochemical smog by Haagen-Smit.

In the 21st century the focus is now shifting again. Atmospheric Chemistry is increasingly studied as one part of the Earth system. Instead of concentrating on atmospheric chemistry in isolation the focus is now on seeing it as one part of a single system with the rest of the atmosphere, biosphere and geosphere. An especially important driver for this is the links between chemistry and climate such as the effects of changing climate on the recovery of the ozone hole and vice versa but also interaction of the composition of the atmosphere with the oceans and terrestrial ecosystems.

Methodology

Observations, lab measurements and modeling are the three central elements in atmospheric chemistry. Progress in atmospheric chemistry is often driven by the interactions between these components and they form an integrated whole. For example observations may tell us that more of a chemical compound exists than previously thought possible. This will stimulate new modelling and laboratory studies which will increase our scientific understanding to a point where the observations can be explained.

Observation

Observations of atmospheric chemistry are essential to our understanding. Routine observations of chemical composition tell us about changes in atmospheric composition over time. One important example of this is the Keeling Curve—a series of measurements from 1958 to today which show a steady rise in of the concentration of carbon dioxide. Observations of atmospheric chemistry are made in observatories such as that on Mauna Loa and on mobile platforms such as aircraft (*e.g.* the UK's Facility for Airborne Atmospheric Measurements), ships and balloons. Observations of atmospheric composition are increasingly made by satellites with important instruments such as GOME and MOPITT giving a global picture of air pollution and chemistry. Surface observations have the advantage that they provide long term records at high time resolution but are limited in the vertical and horizontal space they provide observations from. Some surface based instruments *e.g.* LIDAR can provide concentration profiles of chemical compounds and aerosol but are still restricted in the horizontal region they can cover. Many observations are available on line in Atmospheric Chemistry Observational Databases.

Lab Measurements

Measurements made in the laboratory are essential to our understanding of the sources and sinks of pollutants and naturally occurring compounds. Lab studies tell us which gases react with each other and how fast they react. Measurements of interest include reactions in the gas phase, on surfaces and in water. Also of high importance is photochemistry which quantifies how quickly molecules are split apart by sunlight and what the products are plus thermodynamic data such as Henry's law coefficients.

Modeling

In order to synthesise and test theoretical understanding of atmospheric chemistry, computer models (such as chemical transport models) are used. Numerical models solve the differential equations governing the concentrations of chemicals in the atmosphere. They can be very simple or very complicated. One common trade off in numerical models is between the number of chemical compounds and chemical reactions modelled versus the representation of transport and mixing in the atmosphere. For example, a box model might include hundreds or even thousands of chemical reactions but will only have a very crude representation of mixing in the atmosphere. In contrast, 3D models represent many of the physical processes of the atmosphere but due to constraints on computer resources will have far fewer chemical reactions and compounds. Models can be used to interpret observations, test understanding of chemical reactions and predict future concentrations of chemical compounds in the atmosphere. One important current trend is for atmospheric chemistry modules to become one part of earth system models in which the links between climate, atmospheric composition and the biosphere can be studied.

Some models are constructed by automatic code generators. In this approach a set of constituents are chosen and the automatic code generator will then select the reactions involving those constituents from a set of reaction databases. Once the reactions have been chosen the ordinary differential equations (ODE) that describe their time evolution can be automatically constructed.

6.2 Air Pollution

Air pollution is the human introduction into the atmosphere of chemicals, particulate matter, or biological materials that cause harm or discomfort to humans or other living organisms, or damage the environment. Air pollution causes deaths and respiratory disease. Air pollution is often identified with major stationary sources, but the greatest source of emissions is mobile sources, mainly automobiles. Gases such as carbon dioxide, which contribute to global warming, have recently gained recognition as pollutants by climate scientists, while they also recognize that carbon dioxide is essential for plant life through photosynthesis.

The atmosphere is a complex, dynamic natural gaseous system that is essential to support life on planet Earth. Stratospheric ozone depletion due to air pollution has long been recognized as a threat to human health as well as to the Earth's ecosystems.

6.3 Pollutants

There are many substances in the air which may impair the health of plants and animals (including humans), or reduce visibility. These arise both from natural processes and human activity. Substances not naturally found in the air or at greater concentrations or in different locations from usual are referred to as *pollutants*.

Pollutants can be classified as either primary or secondary. Primary pollutants are substances directly emitted from a process, such as ash from a volcanic eruption or the carbon monoxide gas from a motor vehicle exhaust.

Secondary pollutants are not emitted directly. Rather, they form in the air when primary pollutants react or interact. An important example of a secondary pollutant is ground level ozone—one of the many secondary pollutants that make up photochemical smog.

Note that some pollutants may be both primary and secondary: that is, they are both emitted directly and formed from other primary pollutants.

Major primary pollutants produced by human activity include:

- Sulfur oxides (SO_x) especially sulfur dioxide are emitted from burning of coal and oil.
- Nitrogen oxides (NO_x) especially nitrogen dioxide are emitted from high temperature combustion. Can be seen as the brown haze dome above or plume downwind of cities.
- Carbon monoxide is colourless, odourless, non-irritating but very

poisonous gas. It is a product by incomplete combustion of fuel such as natural gas, coal or wood. Vehicular exhaust is a major source of carbon monoxide.

- Carbon dioxide (CO_2), a greenhouse gas emitted from combustion.
- Volatile organic compounds (VOC), such as hydrocarbon fuel vapors and solvents.
- Particulate matter (PM), measured as smoke and dust. PM_{10} is the fraction of suspended particles 10 micrometers in diameter and smaller that will enter the nasal cavity. $PM_{2.5}$ has a maximum particle size of 2.5 μm and will enter the bronchies and lungs.
- Toxic metals, such as lead, cadmium and copper.
- Chlorofluorocarbons (CFCs), harmful to the ozone layer emitted from products currently banned from use.
- Ammonia (NH_3) emitted from agricultural processes.
- Odors, such as from garbage, sewage, and industrial processes
- Radioactive pollutants produced by nuclear explosions and war explosives, and natural processes such as radon.

Secondary pollutants include:

- Particulate matter formed from gaseous primary pollutants and compounds in photochemical smog, such as nitrogen dioxide.
- Ground level ozone (O_3) formed from NOx and VOCs.
- Peroxyacetyl nitrate (PAN) similarly formed from NOx and VOCs.

Minor air pollutants include:

- A large number of minor hazardous air pollutants. Some of these are regulated in USA under the Clean Air Act and in Europe under the Air Framework Directive.
- A variety of persistent organic pollutants, which can attach to particulate matter.

Sources

Sources of air pollution refer to the various locations, activities or factors which are responsible for the releasing of pollutants in the atmosphere. These sources can be classified into two major categories which are:

Anthropogenic sources (human activity) mostly related to burning different kinds of fuel:

- "Stationary Sources" as smoke stacks of power plants, manufacturing facilities, municipal waste incinerators.
- "Mobile Sources" as motor vehicles, aircraft etc.
- Marine vessels, such as container ships or cruise ships, and related port air pollution.
- Burning wood, fireplaces, stoves, furnaces and incinerators, and industrial activity in general.
- Chemicals, dust and controlled burn practices in agriculture and forestry management, (see Dust Bowl).
- Fumes from paint, hair spray, varnish, aerosol sprays and other solvents.
- Waste deposition in landfills, which generate methane.
- Military, such as nuclear weapons, toxic gases, germ warfare and rocketry.

Natural Sources

- Dust from natural sources, usually large areas of land with little or no vegetation.
- Methane, emitted by the digestion of food by animals, for example cattle.
- Radon gas from radioactive decay within the Earth's crust.
- Smoke and carbon monoxide from wildfires.
- Volcanic activity, which produce sulfur, chlorine, and ash particulates.

Emission Factors

Air pollutant emission factors are representative values that attempt to relate the quantity of a pollutant released to the ambient air with an activity associated with the release of that pollutant. These factors are usually expressed as the weight of pollutant divided by a unit weight, volume, distance, or duration of the activity emitting the pollutant (*e.g.*, kilograms of particulate emitted per megagram of coal burned). Such factors facilitate estimation of emissions from various sources of air pollution. In most cases, these factors are simply averages of all available data of acceptable quality, and are generally assumed to be representative of long-term averages.

The United States Environmental Protection Agency has published a compilation of air pollutant emission factors for a multitude of industrial sources. The United Kingdom, Australia, Canada and other countries have published similar compilations, as has the European Environment Agency.

Indoor Air Quality (IAQ)

A lack of ventilation indoors concentrates air pollution where people often spend the majority of their time. Radon (Rn) gas, a carcinogen, is exuded from the Earth in certain locations and trapped inside houses. Building materials including carpeting and plywood emit formaldehyde (H_2CO) gas. Paint and solvents give off volatile organic compounds (VOCs) as they dry. Lead paint can degenerate into dust and be inhaled. Intentional air pollution is introduced with the use of air fresheners, incense, and other scented items. Controlled wood fires in stoves and fireplaces can add significant amounts of smoke particulates into the air, inside and out. Indoor pollution fatalities may be caused by using pesticides and other chemical sprays indoors without proper ventilation.

Carbon monoxide (CO) poisoning and fatalities are often caused by faulty vents and chimneys, or by the burning of charcoal indoors. Chronic carbon monoxide poisoning can result even from poorly adjusted pilot lights. Traps are built into all domestic plumbing to keep sewer gas, hydrogen sulfide, out of interiors. Clothing emits tetrachloroethylene, or other dry cleaning fluids, for days after dry cleaning.

Though its use has now been banned in many countries, the extensive use of asbestos in industrial and domestic environments in the past has left a potentially very dangerous material in many localities. Asbestosis is a chronic inflammatory medical condition affecting the tissue of the lungs. It occurs after long-term, heavy exposure to asbestos from asbestos-containing materials in structures. Sufferers have severe dyspnea (shortness of breath) and are at an increased risk regarding several different types of lung cancer. As clear explanations are not always stressed in non-technical literature, care should be taken to distinguish between several forms of relevant diseases. According to the World Health Organisation (WHO), these may defined as; asbestosis, *lung cancer*, and *mesothelioma* (generally a very rare form of cancer, when more widespread it is almost always associated with prolonged exposure to asbestos).

Biological sources of air pollution are also found indoors, as gases and airborne particulates. Pets produce dander, people produce dust from minute skin flakes and decomposed hair, dust mites in bedding, carpeting and furniture produce enzymes and micrometre-sized fecal droppings, inhabitants emit methane, mold forms in walls and generates mycotoxins and spores, air conditioning systems can incubate Legionnaires' disease and mold, and houseplants, soil and surrounding gardens can produce pollen, dust, and mold. Indoors, the lack of air circulation allows these airborne pollutants to accumulate more than they would otherwise occur in nature.

Health Effects

The World Health Organization states that 2.4 million people die each year from causes directly attributable to air pollution; with 1.5 million of these deaths attributable to indoor air pollution. A study by the University of Birmingham has shown a strong correlation between pneumonia related deaths and air pollution from motor vehicles. Worldwide more deaths per year are linked to air pollution than to automobile accidents. Published in 2005 suggests that 310,000 Europeans die from air pollution annually. Direct causes of air pollution related deaths include aggravated asthma, bronchitis, emphysema, lung and heart diseases, and respiratory allergies. The US EPA estimates that a proposed set of changes in diesel engine technology (*Tier 2*) could result in 12,000 fewer *premature mortalities*, 15,000 fewer heart attacks, 6,000 fewer emergency room visits by children with asthma, and 8,900 fewer respiratory-related hospital admissions each year in the United States.

The worst short term civilian pollution crisis in India was the 1984 Bhopal Disaster. Leaked industrial vapors from the Union Carbide factory, belonging to Union Carbide, Inc., U.S.A., killed more than 2,000 people outright and injured anywhere from 150,000 to 600,000 others, some 6,000 of whom would later die from their injuries. The United Kingdom suffered its worst air pollution event when the December 4 Great Smog of 1952 formed over London. In six days more than 4,000 died, and 8,000 more died within the following months. An accidental leak of anthrax spores from a biological warfare laboratory in the former USSR in 1979 near Sverdlovsk is believed to have been the cause of hundreds of civilian deaths. The worst single incident of air pollution to occur in the United States of America occurred in Donora, Pennsylvania in late October, 1948, when 20 people died and over 7,000 were injured.

The health effects caused by air pollutants may range from subtle biochemical and physiological changes to difficulty in breathing, wheezing, coughing and aggravation of existing respiratory and cardiac conditions. These effects can result in increased medication use, increased doctor or emergency room visits, more hospital admissions and premature death. The human health effects of poor air quality are far reaching, but principally affect the body's respiratory system and the cardiovascular system. Individual reactions to air pollutants depend on the type of pollutant a person is exposed to, the degree of exposure, the individual's health status and genetics.

Effects on Cystic Fibrosis

A study from 1999 to 2000 by the University of Washington showed that patients near and around particulate matter air pollution had an increased risk of pulmonary exacerbations and decrease in lung function. Patients were examined before the study for amounts of specific pollutants like P. aeruginosa or B. cepacia as well as their socioeconomic standing. Participants involved in the study were located in the United States in close proximity to an Environmental Protection Agency. During the time of the study 117 deaths were associated with air pollution. A trend was noticed that patients living closer or in large metropolitan areas to be close to medical help also had higher level of pollutants found in their system because of more emissions in larger cities. With cystic fibrosis patients already being born with decreased lung function everyday pollutants such as smoke emissions from automobiles, tobacco smoke and improper use of indoor heating devices could add to the dissemination of lung function.

Effects on COPD

Chronic obstructive pulmonary disease (COPD) include diseases such as chronic bronchitis, emphysema, and some forms of asthma. Two researchers Holland and Reid conducted research on 293 male postal workers in London during the time of the 1952 London Fog incident and 477 male postal workers in the rural setting. The amount of the pollutant FEV1 was significantly lower in urban employees however lung function was decreased due to city pollutions such as car fumes and increased amount of cigarette exposure. It is believed that much like cystic fibrosis, by living in a more urban environment serious health hazards become more apparent. Studies have shown that in urban areas patients suffer mucus hypersecretion, lower levels of lung function, and more self diagnosis of chronic bronchitis and emphysema.

London Fog of 1952

In the matter of four days a combination of dense fog and sooty black coal smoke came over the London area. The fog was so dense residents of London could not see in front of them. The extreme reduction in visibility was accompanied by an increase in criminal activity as well as transportation delays and a virtual shut down of the city. During the 4 day period of the fog 12,000 are believed to have been killed.

Effects on Children

Cities around the world with high exposure to air pollutants has the possibility of children living within them to develop asthma, pneumonia and other lower respiratory infections as well as a low initial birth rate. Protective measures to ensure the youths health is being taken in countries such as New Delhi where buses now use compressed natural gas to help eliminate the "pea-soup" fog. Research by the World Health Organization shows there is the greatest concentration of particulate matter particles in countries with low economic world power and high poverty and population rates. Examples of these countries include Egypt, Sudan, Mongolia, and Indonesia. The Clean Air Act was passed in 1970, however in 2002 at least 146 million Americans were living in areas that did not meet at least one of the "criteria pollutants" laid out in the 1997 National Ambient Air Quality Standards. Those pollutants included: ozone, particulate matter, sulfur dioxide, nitrogen dioxide, carbon monoxide, and lead. Because children are outdoors more and have higher minute ventilation they are more susceptible to the dangers of air pollution.

Reduction Efforts

There are various air pollution control technologies and urban planning strategies available to reduce air pollution.

Efforts to reduce pollution from mobile sources includes primary regulation (many developing countries have permissive regulations) expanding regulation to new sources (such as cruise and transport ships, farm equipment, and small gas-powered equipment such as lawn trimmers, chainsaws, and snowmobiles), increased fuel efficiency (such as through the use of hybrid vehicles), conversion to cleaner fuels (such as bioethanol, biodiesel, or conversion to electric vehicles).

Control Devices

The following items are commonly used as pollution control devices by industry or transportation devices. They can either destroy contaminants or remove them from an exhaust stream before it is emitted into the atmosphere.

Particulate Control

- Mechanical collectors (dust cyclones, multicyclones)
- Electrostatic precipitators
- Baghouses
- Particulate scrubbers

Crubbers

- Baffle spray scrubber

- Cyclonic spray scrubber
- Ejector venturi scrubber
- Mechanically aided scrubber
- Spray tower
- Wet scrubber

NOx Control

- Low NOx burners
- Selective catalytic reduction (SCR)
- Selective non-catalytic reduction (SNCR)
- NOx scrubbers
- Exhaust gas recirculation
- Catalytic converter (also for VOC control)

VOC Abatement

- Adsorption systems, such as activated carbon
- Flares
- Thermal oxidizers
- Catalytic oxidizers
- Biofilters
- Absorption (scrubbing)
- Cryogenic condensers
- Vapor recovery systems

Acid Gas/SO_2 Control

- Wet scrubbers
- Dry scrubbers
- Flue gas desulfurization

Mercury Control

- Sorbent Injection Technology
- Electro-Catalytic Oxidation (ECO)
- K-Fuel

Dioxin and Furan Control
Miscellaneous Associated Equipment

- Source capturing systems

- Continuous emissions monitoring systems (CEMS)

Legal Regulations

In general, there are two types of air quality standards. The first class of standards (such as the U.S. National Ambient Air Quality Standards) set maximum atmospheric concentrations for specific pollutants. Environmental agencies enact regulations which are intended to result in attainment of these target levels. The second class (such as the North American Air Quality Index) take the form of a scale with various thresholds, which is used to communicate to the public the relative risk of outdoor activity. The scale may or may not distinguish between different pollutants.

Canada

In Canada, air quality is typically evaluated against standards set by the Canadian Council of Minister for the Environment (CCME), an inter-governmental body of federal, provincial and territorial Ministers responsible for the environment. The CCME has set Canada Wide Standards (CWS). These are:

- CWS for PM2.5 = 30 μg/m³ (24 hour averaging time, by year 2010, based on 98th percentile ambient measurement annually, averaged over 3 consecutive years).
- CWS for *ozone* = 65 ppb (8-hour averaging time, by year 2010, achievement is based on the 4th highest measurement annually, averaged over 3 consecutive years.

Note that there is no consequence in Canada to not achieving these standards. In addition, these only apply to jurisdictions with populations greater than 100,000. Further, provinces and territories may set more stringent standards than those set by the CCME.

European Union

National Emission Ceilings (NEC) for certain atmospheric pollutants are regulated by Directive 2001/81/EC (NECD). As part of the preparatory work associated with the revision of the NECD, the European Commission is assisted by the NECPI working group (National Emission Ceilings— Policy Instruments).

United Kingdom

Air quality targets set by the UK's Department for Environment, Food and Rural Affairs (DEFRA) are mostly aimed at local government representatives responsible for the management of air quality in cities, where air quality management is the most urgent. The UK has established an air quality network where levels of the key air pollutants are published by monitoring centers. Air quality in Oxford, Bath and London is particularly poor. One controversial study performed by the Calor Gas company and published in the Guardian newspaper compared walking in Oxford on an average day to smoking over sixty light cigarettes.

More precise comparisons can be collected from the UK Air Quality Archive which allows the user to compare a cities management of pollutants against the national air quality objectives set by DEFRA in 2000.

Localized peak values are often cited, but average values are also important to human health. The UK National Air Quality Information Archive offers almost real-time monitoring of "current maximum" air pollution measurements for many UK towns and cities. This source offers a wide range of constantly updated data, including:

- Hourly Mean Ozone ($\mu g/m^3$),
- Hourly Mean Nitrogen dioxide ($\mu g/m^3$),
- Maximum 15-Minute Mean Sulphur dioxide ($\mu g/m^3$),
- 8-Hour Mean Carbon monoxide (mg/m^3), and
- 24-Hour Mean PM_{10} ($\mu g/m^3$ Grav Equiv).

DEFRA acknowledges that air pollution has a significant effect on health and has produced a simple banding index system is used to create a daily warning system that is issued by the BBC Weather Service to indicate air pollution levels. DEFRA has published guidelines for people suffering from respiratory and heart diseases.

United States

In the 1960s, 70s, and 90s, the United States Congress enacted a series of Clean Air Acts which significantly strengthened regulation of air pollution. Individual U.S. states, some European nations and eventually the European Union followed these initiatives. The Clean Air Act sets numerical limits on the concentrations of a basic group of air pollutants and provide reporting and enforcement mechanisms.

In 1999, the United States EPA replaced the Pollution Standards Index (PSI) with the Air Quality Index (AQI) to incorporate new PM2.5 and Ozone standards.

The effects of these laws have been very positive. In the United States between 1970 and 2006, citizens enjoyed the following reductions in annual pollution emissions:

- Carbon monoxide emissions fell from 197 million tons to 89 million tons.
- Nitrogen oxide emissions fell from 27 million tons to 19 million tons.
- Sulfur dioxide emissions fell from 31 million tons to 15 million tons.
- Particulate emissions fell by 80 per cent.
- Lead emissions fell by more than 98 per cent.

In an October 2006 letter to EPA, the agency's independent scientific advisors warned that the ozone smog standard "needs to be substantially reduced" and that there is "no scientific justification" for retaining the current, weaker standard. The scientists unanimously recommended a smog threshold of 60 to 70 ppb after they conducted an extensive review of the evidence.

The EPA has proposed, in June 2007, a new threshold of 75 ppb. This falls short of the scientific recommendation, but is an improvement over the current standard.

Polluting industries are lobbying to keep the current (weaker) standards in place. Environmentalists and public health advocates are mobilizing to support compliance with the scientific recommendations.

The National Ambient Air Quality Standards are pollution thresholds which trigger mandatory remediation plans by state and local governments, subject to enforcement by the EPA.

An outpouring of dust layered with man-made sulfates, smog, industrial fumes, carbon grit, and nitrates is crossing the Pacific Ocean on prevailing winds from booming Asian economies in plumes so vast they alter the climate. Almost a third of the air over Los Angeles and San Francisco can be traced directly to Asia. With it comes up to three-quarters of the black carbon particulate pollution that reaches the West Coast.

Air pollution is usually concentrated in densely populated metropolitan areas, especially in developing countries where environmental regulations are generally relatively lax. However, even populated areas in developed countries attain unhealthy levels of pollution.

Statistics

Most Polluted Cities

Most Polluted World Cities by PM	
Particulate matter, µg/m³ (2004)	*City*
169	Cairo, Egypt
150	Delhi, India
128	Kolkata, India (Calcutta)
125	Tianjin, China
123	Chongqing, China
109	Kanpur, India
109	Lucknow, India
104	Jakarta, Indonesia
101	Shenyang, China

Carbon Dioxide Emissions

Total CO_2 emissions

10^6 Tons of CO_2 per year:

- United States: 2,790
- Russia: 661
- Japan: 400
- Australia: 226
- United Kingdom: 212
- China: 2,680
- India: 583
- Germany: 356
- South Africa: 222
- South Korea: 185

Per Capita CO_2 Emissions

Tons of CO_2 per year per capita:

- Australia: 10
- United States: 8.2
- United Kingdom: 3.2
- China: 1.8
- India: 0.5

Atmospheric Dispersion

The basic technology for analyzing air pollution is through the use of a variety of mathematical models for predicting the transport of air pollutants in the lower atmosphere. The principal methodologies are:

- Point source dispersion, used for industrial sources.
- Line source dispersion, used for airport and roadway air dispersion modeling.
- Area source dispersion, used for forest fires or duststorms.
- Photochemical models, used to analyze reactive pollutants that form smog.

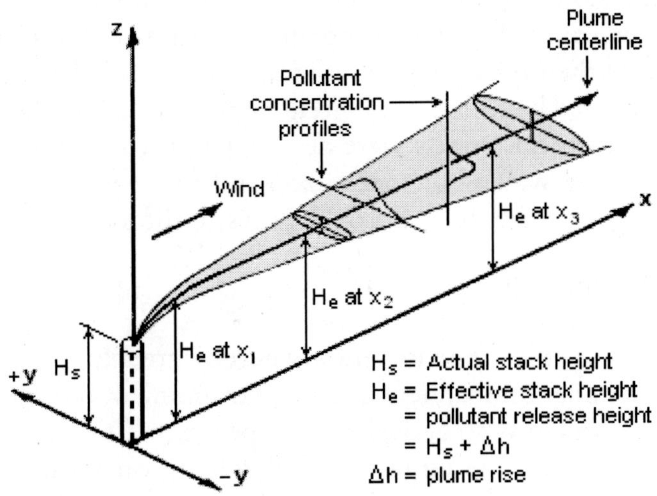

Fig. 6.2: Visualization of a buoyant Gaussian air pollution dispersion plume as used in many atmospheric dispersion models.

The point source problem is the best understood, since it involves simpler mathematics and has been studied for a long period of time, dating back to about the year 1900. It uses a Gaussian dispersion model for buoyant pollution plumes to forecast the air pollution isopleths, with consideration given to wind velocity, stack height, emission rate and stability class (a measure of atmospheric turbulence). This model has been extensively validated and calibrated with experimental data for all sorts of atmospheric conditions.

The roadway air dispersion model was developed starting in the late 1950s and early 1960s in response to requirements of the National Environmental Policy Act and the U.S. Department of Transportation (then known as the Federal Highway Administration) to understand impacts of proposed new highways upon air quality, especially in urban areas. Several research groups were active in this model development, among which were: the Environmental Research and Technology (ERT) group in Lexington, Massachusetts, the ESL Inc. group in Sunnyvale, California and the California Air Resources Board group in Sacramento, California. The research of the ESL group received a boost with a contract award from the United States

Environmental Protection Agency to validate a line source model using sulfur hexafluoride as a tracer gas. This program was successful in validating the line source model developed by ESL inc. Some of the earliest uses of the model were in court cases involving highway air pollution, the Arlington, Virginia portion of Interstate 66 and the New Jersey Turnpike widening project through East Brunswick, New Jersey.

Area source models were developed in 1971 through 1974 by the ERT and ESL groups, but addressed a smaller fraction of total air pollution emissions, so that their use and need was not as widespread as the line source model, which enjoyed hundreds of different applications as early as the 1970s. Similarly photochemical models were developed primarily in the 1960s and 1970s, but their use was more specialized and for regional needs, such as understanding smog formation in Los Angeles, California.

Environmental Impacts

The greenhouse effect is a phenomenon whereby greenhouse gases create a condition in the upper atmosphere causing a trapping of heat and leading to increased surface and lower tropospheric temperatures. It shares this property with many other gases, the largest overall forcing on Earth coming from water vapour. Other greenhouse gases include methane, hydrofluorocarbons, perfluorocarbons, chlorofluorocarbons, NOx, and ozone. Many greenhouse gases, contain carbon, and some of that from fossil fuels.

This effect has been understood by scientists for about a century, and technological advancements during this period have helped increase the breadth and depth of data relating to the phenomenon. Currently, scientists are studying the role of changes in composition of greenhouse gases from natural and anthropogenic sources for the effect on climate change.

A number of studies have also investigated the potential for long-term rising levels of atmospheric carbon dioxide to cause slight increases in the acidity of ocean waters and the possible effects of this on marine ecosystems. However, carbonic acid is a very weak acid, and is utilized by marine organisms during photosynthesis.

6.4 Causes and Effects of Air Pollution

I. Introduction

Children learn in the earliest science experience that problem-solving is an essential part of the learning process. Problems probe one's depth of understanding. Students should be stimulated to delve into areas that are not explicitly covered by this unit. While concentrating on the physical reasoning of air pollution, students should be able to develop skills in reading, language, mathematics, science, and social studies. This unit will be helpful to the self-contained classroom teacher as well as the departmentalized subject teacher. It is geared toward the middle school student from grades five through eight. The time span would vary according to the depth desired by individual teachers. A recommended period of four to eight weeks would allow student from below average range to be successful. Students will be able to:

1. Classify air pollutants.
2. Explore the adverse effects of pollution.
3. Determine the effects of smokestacks.
4. Determine the effectiveness of government control on pollution.
5. Understand the main idea of a paragraph.
6. Read for high comprehension and supporting ideas.
7. Interpret information from graphs and charts.
8. Use pictures to support main ideas of a subject.
9. Use the newspaper to study pollution problems in their own environment.

Pollution has become a major problem. It is not a future risk. Pollution is killing and destroying the health of people right now. It is impossible to escape. We have become so accustomed to low levels of exposure that it is hardly noticed.

Eliminating pollution from the environment has not proved as easy as eliminating it from the pages of a book. Industry is now spending several billion dollars a year on pollution control.

The first problem in understanding air pollution is to decide what is and what is not an air pollutant. Many of the things generally considered pollutants are present in the natural air. The amount of a substance locally present in the air is clearly important in defining a pollutant. Also the amount of harm or inconvenience caused by the substance and how long it remains in the atmosphere. These three factors are known as the three T's (tonnage, toxicity, and time in the atmosphere).

A great deal of power is needed to run the factories of modern industrial nations. Automobiles, trains, planes, and buses need power too. Nearly all of this power is produced in the same way—by burning fuels. The burning produces wastes. Some of the wastes get into the air, causing air pollution. The eventual fate of air pollution is to be wasted out of air.

A smokestack with a billowing black plume, for years the proud symbol of America's industrial wealth and technological prowess, has in the last decade acquired another meaning. The puffing smokestack has come to signify the Achilles heel, rather than its strength.

II. Types of Air Pollution

Particulates of both natural and human origin also cause pollution. Smoke both natural from fires and human activities cause from industries and other sources are major and cause much damage. The chemicals that are most trouble-some in air pollution are formed in the atmosphere by gases. The pollutant introduced into the atmosphere in the largest quantities by human activity is carbon monoxide. It is the product of incomplete combustion and the largest contribution comes from exhaust. Carbon monoxide is a colorless, odorless, tasteless gas, with the formula CO.

Carbon dioxide (CO_2) is also a product of combustion of fossil fuels. It is a minor constituent of natural air (about 0.03%), but the increased use of fossil fuels may cause an increase in the amount of carbon dioxide in the atmosphere.

The gases given off by engine exhaust are the oxides of nitrogen and the unburned hydrocarbons. The energy that causes these gases to react to form new compounds, comes from the sun. This reaction is called a photochemical reaction. The "air" that is exhausted from diesel engines is also too poor to breathe, although it contains more oxygen per cubic foot, but more particles.

Ozone is a chemically reactive substance (03) that is sometimes used to deodorizing exhausts by oxiding them to less objectionable odorous products before they are released to the atmosphere.

Industrial or gray smog is considered the most serious type of air pollution. Smoke and oxides of that are released by burning coal and oil containing minor amounts of sulfur is the cause.

The oxides of sulfur form sulfuric acid in the atmosphere which is both toxic to life and damaging to many materials. The smoke gives the air a gray color.

Industrial smog has been known to cause air pollution disasters. One of the worst occurred in London in December of 1952. Five days of stagnant air brought about high pressure systems caused between 3,500 and 4,000 deaths. In Donora, Pennsylvania 20 died and 6,000 became ill in 1948 because of a similar instance.

Photochemical smog also know as brown smog is largely caused by exhaust gases. It is common in warm cities in dry areas with lots of sunshine, such as Los Angeles, Denver, and Salt Lake City. This type of smog can obscure vision, cause plant damage, and irritate eyes.

III. Effects of Air Pollution on the Lungs

One study by Ishikawa et al. provided evidence that air pollution may cause or contribute to emphysema. A comparison was made of autopsy lung material from residents of two cities, Winnipeg Manitoba and St. Louis, Missouri. The Canadian city has a relatively low level of air pollution, whereas the American city characteristically has high levels of industrial contaminants. Emphysema was found to be seven times more common in St. Louis for ages 20-49 and twice as common for ages over 60. Lets look at a comparison. Smoking was significant but not an isolated factor.

A 1960-66 post mortem examination of lungs of 300 residents of St. Louis, Missouri, and an equal number from Winnipeg, Canada. The subjects were matched by sex, occupation, socio-economic status, length of residence, smoking habits, and age at death.

IV. Damages

Throughout the world the damage cause by air pollution is enormous. In money alone it represents a loss of billions of dollars each year. Many flower and vegetable crops suffer ill effects from air pollution caused by exhaust gases. Trees have been killed by pollution. Cattle have been poisoned. Air pollution causes rubber tires on automobiles to crack and become porous. Fine buildings become shabby, their walls blackened with soot as a result of the pollution that has settled on building stones and surfaces for years.

V. Cost of Treating the Sick

The high cost of air pollution is strikingly illustrated in its damaging effects on the human body. Besides the unpleasantness of irritated eyes and scratchy throats, it presents a threat to the respiratory tract, contributing to a number of serious diseases. In both the United States and Europe, episodes of high levels of air pollution were implicated in a large number of deaths.

VI. Effects of Airplanes

The jet airplane has revolutionized travel since about 1960. It has brought people and cultures closer together. It has created environmental problems. Harmful chemicals sift down from the smoky trails of low-flying jets. The scream of jet engines is constantly heard by people who live near big-city airports. Jet aircraft, particularly the supersonic transport (SST) could engender stratospheric air pollution with consequent changes in climate. Jet exhaust contains water, CO2 oxides of nitrogen, and particulate matter. It is speculative just how harmful these pollutants can be. For example, it is estimated that a fleet of five hundred SST's over a period of years could increase the water content of the stratosphere by 50 to 100 percent, which could result in a rise in average. Temperature of the surface of the earth of about 0.2 Celsius degrees and could cause destruction of some of the stratospheric ozone that protects the earth from ultraviolet radiation.

It is a fact that many people prefer air travel rather than ground or water transportation. This has prompted a critical look at safety and quality control. Contributions to air pollution is a chief concern because of this revolutionary change in public transportation in the United States and around the world. The government must also establish standards for exhaust emissions. Thus manufacturers are forced to develop low-pollutant engines. Contact with government agencies will give a greater insight of this subject. Information on how to contact environmental groups in New England can be found on the reference page of this unit.

This graph shows the dramatic increases in the air traffic since 1950, including passenger miles of U.S. airlines. (Aviation Facts and Figures 1953. Washington, DC: Lincoln Press.)

An airplane needs an energy supply and an engine, for propulsion, that will function whenever they are needed. These internal combustion engines have discharged pollutants into the air. The combustion exhaust must be dealt with. Modification of air/fuel ratios can provide a partial solution. The problem of air pollution from airplanes involve a complex set of interactions among technical, social, and economic factors.

Emission from jet aircraft, particularly on landing and take-offs, are a source of bitter complaints from nearby residents. In a few airports visibility has been dangerously restricted by particulate emissions and photochemical smog. Airlines have a considerable expense in cleaning the obnoxious odors of unburned fuel from aircraft air conditioning systems. Most pilots prefer exhaust plumes, because aircrafts are made more visible.

In a jet engine, air enters through the front and is compressed by rotating vanes as it is forced into combustion chambers arranged around the circumference. Fuel is steadily sprayed into the leading end of each chamber where it ignites in the not compressed air, burns and causes the air to expand. The burning gases push toward the rear, striking turbine blades whose rotation drives the compressor they are connected to. The burning gases are further compressed at the exhaust nozzle to provide a high-velocity exhaust. This provides forward thrust to the aircraft. The diagram below illustrates this operation.

Reliable data on engine exhausts are difficult to obtain because engines operate under many different conditions. Engine design plays an important role in reducing pollution emissions. One engine may emit more than a comparable engine. Older aircraft have experienced a substantial reduction in hydrocarbon, carbon monoxide, and particulate emission because the fuel is more completely burned with the installation of "clean burner cans".

VII. Altitude

The temperature of the atmosphere is not uniform, but decreases with altitude up to 20 kilometers where it is found to be only 220°K. At altitudes above 20 km, the temperature is observed to increase with altitude as a result of the absorption of solar radiation by ozone in the upper atmosphere. At much higher altitudes the temperature decreases again. Any mountain climber knows there is a decrease of air density with altitude. At the highest permanently inhabited village in the Peruvian Andes, located at an altitude of 5.3 km, the air density is about half of the sea level density. In adapting to these conditions the inhabitants have developed unusually large lung capacities. Because of the exponential decreased of density with altitude, most of the atmospheric mass is beneath an altitude of 33 km, about three times the altitude of Mt. Everest.

VIII. Plumes of Smokestacks

A plume from an elevated source such as a tall exhaust stack mixes vertically and horizontally with ambient air as it drifts downwind, Vertical mixing is determined largely by the degree of instability of the lower troposphere. This depends upon the temperature profile. Horizontal mixture can also be influenced indirectly by the elapse rate. For a movement of air in the vertical direction cannot proceed without horizontal movement somewhere.

Examples

(a) *Looping*—large-scale turbulent eddies cause sizable parcels of air, together with portions of the plume, to deviate from a straight downwind direction

(b) *Coning*—the shape of the plume is commonly vertically symmetrical about what is call the plume line.

(c) *Fanning*—suppressed vertical mixing, but not horizontal mixing entirely causes the plume to spread only parallel to the ground and appears to take on the shape of a fan as seen from below.

(d) *Lofting*—the lapse rate in the upper portion of the plume is unstable and in the lower it is stable. Mixing is vigorous in the upward direction.

(d) *Umigation*—poses a potentially serious air pollution situation. Here the plume is released just under an elevated inversion layer. When the low-level unstable lapse rate reaches the plume, the effluent suddenly mixes downward toward the ground.

In the effluent from a smelter in a valley is trapped in a radiational inversion, diffusing neither upward or downward, but drifting down the valley, the ground level concentration will be highest. The daily uniform warming of the valley floor erodes the inversion from beneath, and when the layer containing pollutants becomes unstable widespread fumigation occurs along a great length of the valley.

PLUME TRAPPED WITHIN AN INVERSION DRIFTS INLAND UNTIL IT EMERGES FROM THE UPWARD-SLOPING BASE OF THE INVERSION. THE VENTING OF AIR UP A MOUNTAIN SLOPE IS KNOWN AS THE CHIMNEY EFFECT.

IX. Air Pollution Control

In an effort to control the concentration of air pollutants at ground level, some companies have built very tall smokestacks—up to a thousand feet high. Tall stacks along with, "intermittent controls", a term used to describe the practice of cutting back production when weather conditions threaten to raise local pollution levels above ambient air standards. What can be done to avoid air pollution? This question does not have a simple answer. Every industrial process brings its own set of contaminants to the air. Cooking a meal or simply heating your home contributes to this problem. In fact there are very few activities that so not release some type of contaminant into the atmosphere. It is evident that air pollution will never be completely eliminated. So a more realistic question might be: How can air pollution be reduced to a less harmful level?

Legal actions to place control over the emission of air pollutants have been instituted in several ways. One is in the form of a *public nuisance* law. This is when conditions cause discomfort, inconvenience, damage to property, or injury. A court injunction can be placed against the person or corporation responsible. In a case of community smog it would be pretty impossible to identify who is responsible. So the law governing public nuisances is not very effective.

Private litigation may be sought in cases of damages for individuals. The individual must clearly link the damage to the pollutant emitted. Thus the burden of proof is on the complainant. This can be very expensive. Often the court will weigh the costs of improving conditions against the benefits.

The government has also intervened in the protection of the public. As a result of much research; devices for pollution control have been developed, guidelines for air quality were established, tax incentives were introduced, and most importantly, enforcement of ordinances for restricting the emission of contaminants—*prescribed emission standards*.

In 1970, Congress passed the Clean Air Act, the first comprehensive legislation to reduce air pollution in the United States. This was complemented in 1972 by the similarly aimed Water Pollution Control Act. Both dealt with industrial sources of pollution.

X. Atmospheric Structure

(a) *Troposphere*—This layer is nearest the surface of the earth and most "weather" occurs here. It extends up about eight miles. In the troposphere the air mixed by vertical circulation. Therefore it is affected by the conditions near the surface. The air in the troposphere becomes warmer or cooler, drier or moister, as surface conditions change from day to day. The temperature falls between 0.65°C per 100 meters (3.5°F per 1000 feet).

(b) *Stratosphere*—This layer extends from the tropopause (the top of the troposphere) to about 50 kilometers (30 miles). Here the air moves up and down very little. It is in layers or strata, so it is said to be stratified. The temperature is constant up to about 20 kilometers (12 miles) and then increases to stratopause.

(c) *Ozonosphere*—The ozone layer is a part of the stratosphere and extends from about 10 to 50 kilometers (6 to 30 miles), and the maximum concentration is about 5 parts per million at about 30 kilometers (18 miles). This tiny amount of ozone, which it itself lethal in higher concentrations, shields us from lethal ultraviolet radiation form the sun.

(d) *Mesosphere*—Above the stratopause is the mesosphere another zone of falling temperature that extends to about 85 kilometers (50 miles). The temperature at the mesosphere is about -85°C (-120°F).

(e) *Thermosphere*—A zone of increasing temperature that extends several hundred kilometers. Temperatures reach very high, but has little meaning in this zone of very thin atmosphere.

During December, 1952 episode in London the death rate is closely correlated the change in the atmosphere. There was an increase in mortality in all age groups but the highest increase was among the elderly. The main causes were chronic bronchitis, bronchopneumonia, and heart disease. It is quite evident that pollution causes acute respiratory and heart disease.

XI. Greenhouse Effect

The short wavelength radiation from the sun can penetrate the atmosphere, but the long wavelength radiation from the earth cannot and so is trapped heating the atmosphere. This is similar to the way the glass in a greenhouse was thought to let the short-wavelength radiation from the interior of the warm greenhouse.

The greenhouse effect keeps the average temperature of the earth's surface about 35°C (63°F) warmer than it would be without it. It is also reason the troposphere is heated from below.

XII. Resource Agencies in New England Relating to Air Pollution

AIR COMPLIANCE UNIT
DEPARTMENT OF ENVIRONMENTAL PROTECTION
STATE OFFICE BUILDING
HARTFORD, CT 06115
(203) 566-4030

CONNECTICUT AIR CONSERVATION COMMITTEE
45 ASH STREET
EAST HARTFORD, CT 06108
(203) 528-9437
OFFICE OF PUBLIC AFFAIRS, REGION 1

UNITED STATES FEDERAL PROTECTION AGENCY
ROOM 2203, JOHN F. KENNEDY FEDERAL BUILDING
BOSTON, MA 02203.

SOUTH CENTRAL BRANCH
AMERICAN LUNG ASSOCIATION OF CONNECTICUT
364 WHITNEY AVENUE
NEW HAVEN, CT 06511
(203) 777-6821

Note: Available at the American Lung Association are many brochures, education films (on loan), and Environmental Health catalogs and order forms.

It will be necessary to pursue in more details subjects covered in this unit. The references listed and the bibliography of this unit should be consulted. These references are important because this unit is not intended to cover all aspects of air pollution. Many contaminants will not be discussed in full detail. The basic principles gained from this unit should lead to a good position to consult more inclusive and most recent literature, depending on the time of use. Information constantly change and must be updated.

Understand that the sample lesson plans can be expanded according to the depth in which the individual desires to venture and according to the needs of the students. Many activities can derive from these when it is appropriate. Also the order in which they appear is not necessarily the suggested order they are performed.

XIII. Activities

Activity I

Reading

OBJECTIVE: Students will identify the main idea of a paragraph.

WHAT TO DO: Underline the sentence that states the main idea of each paragraph.

1. The idea that polluted air can be harmful to man dates back to the Middle Ages. Direct evidence of bad effects from polluted air began to accumulate after the first use of coal. This was noticed around the beginning of the fourteenth century. The dark smoke, the unpleasant odors, the blackening of buildings and monuments were clear results.
2. Air pollution has caused widespread damage to trees, fruit, vegetables, and ornamental flowers. The total annual cost of plant damage in the United States has been estimated at close to one billion dollars. The most dramatic instance of such effects were seen in the total destruction of vegetation by sulfur dioxide in the areas surrounding smelters.

Activity II

Reading
Science
Social Studies
Grades 5-8
Below Average To Above Average Levels

OBJECTIVES: Students will identify pollution problems in their immediate environment.

Students will identify the main idea of an article.

WHAT TO DO: Students will use newspapers and magazines to find articles on pollution. Each student must bring in one article per week for four weeks. Have them mount each article on construction paper and write one sentence that give the main idea of the article.

Activity III

Reading
Science
Social Studies
Grades 5-8
Below Average to above Average Levels

OBJECTIVE: Students will obtain and organize relevant data on a particular subject or theme.

WHAT TO DO:

1. Each student should receive a folder or material to make a folder to keep materials together and organized.
2. Each student will research and write a composition.

   (a) FIFTH AND SIXTH graders will use encyclopedias and dictionaries to write a composition entitled "What Is Air Pollution?" (minimum of 1 page)
   (b) SEVENTH graders will use encyclopedias, dictionaries, and other reference materials to write an essay called, "The Effects of Air Pollution" (minimum of 3 pages).
   (c) EIGHTH graders will write a research paper geared toward the cause and effects of air pollution. Students may choose their own titles. This paper should be four to five pages in length. (Teachers may use this opportunity to introduce footnotes and bibliographies if they have not been taught. Teachers may also wish to monitor students' note taking skills.)

Activity IV

Reading
Social Studies

Grades 5-8
Below Average To Above Average Levels

OBJECTIVE: Students will develop a visual study or outline that will support the main idea of a subject with supportive details.

WHAT TO DO: Have students make a collage using the newspaper, stressing the fact that all sections may be used—cartoons, pictures, ads, printed matter. They will concentrate on the subject of air pollution. Examples may be done by the teacher and shared with students to help them get started.

On completion of the work, each student will present his or her project orally specifying the main idea and showing how each part of the collage supports the main idea. The class will be looking for irrelevant details.

Activity V

Social Studies
Grades 7-8
Below Average To Above Average Levels

OBJECTIVE: Students will be able to develop plans and solutions of social problem solving.

WHAT TO DO: Separate the class into groups of four or five. Give each group the same problem.

PROBLEM: You are a member of a planning commission which has the responsibility for selecting a site which will be zoned for heavy industry.

TASK: List criteria concerning aspects of air pollution which would bear on the matter. Present relevant information.

Activity VI

Reading
Grades 7-8
Below Average To Above Average Levels

OBJECTIVE: Students will read for high comprehension and supporting details.

WHAT TO DO: Write the following HINTS TO SUCCESSFUL CRITICAL READING on the board. Go over them with the students and allow them to comment and/or ask questions.

1. Read for high comprehension of ideas and supporting details.

2. Re-read if necessary to insure full understanding.
3. Look for author's purpose.
4. Note any inaccuracies.
5. Determine if the facts justify the conclusion.
6. Find the basic thoughts
7. Be sure you have a clear overview.
8. Relate the new material to what you already know
9. Remember key words as clues.
10. Define words that are not familiar.

Now give students selected reading material to read independently. Then ask students to write three important details they remember. Make sure the reading has lots of details and is interesting.

Activity VII

Social Studies
Grades 5-8
Below Average To Above Average Levels

OBJECTIVE: Students will support an idea and use skill of persuasion to substantiate their view.

WHAT TO DO: Divide the class in two teams. One team for and the other against a statement. Each team will develop an argument in favor of their view point. The teams will select 5 students to represent their team. The remaining students will serve as a support group. Rules for alternates can be established by teachers.

PROBLEM: Tall chimneys do not collect or destroy anything, all they do is protect the nearby area at the expense of more distant places which will eventually get all the pollutants anyway. Are tall smokestacks an asset in control of air pollution?

Activity VIII

Math
Grades 5-8
Below Average To Above Average Levels

OBJECTIVES: Students will be able to read scales.
Students will be able to obtain information from scales.

Students will be able to compare Celsius and Fahrenheit temperature scales.

WHAT TO DO: Have students study the thermometers and fill in the missing temperatures. Then have them answer the questions.

1. What is the boiling point of water? ___ C, ___ F
2. What is the normal body temperature? ___ C, ___ F
3. What is normal room temperature? ___ C, ___ F
5. What is the difference between the freezing point of water and the boiling point? ___ C, ___ F

7

Geochemistry, Soil Chemistry and Soil Pollution

7.1 Geochemistry

A field that encompasses the investigation of the chemical composition of the Earth, other planets, and the solar system and universe as a whole, as well as the chemical processes that occur within them. The discipline is large and very important because basic knowledge about the chemical processes involved is critical for understanding subjects as diverse as the formation of economically valuable ore deposits, safe disposal of toxic wastes, and variations in the Earth's climate.

Isotope geochemistry is based on the fact that the isotopic compositions of various chemical elements may reveal information about the age, history, and origin of terrestrial and extraterrestrial materials. Isotopes of an element share the same chemical properties but have slightly different nuclear makeup's and therefore different masses. Some naturally occurring isotopes are radioactive and decay at known rates to form daughter isotopes of another element; for example, radioactive uranium isotopes decay to stable isotopes of lead. Radioactive decay is the basis of geochronology, or age determination: the age of a sample can be found by measuring its content of the daughter isotope. Both radioactive decay and the processes that enrich or deplete materials in certain isotopes cause different parts of the Earth and solar system to have different, characteristic isotopic compositions for some elements. These differences serve as fingerprints for tracing the origins of, and characterizing the interactions between, various geochemical reservoirs.

Cosmo chemistry deals with no earthly materials. Typically, Cosmo chemists use the same kinds of analytical and theoretical approaches as other geochemists but apply them to problems involving the origin and history of meteorites, the formation of the solar system, the chemical processes on other planets, and the ultimate origin of the elements themselves in stars.

Organic geochemistry deals with carbon-containing compounds, largely those produced by living organisms. These are widely dispersed in the outer part of the Earth" in the oceans, the atmosphere, soil, and sedimentary rocks. Organic geochemistry is important for understanding many of the chemical cycles that occur on Earth because biology often plays a major role. Organic geochemists are also active in investigating such areas as the origin of life, the formation of some types of ore deposits that may be biologically mediated, and the origin of coal, petroleum, and natural gas.

In recent years there has been widespread application of geochemical techniques to problems in pale climatology and pale oceanography. In this approach, ocean sediments, sedimentary rocks on land, ice cores, and other continuous records of the Earth's history are analyzed for fossil chemical evidence of past climates or seawater composition. As in most areas of geochemistry, precise and accurate analytical methods for determining the isotopic and elemental composition of the samples are critical.

Study of the chemical changes on the earth. More specifically, it is the study of the absolute and relative abundances of chemical elements in the minerals, soils, ores, rocks, water, and atmosphere of the earth and the distribution and movement of these elements from one place to another as a result of their chemical and physical properties. Geochemical studies also include the study of isotopes of chemical elements, especially their abundance and stability in the universe. Geochemistry provides a theoretical basis for ore prospecting and has refined and improved the methods of determining the age of rocks including the use of radioactive isotopes to date the rock. Chemical studies of ancient sedimentary rocks and the fluids contained in them have provided insights into the evolution of the oceans and the atmosphere. Experiments have been conducted with gases that recreate the primordial atmosphere. Today, important work in geochemistry involves the study of geochemical cycles in the atmosphere; marine and estuarine waters; and the earth's crust. There are many studies in relation to the effects of massive amounts of pollutants on the environment.

The field of *geochemistry* involves study of the chemical composition of the Earth and other planets, chemical processes and reactions that govern the composition of rocks and soils, and the cycles of matter and energy that transport the Earth's chemical components in time and space, and their interaction with the hydrosphere and the atmosphere.

The most important fields of geochemistry are:

1. *Isotope geochemistry*: Determination of the relative and absolute concentrations of the elements and their isotopes in the earth and on earth's surface.
2. Examination of the distribution and movements of elements in different parts of the earth (crust, mantle, hydrosphere etc.) and in minerals with the goal to determine the underlying system of distribution and movement.
3. *Cosmo chemistry*: Analysis of the distribution of elements and their isotopes in the cosmos.
4. *Organic geochemistry*: A study of the role of processes and compounds that are derived from living or once-living organisms.
5. Applications to environmental, hydrological and mineral exploration studies.

The man considered by most to be the father of modern geochemistry was Victor Goldschmidt, and the ideas of the subject were formed by him in a series of publications from 1922 under the title 'Geochemist Verteilungsgesetze der Element'™.

Chemical Characteristics

The more common rock constituents are nearly all oxides; chlorine, sulfur and fluorine are the only important exceptions to this and their total amount in any rock is usually much less than 1 per cent. F. W. Clarke has calculated that a little more than 47 per cent of the earth's crust consists of oxygen. It occurs principally in combination as oxides, of which the chief are silica, alumina, iron oxides, lime, magnesia, potash and soda. The silica functions principally as an acid, forming silicates, and all the commonest minerals of igneous rocks are of this nature. From a computation based on 1672 analyses of all kinds of rocks Clarke arrived at the following as the average percentage composition: SiO_2 = 59.71, Al_2O_3 = 15.41, Fe_2O_3 = 2.63, FeO = 3.52, MgO = 4.36, CaO = 4.90, Na_2O = 3.55, K_2O = 2.80, H_2O = 1.52, TiO_2 = 0.60, P_2O_5 = 0.22, total 99.22%). All the other constituents occur only in very small quantities, usually much less than 1 per cent.

These oxides do not combine in a haphazard way. The potash and soda, for example, combine to produce feldspars. In some cases they may take other forms, such as nepheline, leucite and muscovite, but in the great majority of instances they are found as felspar. The phosphoric acid with lime forms apatite. The titanium dioxide with ferrous oxide gives rise to limonite. Part of the lime forms lime felspar. Magnesia and iron oxides with silica crystallize as olivine or enstatite, or with alumina and lime form the complex ferromagnesian silicates of which the pyroxenes, amphiboles and biotitic are the chief. Any excess of silica above what is required to neutralize the bases will separate out as quartz; excess of alumina crystallizes as corundum. These must be regarded only as general tendencies. It is possible by inspection of a rock analysis to say approximately what minerals the rock will contain, but there are numerous exceptions to any rule which can be laid down.

Mineral Constitution

Hence we may say that except in acid or siliceous rocks containing 66 per cent of silica and over, quartz will not be abundant. In basic rocks (containing 60% of silica or less) it is rare and accidental. If magnesia and iron be above the average while silica is low olivine may be expected; where silica is present in greater quantity over ferromagnesian minerals, such as augite, hornblende, enstatite or biotite, occur rather than olivine. Unless potash is high and silica relatively low leucite will not be present, for leucite does not occur with free quartz. Nepheline, likewise, is usually found in rocks with much soda and comparatively little silica. With high alcalis soda-bearing pyroxenes and amphiboles may be present. The lower the percentage of silica and the alkalis the greater is the prevalence of t lime felspar as contracted with soda or potash felspar. Clarke has calculated the relative abundance of the principal rock-forming minerals with the following results: Apatite = 0.6, titanium minerals = 1.5, quartz = 12.0, felspars = 59.5, biotite = 3.8, hornblende and pyroxene = 16.8, total = 94.2 per cent. This, however, can only be a rough approximation. The other determining factor, namely the physical conditions attending consolidation, plays on the whole a smaller part, yet is by no means negligible, as a few instances will prove. There are certain minerals which are practically confined to deep-seated intrusive rocks, *e.g.* microcline, muscovite, diallage. Leucite is very rare in plutonic masses; many minerals have special peculiarities in microscopic character according to whether they crystallized in depth or near the surface, *e.g.* hypersthene, orthoclase, quartz. There are some curious instances of rocks having the same chemical composition but consisting of entirely different minerals, *e.g.* the hornblendite of Gran, in Norway, containing only hornblende, has the same composition as some of the camptonites of the same locality which contain felspar and hornblende of a different variety. In this connection we may repeat what has been said above about the corrosion of porphyritic minerals in igneous rocks. In rhyolites and trachytes early crystals of hornblende and biotite may be found in great numbers partially converted into augite and magnetite. The hornblende and biotite were stable under the pressures and other conditions which obtained below the surface, but unstable at higher levels. In the ground-mass of these rocks augite is almost universally present. But the plutonic representatives of the same magma, granite and syenite contain biotite and hornblende far more commonly than augite.

Acid, Intermediate and Basic Igneous Rocks

Those rocks which contain most silica and on crystallizing yield free quartz are erected into a group generally designated the "acid" rocks. Those again which contain least silica and most magnesia and iron, so that quartz is absent while olivine is usually abundant, form the "basic" group. The "intermediate" rocks include those which are characterized by the general absence of both quartz and olivine. An important subdivision of these contains a very high percentage of alkalis, especially soda, and consequently has minerals such as nepheline and leucite not common in other rocks. It is often separated from the others as the "alkali" or "soda" rocks, and there is a corresponding series of basic rocks. Lastly a small sub-group rich in olivine and without felspar has been called the "ultrabasic" rocks. They have very low percentages of silica but much iron and magnesia.

Except these last practically all rocks contain felspars or felspathoid minerals. In the acid rocks the common felspars are orthoclase, which perthite, microcline, oligoclase, all having much silica and alkalis. In the basic rocks labradorite, anorthite and bytownite prevail, being rich in lime and poor in silica, potash and soda. Augite is the commonest ferro-magnesian of the basic rocks, but biotite and hornblende are on the whole more frequent in the acid.

	Acid	*Intermediate*		*Basic*	*Ultrabasic*
Commonest Minerals	*Quartz Orthoclase (and Oligoclase), Mica, Hornblende, Augite*	*Little or no Quartz: Orthoclase hornblende, Augite, Biotite*	*Little or no Quartz: Plagioclase Hornblende, Augite, Biotite*	*No Quartz Plagioclase Augite, Olivine*	*No Felspar Augite, Hornblende, Olivine*
Plutonic or Abyssal type	Granite	Syenite	Diorite	Gabbro	Peridotite
Intrusive or Hypabyssal type	Quartz-porphyry	Orthoclase-porphyry	Porphyrite	Dolerite	Picrite
Lavas or Effusive type	Rhyolite, Obsidian	Trachyte	Andesite	Basalt	Limburgite

The rocks which contain leucite or nepheline, either partly or wholly replacing felspar are not included in this table. They are essentially of intermediate or of basic character. We might in consequence regard them as varieties of syenite, diorite, gabbro, etc., in which felspathoid minerals occur, and indeed there are many transitions between syenites of ordinary type and nepheline â€" or leucite â€" syenite, and between gabbro or dolerite and theralite or essonite. But as many minerals develop in these "alcali" rocks which are uncommon elsewhere, it is convenient in a purely formal classification like that which is outlined here to treat the whole assemblage as a distinct series.

Nepheline and Leucite-bearing Rocks

Commonest Minerals	*Alkali Felspar, Nepheline or Leucite, Augite, Hornblend, Biotite*	*Soda Lime Felspar, Nepheline or Leucite, Augite, Hornblende (Olivine)*	*Nepheline or Leucite, Augite, Hornblende, Olivine*
Plutonic type	Nepheline-syenite, Leucite-syenite, Nepheline-porphyry	Essexite and Theralite	Ijolite and Missourite
Effusive type or Lavas	Phonolite, Leucitophyre	Tephrite and Basanite	Nepheline-basalt, Leucite-basalt

This classification is based essentially on the mineralogical constitution of the igneous rocks. Any chemical distinctions between the different groups, though implied, are relegated to a subordinate position. It is admittedly artificial by it has grown up with the grown of the science and is still adopted as the basis on which more minute subdivisions are erected. The subdivisions are by no means of equal value. The syenites, for example, and the peridotites, are far less important than the granites, diorites and gabbros. Moreover, the effusive andesites do not always correspond to the plutonic diorites but partly also to the gabbros. As the different kinds of rock, regarded as aggregates of minerals, pass gradually into one another, transitional types are very common and are often so important as to receive special names. The quartz-syenites and nordmarkites may be interposed between granite and syenite, the tonalities and adamellites between granite and diorite, the monzoaites between syenite and diorite, norites and hyperites between diorite and gabbro, and so on.

7.2 Soil Chemistry

Soil chemistry studies the chemical characteristics of soil. Soil chemistry is affected by mineral composition, organic matter and environmental factors.

Overview

Until the late 1960s, soil chemistry focused primarily on chemical reactions in the soil that contribute to pedogenesis or that affect plant growth. Since then concerns have grown about environmental pollution, organic and inorganic soil contamination and potential ecological health and environmental health risks. Consequently, the emphasis in soil chemistry has shifted from pedology and agricultural soil science to an emphasis on environmental soil science.

A knowledge of environmental soil chemistry is paramount to predicting the fate, mobility and potential toxicity of contaminants in the environment. The vast majority of environmental contaminants are initially released to the soil. Once a chemical is exposed to the soil environment a myriad of chemical reactions can occur that may increase/decrease contaminant toxicity. These reactions include adsorption/desorption, pcipitation, polymerization, dissolution, complexation, and oxidation/reduction. These reactions are often disregarded by scientists and engineers involved with environmental remediation. Understanding these processes enable us to better predict the fate and toxicity of contaminants and provide the knowledge to develop scientifically correct, and cost-effective remediation strategies.

Concepts

- ❖ Anion and cation exchange capacity
- ❖ Soil pH
- ❖ Mineral formation and transformation processes
- ❖ Clay mineralogy
- ❖ Sorption and precipitation reactions in soil
- ❖ Oxidation-reduction reactions
- ❖ Chemistry of problem soils

7.3 Soil Contamination

Soil contamination is caused by the presence of man-made chemicals or other alteration in the natural soil environment. This type of contamination typically arises from the rupture of underground storage tanks, application of pesticides, percolation of contaminated surface water to subsurface strata, oil and fuel dumping, leaching of wastes from landfills or direct discharge of industrial wastes to the soil. The most common chemicals involved are petroleum hydrocarbons, solvents, pesticides, lead and other heavy metals. This occurrence of this phenomenon is correlated with the degree of industrialization and intensity of chemical usage.

The concern over soil contamination stems primarily from health risks, both of direct contact and from secondary contamination of water supplies Mapping of contaminated soil sites and the resulting cleanup are time consuming and expensive tasks, requiring extensive amounts of geology, hydrology, chemistry and computer modeling skills.

It is in North America and Western Europe that the extent of contaminated land is most well known, with many of countries in these areas having a legal framework to identify and deal with this environmental problem; this however may well be just the tip of the iceberg with developing countries very likely to be the next generation of new soil contamination cases.

The immense and sustained growth of the People's Republic of China since the 1970s has exacted a price from the land in increased soil pollution. The State Environmental Protection Administration believes it to be a threat to the environment, to food safety and to sustainable agriculture. According to a scientific sampling, 150 million mi (100,000 square kilometres) of China's cultivated land have been polluted, with contaminated water being used to irrigate a further 32.5 million mi (21,670 square kilometres) and another 2 million mi (1,300 square kilometres) covered or destroyed by solid waste. In total, the area accounts for one-tenth of China's cultivatable land, and is mostly in economically developed areas. An estimated 12 million tonnes of grain are contaminated by heavy metals every year, causing direct losses of 20 billion yuan (US$2.57 billion).

The United States, while having some of the most widespread soil contamination, has actually been a leader in defining and implementing standards for cleanup· Other industrialized countries have a large number of contaminated sites, but lag the U.S. in executing remediation. Developing countries may be leading in the next generation of new soil contamination cases.

Each year in the U.S., thousands of sites complete soil contamination cleanup, some by using microbes that "eat up" toxic chemicals in soil, many others by simple excavation and others by more expensive high-tech soil vapor extraction or air stripping. At the same time, efforts proceed worldwide in creating and identifying new sites of soil contamination, particularly in industrial countries other than the U.S., and in developing countries which lack the money and the technology to adequately protect soil resources.

Microanalysis of Soil Contamination

To understand the fundamental nature of soil contamination, it is necessary to envision the variety of mechanisms for pollutants to become entrained in soil. Soil particulates may be composed of a gamut of organic and inorganic chemicals with variations in cation exchange capacity, buffering capacity, and redox poise. For example, at the extremes, one has a sand component, a coarse grained, inert, and totally inorganic substance; whereas peat soils are dominated by a fine organic material, made of decomposing organic material and highly active. Most soils are mixtures of soil subtypes and thus have quite complex characteristics. There is also a great diversity of soil porosity, ranging from gravels to sands to silt to clay (in increasing order of porosity), pore size, and pore tortuosity (both in decreasing order). Finally there is a wide spectrum of chemical bonding or adhesion characteristics: each contaminant has a different interaction or bonding mechanism with a given soil type.

On balance, some contaminants may literally drain through soils such as sand and gravel and move to other soils or deeper aquifers, while polar or organic chemicals discharged into a clay soil will have a very high adsorption. Thus most soil contamination is the result of pollutants adhering to the soil particle surface, or lodging in interstices of a soil matrix. Clearly the equilibrium reached is a dynamic one, where new pollutants may lodge on new soil particles and the action of groundwater movement may over time transport some of the soil contaminants to other locations or depths.

Soil contamination results when hazardous substances are either spilled or buried directly in the soil or migrate to the soil from a spill that has occurred elsewhere. For example, soil can become contaminated when small particles containing hazardous substances are released from a smokestack and are deposited on the surrounding soil as they fall out of the air. Another source of soil contamination could be water that washes contamination from an area containing hazardous substances and deposits the contamination in the soil as it flows over or through it.

Health Effects

The major concern is that there are many sensitive land uses where people are in direct contact with soils such as residences, parks, schools and playgrounds. Other contact mechanisms include contamination of drinking water or inhalation of soil contaminants which have vaporized. There is a very large set of health consequences from exposure to soil contamination depending on pollutant type, pathway of attack and vulnerability of the exposed population. Chromium and obsolete pesticide formulations are carcinogenic to populations. Lead is especially hazardous to young children, in which group there is a high risk of developmental damage to the brain, while to all populations kidney damage is a risk.

Chronic exposure to at sufficient concentrations is known to be associated with higher incidence of leukemia. Obsolete pesticides such as mercury and cyclodienes are known to induce higher incidences of kidney damage, some irreversible; cyclodienes are linked to liver toxicity. Organophosphates and carbamates can induce a chain of responses leading to neuromuscular blockage. Many chlorinated solvents induce liver changes, kidney changes and depression of the central nervous system. There is an entire spectrum of further health effects such as headache, nausea, fatigue (physical), eye irritation and skin rash for the above cited and other chemicals.

Ecosystem Effects

Not unexpectedly, soil contaminants can have significant deleterious consequences for ecosystems· There are radical soil chemistry changes which can arise from the presence of many hazardous chemicals even at low concentration of the contaminant species. These changes can manifest in the alteration of metabolism of endemic microorganisms and arthropods resident in a given soil environment. The result can be virtual eradication of some of the primary food chain, which in turn have major consequences for predator or consumer species. Even if the chemical effect on lower life forms is small, the lower pyramid levels of the food chain may ingest alien chemicals, which normally become more concentrated for each consuming rung of the food chain. Many of these effects are now well known, such as the concentration of persistent DDT materials for avian consumers, leading to weakening of egg shells, increased chick mortality and potentially species extinction.

Effects occur to agricultural lands which have certain types of soil contamination. Contaminants typically alter plant metabolism, most commonly to reduce crop yields. This has a secondary effect upon soil conservation, since the languishing crops cannot shield the earth's soil mantle from erosion phenomena. Some of these chemical contaminants have long half-lives and in other cases derivative chemicals are formed from decay of primary soil contaminants.

Regulatory Framework

United States of America

Until about 1970 there was little widespread awareness of the worldwide scope of soil contamination or its health risks. In fact, areas of concern were often viewed as unusual or isolated incidents. Since then, the U.S. has established guidelines for handling hazardous waste and the cleanup of soil pollution. In 1980 the U.S. Superfund/CERCLA established strict rules on legal liability for soil contamination. Not only did CERCLA stimulate identification and cleanup of thousands of sites, but it raised awareness of property buyers and sellers to make soil pollution a focal issue of land use and management practices.

While estimates of remaining soil cleanup in the U.S. may exceed 200,000 sites, in other industrialized countries there is a lag of identification and cleanup functions. Even though their use of chemicals is lower than industrialized countries, often their controls and regulatory framework is quite weak. For example, some persistent pesticides that have been banned in the U.S. are in widespread uncontrolled use in developing countries. It is worth noting that the cost of cleaning up a soil contaminated site can range from as little as about $10,000 for a small spill, which can be simply excavated, to millions of dollars for a widespread event, especially for a chemical that is very mobile such as perchloroethylene.

China

China, an economy that regularly records double digit annual economic growth, has little or no legislation to protect the environment. Currently, given the amount of land in question (up to one-tenth of China's cultivatable land may be polluted), the degree of the pollution in specific locations is unclear, making both prevention and remedy difficult. There are no laws or environmental standards regarding soil. Funding is limited, too, so there is little advanced scientific study of China's soil taking place. The severity of the pollution is not known by the public or business population, and the situation is most likely worsening as a result.

United Kingdom

Generic guidance commonly used in the UK are the Soil Guideline Values published by DEFRA and the Environment Agency. These are screening values that demonstrate the minimal acceptable level of a substance. Above this there can be no assurances in terms of significant risk of harm to human health. These have been derived using the Contaminated Land Exposure Asseeement Model (CLEA UK). Certain input parameters such as Health Criteria Values, age and land use are fed into CLEA UK to obtain a probablistic output.

Guidance by the Inter Departmental Committee for the Redevelopment of Contaminated Land (ICRCL) has been formally withdrawn by the Department for Environment, Food and Rural Affairs (DEFRA), for use as a prescriptive document to determine the potential need for remediation or further assessment. Therefore, no further reference is made to these former guideline values.

Other generic guidance that exists (to put the concentration of a particular contaminant in context), includes the United States EPA Region 9 Preliminary Remediation Goals (US PRGs), the US EPA Region 3 Risk Based Concentrations (US EPA RBCs) and National Environment Protection Council of Australia Guideline on Investigation Levels in Soil and Groundwater.

However international guidance should only be used in the UK with clear justification. This is because foreign standards are usually particular to that country due to drivers such as political policy, geology, flood regime and epidemiology. It is generally accepted by UK regulators that only robust scientific methods that relate to the UK should be used.

The CLEA model published by DEFRA and the Environment Agency (EA) in March 2002 sets a framework for the appropriate assessment of risks to human health from contaminated land, as required by Part IIA of the Environmental Protection Act 1990. As part of this framework, generic Soil Guideline Values (SGVs) have currently been derived for ten contaminants to be used as "intervention values". These values should not be considered as remedial targets but values above which further detailed assessment should be considered.

Three sets of CLEA SGVs have been produced for three different land uses, namely

- Residential (with and without plant uptake)
- Allotments
- Commercial/industrial

It is intended that the SGVs replace the former ICRCL values. It should be noted that the CLEA SGVs relate to assessing chronic (long term) risks to human health and do not apply to the protection of ground workers during construction, or other potential receptors such as groundwater, buildings, plants or other ecosystems. The CLEA SGVs are not directly applicable to a site completely covered in hardstanding, as there is no direct exposure route to contaminated soils.

To date, the first ten of fifty-five contaminant SGVs have been published, for the following: arsenic, cadmium, chromium, lead, inorganic mercury, nickel, selenium ethyl benzene, phenol and toluene. Draft SGVs for benzene, naphthalene and xylene have been produced but their publication is on hold.

Toxicological data (Tox) has been published for each of these contaminants as well as for benzo[a]pyrene, benzene, dioxins, furans and dioxin-like PCBs, naphthalene, vinyl chloride, 1,1,2,2 tetrachloroethane and 1,1,1,2 tetrachloroethane, 1,1,1 trichloroethane, tetrachloroethene, carbon tetrachloride, 1,2-dichloroethane, trichloroethene and xylene. The SGVs for ethyl benzene, phenol and toluene are dependent on the soil organic matter (SOM) content (which can be calculated from the total organic carbon (TOC) content). As an initial screen the SGVs for 1 per cent SOM are considered to be appropriate.

Cleanup Options

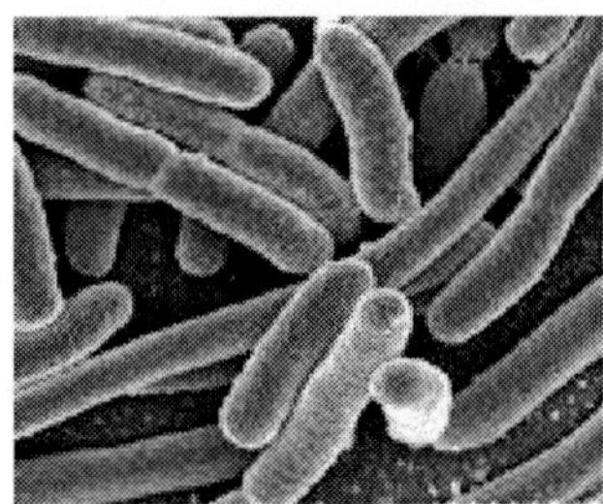

Fig. 7.1: Microbes can be used in soil cleanup.

Cleanup or remediation is analyzed by environmental scientists who utilize field measurement of soil chemicals and also apply computer models for analyzing transport and fate of soil chemicals. Thousands of soil contamination cases are currently in active cleanup across the U.S. as of 2006. There are several principal strategies for remediation:

- Excavate soil and remove it to a disposal site away from ready pathways for human or sensitive ecosystem contact. This technique also applies to dredging of bay muds containing toxins.
- Aeration of soils at the contaminated site (with attendant risk of creating air pollution)
- Bioremediation, involving microbial digestion of certain organic chemicals. Techniques used in bioremediation include landfarming, biostimulation and bioaugmentation soil biota with commercially available microflora.
- Extraction of groundwater or soil vapor with an active electromechanical system, with subsequent stripping of the contaminants from the extract.
- Containment of the soil contaminants (such as by capping or paving

8

Aquatic Chemistry, Water Pollution and Wastewater Treatment

8.1 Aquatic Chemistry

Aquatic chemistry is the study of chemical reactions in aqueous solutions, including acid-base reactions, redox reactions, precipitation reactions, and dissolution reactions.

8.2 Chemical Hydrology

Chemical hydrology or *hydrochemistry* is the subdivion of hydrology that deals with the chemical characteristics of water.

Hydrology (from Greek: Ydwr, *hudôr*, "water"; and lügoV, *logos*, "study") is the study of the movement, distribution, and quality of water throughout the Earth, and thus addresses both the hydrologic cycle and water resources. A practitioner of hydrology is a hydrologist, working within the fields of either earth or environmental science, physical geography or civil and environmental engineering.

Domains of hydrology include hydrometeorology, surface hydrology, hydrogeology, drainage basin management and water quality, where water plays the central role. Oceanography and meteorology are not included because water is only one of many important aspects.

Hydrological research is useful in that it allows us to better understand the world in which we live, and also provides insight for environmental engineering, policy and planning.

History of Hydrology

Hydrology has been a subject of investigation and engineering for millennia. For example, in about 4000 B.C. the Nile was dammed to improve agricultural productivity of previously barren lands. Mesopotamian towns were protected from flooding with high earthen walls. Aqueducts were built by the Greeks and Ancient Romans, while the History of China shows they built irrigation and flood control works. The ancient Sinhalese used hydrology to build complex irrigation Works in Sri Lanka, also known for invention of the Valve Pit which allowed construction of large reservoirs, anicuts and canals which still function.

Marcus Vitruvius, in the first century B.C., described a philosophical theory of the hydrologic cycle, in which precipitation falling in the mountains infiltrated the earth's surface and led to streams and springs in the lowlands. With adoption of a more scientific approach, Leonardo da Vinci and Bernard Palissy independently reached an accurate representation of the hydrologic cycle. It was not until the 17th century that hydrologic variables began to be quantified.

Pioneers of the modern science of hydrology include Pierre Perrault, Edme Mariotte and Edmund Halley. By measuring rainfall, runoff, and drainage area, Perrault showed that rainfall was sufficient to account for flow of the Seine. Marriotte combined velocity and river cross-section measurements to obtain discharge, again in the Seine. Halley showed that the evaporation from the Mediterranean Sea was sufficient to account for the outflow of rivers flowing into the sea.

Advances in the 18th century included the Bernoulli piezometer and Bernoulli's equation, by Daniel Bernoulli, the Pitot tube. The 19th century saw development in groundwater hydrology, including Darcy's law, the Dupuit-Thiem well formula, and Hagen-Poiseuille's capillary flow equation.

Rational analyses began to replace empiricism in the 20th century, while governmental agencies began their own hydrological research programs. Of particular importance were Leroy Sherman's unit hydrograph, the infiltration theory of Robert E. Horton, and C.V. Theis's Aquifer test/equation describing well hydraulics.

Since the 1950's, hydrology has been approached with a more theoretical basis than in the past, facilitated by advances in the physical understanding of hydrological processes and by the advent of computers and especially Geographic Information Systems (GIS).

Hydrologic Cycle

The central theme of hydrology is that water moves throughout the Earth through different pathways and at different rates. The most vivid image of this is in the evaporation of water from the ocean, which forms clouds. These clouds drift over the land and produce rain. The rainwater flows into lakes, rivers, or aquifers. The water in lakes, rivers, and aquifers then either evaporates back to the atmosphere or eventually flows back to the ocean, completing a cycle.

Branches of Hydrology

- *Chemical hydrology* is the study of the chemical characteristics of water.
- *Ecohydrology* is the study of interactions between organisms and the hydrologic cycle.
- *Hydrogeology* is the study of the presence and movement of water in aquifers.
- *Hydroinformatics* is the adaptation of information technology to hydrology and water resources applications.
- *Hydrometeorology* is the study of the transfer of water and energy between land and water body surfaces and the lower atmosphere.
- *Isotope hydrology* is the study of the isotopic signatures of water.
- *Surface hydrology* is the study of hydrologic processes that operate at or near the Earth's surface.

Related Fields

- Aquatic chemistry
- Civil engineering
- Climatology
- Environmental engineering
- Geomorphology
- Hydrography
- Hydraulic engineering
- Limnology
- Oceanography
- Physical geography

Hydrologic Measurements

The movement of water through the Earth can be measured in a number of ways. This information is important for both assessing water resources and understanding the processes involved in the hydrologic cycle. Following is a list of devices used by hydrologists and what they are used to measure.

- Capacitance probe-soil moisture,

- Disdrometer—precipitation characteristics,
- Evaporation—Symon's evaporation pan,
- Infiltrometer—infiltration,
- Piezometer—groundwater pressure and, by inference, groundwater depth (see: aquifer test),
- Radar—cloud properties, rain rate estimation, hail and snow detection
- Rain gauge—rain and snowfall,
- Satellite—rainy area identification, rain rate estimation, land-cover/land-use, soil moisture,
- Sling psychrometer—humidity,
- Stream gauge—stream flow,
- Tensiometer—soil moisture, and
- Time domain reflectometer—soil moisture.

Hydrologic Prediction

Observations of hydrologic processes are used to make predictions of the future behaviour of hydrologic systems (water flow, water quality). One of the major current concerns in hydrologic research is the Prediction in Ungauged Basins (PUB), *i.e.* in basins where no or only very few data exist.

Statistical Hydrology

By analysing the statistical properties of hydrologic records, such as rainfall or river flow, hydrologists can estimate future hydrologic phenomena. This, however, assumes the characteristics of the processes remain unchanged.

These estimates are important for engineers and economists so that proper risk analysis can be performed to influence investment decisions in future infrastructure and to determine the yield reliability characteristics of water supply systems. Statistical information is utilised to formulate operating rules for large dams forming part of systems which include agricultural, industrial and residential demands.

Hydrologic Modeling

Hydrologic models are simplified, conceptual representations of a part of the hydrologic cycle. They are primarily used for hydrologic prediction and for understanding hydrologic processes. Two major types of hydrologic models can be distinguished:

- *Models based on data.* These models are black box systems, using mathematical and statistical concepts to link a certain input (for instance rainfall) to the model output (for instance runoff). Commonly used techniques are regression, transfer functions, neural networks and system identification. These models are known as stochastic hydrology models.

- *Models based on process descriptions.* These models try to represent the physical processes observed in the real world. Typically, such models contain representations of surface runoff, subsurface flow, evapotranspiration, and channel flow, but they can be far more complicated. These models are known as deterministic hydrology models. Deterministic hydrology models can be subdivided into single-event models and continuous simulation models.

Recent research in hydrologic modeling tries to have a more global approach to the understanding of the behaviour of hydrologic systems to make better predictions and to face the major challenges in water resources management.

Hydrologic Transport

Water movement is a significant means by which other material, such as soil or pollutants, are transported from place to place. Initial input to receiving waters may arise from a point source discharge or a line source or area source, such as surface runoff. Since the 1960s rather complex mathematical models have been developed, facilitated by the availability of high speed computers. The most common pollutant classes analyzed are nutrients, pesticides, total dissolved solids and sediment.

Applications of Hydrology

- Determining the water balance of a region.
- Designing riparian restoration projects.
- Mitigating and predicting flood, landslide and drought risk.
- Real-time flood forecasting and flood warning.
- Designing irrigation schemes and managing agricultural productivity.
- Part of the hazard module in catastrophe modeling.
- Providing drinking water.
- Designing dams for water supply or hydroelectric power generation.
- Designing bridges.
- Designing sewers and urban drainage system.

- Analyzing the impacts of antecedent moisture on sanitary sewer systems.
- Predicting geomorphological changes, such as erosion or sedimentation.
- Assessing the impacts of natural and anthropogenic environmental change on water resources.
- Assessing contaminant transport risk and establishing environmental policy guidelines.

8.3 Water Pollution

Water pollution is the contamination of water bodies such as lakes, rivers, oceans, and groundwater caused by human activities, which can be harmful to organisms and plants which live in these water bodies.

Although natural phenomena such as volcanoes, algae blooms, storms, and earthquakes also cause major changes in water quality and the ecological status of water, water is typically referred to as polluted when it impaired by anthropogenic contaminants and either does not support a human use (like serving as drinking water) or undergoes a marked shift in its ability to support its constituent biotic communities. Water pollution has many causes and characteristics. The primary sources of water pollution are generally grouped into two categories based on their point of origin. Point-source pollution refers to contaminants that enter a waterway through a discrete "point source". Examples of this category include discharges from a wastewater treatment plant, outfalls from a factory, leaking underground tanks, etc. The

Fig. 8.1: Raw sewage and industrial waste flows into the U.S. from Mexico as the New River passes from Mexicali, Baja California to Calexico, California

second primary category, non-point source pollution, refers to contamination that, as its name suggests, does not originate from a single discrete source. Non-point source pollution is often a cumulative effect of small amounts of contaminants gathered from a large area. Nutrient runoff in stormwater from sheet flow over an agricultural field, or metals and hydrocarbons from an area with high impervious surfaces and vehicular traffic are examples of non-point source pollution. The primary focus of legislation and efforts to curb water pollution for the past several decades was first aimed at point sources. As point sources have been effectively regulated, greater attention has come to be placed on non-point source contributions, especially in rapidly urbanizing/suburbanizing or developing areas.

The specific contaminants leading to pollution in water include a wide spectrum of chemicals, pathogens, and physical or sensory changes. While many of the chemicals and substances that are regulated may be naturally occurring (iron, manganese, etc) the concentration is often the key in determining what is a natural component of water, and what is a contaminant. Many of the chemical substances are toxic. Pathogens can produce waterborne diseases in either human or animal hosts. Alteration of water's physical chemistry include acidity, electrical conductivity, temperature, and eutrophication. Eutrophication is the fertilisation of surface water by nutrients that were previously scarce. Water pollution is a major problem in the global context. It has been suggested that it is the leading worldwide cause of deaths and diseases, and that it accounts for the deaths of more than 14,000 people daily.

Contaminants

Contaminants may include organic and inorganic substances.

Some organic water pollutants are:

- Insecticides and herbicides, a huge range of organohalide and other chemicals;
- Bacteria, often is from sewage or livestock operations;
- Food processing waste, including pathogens;
- Tree and brush debris from logging operations;
- VOCs (volatile organic compounds), such as industrial solvents, from improper storage;
- DNAPLs (dense non-aqueous phase liquids), such as chlorinated solvents, which may fall at the bottom of reservoirs, since they don't mix well with water and are more dense;
- Petroleum Hydrocarbons including fuels (gasoline, diesel, jet fuels, and fuel oils) and lubricants (motor oil) from oil field operations, refineries,

pipelines, retail service station's underground storage tanks, and transfer operations. Note: VOCs include gasoline-range hydrocarbons;
- Detergents;
- Various chemical compounds found in personal hygiene and cosmetic products; and
- Disinfection by-products (DBPs) found in chemically disinfected drinking water.

Some inorganic water pollutants include:

- Heavy metals including acid mine drainage,
- Acidity caused by industrial discharges (especially sulfur dioxide from power plants),
- Pre-production industrial raw resin pellets, an industrial pollutant,
- Chemical waste as industrial by products,
- Fertilizers, in runoff from agriculture including nitrates and phosphates,
- Silt in surface runoff from construction sites, logging, slash and burn practices or land clearing sites.

Transport and Chemical Reactions of Water Pollutants

Most water pollutants are eventually carried by the rivers into the oceans. In some areas of the world the influence can be traced hundred miles from the mouth by studies using hydrology transport models. Advanced computer models such as SWMM or the DSSAM Model have been used in many locations worldwide to examine the fate of pollutants in aquatic systems. Indicator filter feeding species such as copepods have also been used to study pollutant fates in the New York Bight, for example. The highest toxin loads are not directly at the mouth of the Hudson River, but 100 kilometers south, since several days are required for incorporation into planktonic tissue. The Hudson discharge flows south along the coast due to coriolis force. Further south then are areas of oxygen depletion, caused by chemicals using up oxygen and by algae blooms, caused by excess nutrients from algal cell death and decomposition. Fish and shellfish kills have been reported, because toxins climb the foodchain after small fish consume copepods, then large fish eat smaller fish, etc. Each successive step up the food chain causes a stepwise concentration of pollutants such as heavy metals (*e.g.* mercury) and persistent organic pollutants such as DDT. This is known as biomagnification which is occasionally used interchangeably with bioaccumulation.

The big gyres in the oceans trap floating plastic debris. The North Pacific

Gyre for example has collected the so-called "Great Pacific Garbage Patch" that is now estimated at two times the size of Texas. Many of these long-lasting pieces wind up in the stomachs of marine birds and animals. This results in obstruction of digestive pathways which leads to reduced appetite or even starvation.

Many chemicals undergo reactive decay or chemically change especially over long periods of time in groundwater reservoirs. A noteworthy class of such chemicals are the chlorinated hydrocarbons such as trichloroethylene (used in industrial metal degreasing and electronics manufacturing) and tetrachloroethylene used in the dry cleaning industry (note latest advances in liquid carbon dioxide in dry cleaning that avoids all use of chemicals). Both of these chemicals, which are carcinogens themselves, undergo partial decomposition reactions, leading to new hazardous chemicals (including dichloroethylene and vinyl chloride).

Groundwater pollution is much more difficult to abate than surface pollution because groundwater can move great distances through unseen aquifers. Non-porous aquifers such as clays partially purify water of bacteria by simple filtration (adsorption and absorption), dilution, and, in some cases, chemical reactions and biological activity: however, in some cases, the pollutants merely transform to soil contaminants. Groundwater that moves through cracks and caverns is not filtered and can be transported as easily as surface water. In fact, this can be aggravated by the human tendency to use natural sinkholes as dumps in areas of Karst topography.

There are a variety of secondary effects stemming not from the original pollutant, but a derivative condition. Some of these secondary impacts are:

- ❖ Silt bearing surface runoff from can inhibit the penetration of sunlight through the water column, hampering photosynthesis in aquatic plants.
- ❖ Thermal pollution can induce fish kills and invasion by new thermophilic species. This can cause further problems to existing wildlife.

Sampling & Monitoring

Sampling water can take several forms depending on the accuracy needed and the characteristics of the contaminant. Many contamination events are temporal and most commonly in association with rain events. For this reason 'grab' samples can be used as indicators, but are often inadequate for fully accessing contaminant concerns in a water body. Scientists gathering this type of data often employ auto-sampler devices that pump increments of water at either time or discharge intervals.

Regulatory Framework

In the UK there are common law rights (civil rights) to protect the passage of water across land unfettered in either quality of quantity. Criminal laws dating back to the 16th century exercised some control over water pollution but it was not until the *River (Prevention of pollution) Acts 1951—1961* were enacted that any systematic control over water pollution was established. These laws were strengthened and extended in the *Control of Pollution Act 1984* which has since been updated and modified by a series of further acts. It is a criminal offense to either pollute a lake, river, groundwater or the sea or to discharge any liquid into such water bodies without proper authority. In England and Wales such permission can only be issued by the Environment Agency and in Scotland by SEPA.

In the USA, concern over water pollution resulted in the enactment of state anti-pollution laws in the latter half of the 19th century, and federal legislation enacted in 1899. The Refuse Act of the federal Rivers and Harbors Act of 1899 prohibits the disposal of any refuse matter from into either the nation's navigable rivers, lakes, streams, and other navigable bodies of water, or any tributary to such waters, unless one has first obtained a permit. The Water Pollution Control Act, passed in 1948, gave authority to the Surgeon General to reduce water pollution.

Growing public awareness and concern for controlling water pollution led to enactment of the Federal Water Pollution Control Act Amendments of 1972. As amended in 1977, this law became commonly known as the Clean Water Act. The Act established the basic mechanisms for regulating contaminant discharge. It established the authority for the United States Environmental Protection Agency to implement wastewater standards for industry. The Clean Water Act also continued requirements to set water quality standards for all contaminants in surface waters. Further amplification of the Act continued including the enactment of the Great Lakes Legacy Act of 2002.

8.4 Water Pollution and Society

Introduction

Comprising over 70 per cent of the Earth's surface, water is undoubtedly the most precious natural resource that exists on our planet. Without the seemingly invaluable compound comprised of hydrogen and oxygen, life on Earth would be non-existent: it is essential for everything on our planet to grow and prosper. Although we as humans recognize this fact, we disregard it by polluting our rivers, lakes, and oceans. Subsequently, we are slowly but surely harming our planet to the point where organisms are dying at a very alarming rate. In addition to innocent organisms dying off, our drinking water has become greatly affected as is our ability to use water for recreational purposes. In order to combat water pollution, we must understand the problems and become part of the solution.

Point and Nonpoint Sources

According to the American College Dictionary, pollution is defined as: 'to make foul or unclean; dirty.' Water pollution occurs when a body of water is adversely affected due to the addition of large amounts of materials to the water. When it is unfit for its intended use, water is considered polluted. Two types of water pollutants exist; point source and nonpoint source. Point sources of pollution occur when harmful substances are emitted directly into a body of water. The Exxon Valdez oil spill best illustrates a point source water pollution. A nonpoint source delivers pollutants indirectly through environmental changes. An example of this type of water pollution is when fertilizer from a field is carried into a stream by rain, in the form of run-off which in turn effects aquatic life. The technology exists for point sources of pollution to be monitored and regulated, although political factors may complicate matters. Nonpoint sources are much more difficult to control. Pollution arising from nonpoint sources accounts for a majority of the contaminants in streams and lakes.

Causes of Pollution

Many causes of pollution including sewage and fertilizers contain nutrients such as nitrates and phosphates. In excess levels, nutrients over stimulate the growth of aquatic plants and algae. Excessive growth of these types of organisms consequently clogs our waterways, use up dissolved oxygen as they decompose, and block light to deeper waters.

This, in turn, proves very harmful to aquatic organisms as it affects the respiration ability or fish and other invertebrates that reside in water.

Pollution is also caused when silt and other suspended solids, such as soil, wash off plowed fields, construction and logging sites, urban areas, and eroded river banks when it rains. Under natural conditions, lakes, rivers, and other water bodies undergo Eutrophication, an aging process that slowly fills in the water body with sediment and organic matter. When these sediments enter various bodies of water, fish respiration becomes impaired, plant productivity and water depth become reduced, and aquatic organisms and their environments become suffocated. Pollution in the form of organic material enters waterways in many different forms as sewage, as leaves and grass clippings, or as runoff from livestock feedlots and pastures. When natural bacteria and protozoan in the water break down this organic material, they begin to use up the oxygen dissolved in the water. Many types of fish and bottom-dwelling animals cannot survive when levels of dissolved oxygen drop below two to five parts per million. When this occurs, it kills aquatic organisms in large numbers which leads to disruptions in the food chain.

The pollution of rivers and streams with chemical contaminants has become one of the most curtail environmental problems within the 20th century. Waterborne chemical pollution entering rivers and streams cause tremendous amounts of destruction.

Pathogens are another type of pollution that prove very harmful. They can cause many illnesses that range from typhoid and dysentery to minor respiratory and skin diseases. Pathogens include such organisms as bacteria, viruses, and protozoan. These pollutants enter waterways through untreated

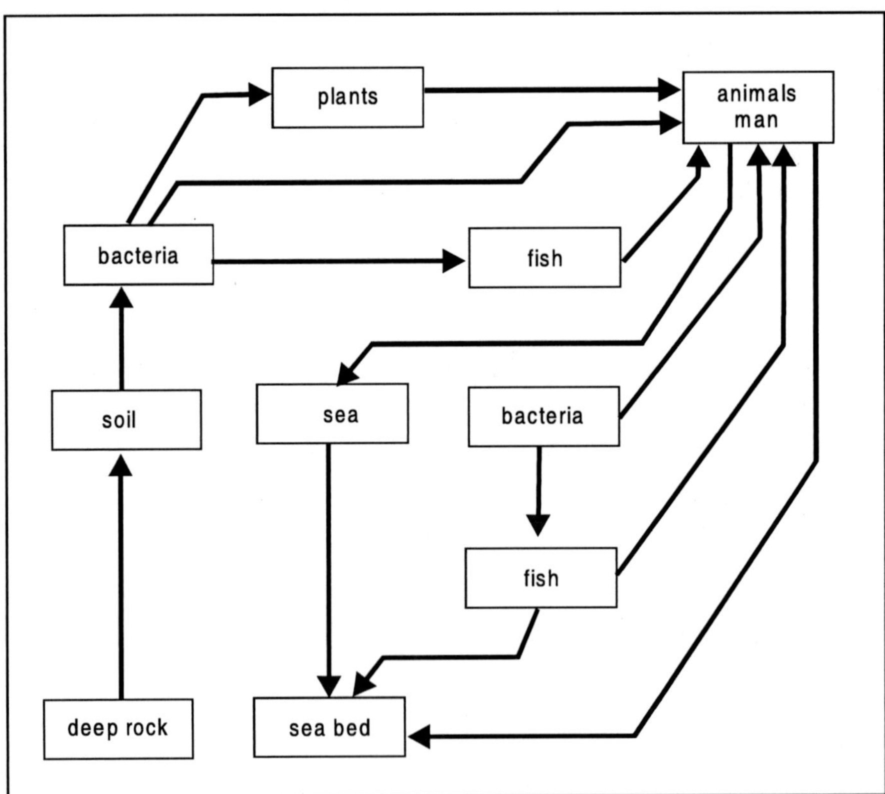

Fig. 8.2: The pathway of contamination.

sewage, storm drains, septic tanks, runoff from farms, and particularly boats that dump sewage. Though microscopic, these pollutants have a tremendous effect evidenced by their ability to cause sickness.

Additional Forms of Water Pollution

Three last forms of water pollution exist in the forms of petroleum, radioactive substances, and heat. Petroleum often pollutes water bodies in the form of oil, resulting from oil spills. The previously mentioned Exxon Valdez is an example of this type of water pollution. These large-scale accidental discharges of petroleum are an important cause of pollution along shore lines. Besides the supertankers, off-shore drilling operations contribute a large share of pollution. One estimate is that one ton of oil is spilled for every million tons of oil transported. This is equal to about 0.0001 per cent. Radioactive substances are produced in the form of waste from nuclear power plants, and from the industrial, medical, and scientific use of radioactive materials. Specific forms of waste are uranium and thorium mining and refining. The last form of water pollution is heat. Heat is a pollutant because increased temperatures result in the deaths of many aquatic organisms. These decreases in temperatures are caused when a discharge of cooling water by factories and power plants occurs.

Demonstrators Protest Drilling

Oil pollution is a growing problem, particularly devastating to coastal wildlife. Small quantities of oil spread rapidly across long distances to form deadly oil slicks. In this picture, demonstrators with "oil-covered" plastic animals protest a potential drilling project in Key Largo, Florida. Whether or not accidental spills occur during the project, its impact on the delicate marine ecosystem of the coral reefs could be devastating.

Oil Spill Clean-up

Workers use special nets to clean up a California beach after an oil tanker spill. Tanker spills are an increasing environmental problem because once oil has spilled, it is virtually impossible to completely remove or contain it. Even small amounts spread rapidly across large areas of water. Because oil and water do not mix, the oil floats on the water and then washes up on broad expanses of shoreline. Attempts to chemically treat or sink the oil may further disrupt marine and beach ecosystems.

Classifying Water Pollution

The major sources of water pollution can be classified as municipal, industrial, and agricultural. Municipal water pollution consists of waste water from homes and commercial establishments. For many years, the main goal of treating municipal wastewater was simply to reduce its content of suspended solids, oxygen-demanding materials, dissolved inorganic compounds, and harmful bacteria. In recent years, however, more stress has been placed on improving means of disposal of the solid residues from the municipal treatment processes. The basic methods of treating municipal wastewater fall into three stages: primary treatment, including grit removal, screening, grinding, and sedimentation; secondary treatment, which entails oxidation of dissolved organic matter by means of using biologically active sludge, which is then filtered off; and tertiary treatment, in which advanced biological methods of nitrogen removal and chemical and physical methods such as granular filtration and activated carbon absorption are employed. The handling and disposal of solid residues can account for 25 to 50 percent of the capital and operational costs of a treatment plant. The characteristics of industrial waste waters can differ considerably both within and among industries. The impact of industrial discharges depends not only on their collective characteristics, such as biochemical oxygen demand and the amount of suspended solids, but also on their content of specific inorganic and organic substances. Three options are available in controlling industrial wastewater. Control can take place at the point of generation in the plant; wastewater can be pretreated for discharge to municipal treatment sources; or wastewater can be treated completely at the plant and either reused or discharged directly into receiving waters.

Wastewater Treatment

Raw sewage includes waste from sinks, toilets, and industrial processes. Treatment of the sewage is required before it can be safely buried, used, or released back into local water systems. In a treatment plant, the waste is passed through a series of screens, chambers, and chemical processes to reduce its bulk and toxicity. The three general phases of treatment are primary, secondary, and tertiary. During primary treatment, a large percentage of the suspended solids and inorganic material is removed from the sewage. The focus of secondary treatment is reducing organic material by accelerating natural biological processes. Tertiary treatment is necessary when the water will be reused; 99 per cent of solids are removed and various chemical processes are used to ensure the water is as free from impurity as possible.

Agriculture, including commercial livestock and poultry farming, is the source of many organic and inorganic pollutants in surface waters and groundwater. These contaminants include both sediment from erosion cropland and compounds of phosphorus and nitrogen that partly originate in animal wastes and commercial fertilizers. Animal wastes are high in oxygen demanding material, nitrogen and phosphorus, and they often harbor pathogenic organisms. Wastes from commercial feeders are contained and disposed of on land; their main threat to natural waters, therefore, is from runoff and leaching. Control may involve settling basins for liquids, limited biological treatment in aerobic or anaerobic lagoons, and a variety of other methods.

Ground Water

Ninety-five percent of all fresh water on earth is ground water. Ground water is found in natural rock formations. These formations, called aquifers, are a vital natural resource with many uses. Nationally, 53 per cent of the population relies on ground water as a source of drinking water. In rural areas this figure is even higher. Eighty one percent of community water is dependent on ground water. Although the 1992 Section 305(b) State Water Quality Reports indicate that, overall, the Nation's ground water quality is good to

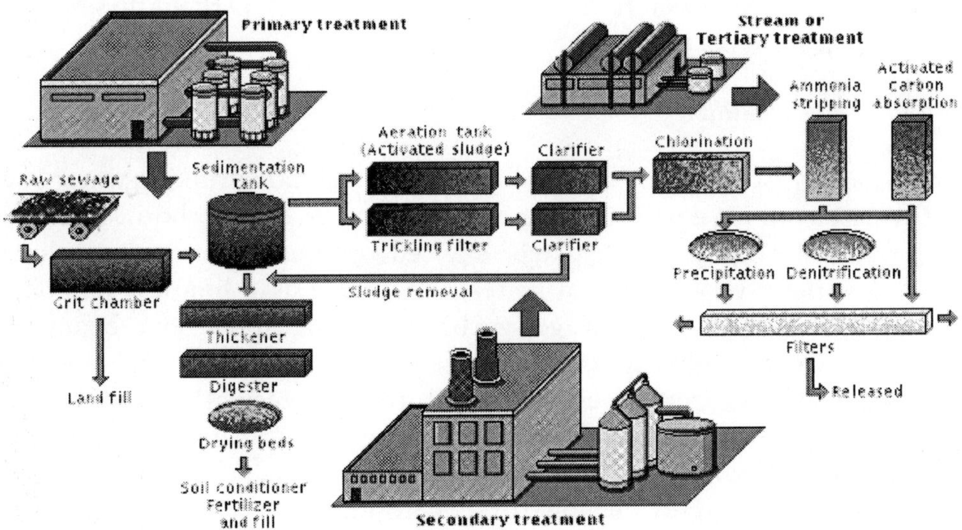

Fig. 8.3: Wastewater treatment.

excellent, many local areas have experienced significant ground water contamination.

Some examples are leaking underground storage tanks and municipal landfills.

Legislation

Several forms of legislation have been passed in recent decades to try to control water pollution. In 1970, the Clean Water Act provided 50 billion dollars to cities and states to build wastewater facilities. This has helped control surface water pollution from industrial and municipal sources throughout the United States. When congress passed the Clean Water Act in 1972, states were given primary authority to set their own standards for their water. In addition to these standards, the act required that all state beneficial uses and their criteria must comply with the 'fishable and swimmable' goals of the act. This essentially means that state beneficial uses must be able to support aquatic life and recreational use. Because it is impossible to test water for every type of disease-causing organism, states usually look to identify indicator bacteria. One for a example is a bacteria known as fecal coliforms. These indicator bacteria suggest that a certain selection of water may be contaminated with untreated sewage and that other, more dangerous, organisms are present. These legislations are an important part in the fight against water pollution. They are useful in preventing Environmental catastrophes. The graph shows reported pollution incidents since 1989-1994. If stronger legislations existed, perhaps these events would never have occurred.

Global Water Pollution

Estimates suggest that nearly 1.5 billion people lack safe drinking water and that at least 5 million deaths per year can be attributed to waterborne diseases. With over 70 percent of the planet covered by oceans, people have long acted as if these very bodies of water could serve as a limitless dumping ground for wastes. Raw sewage, garbage, and oil spills have begun to overwhelm the diluting capabilities of the oceans, and most coastal waters are now polluted. Beaches around the world are closed regularly, often because of high amounts of bacteria from sewage disposal, and marine wildlife is beginning to suffer.

Perhaps the biggest reason for developing a worldwide effort to monitor and restrict global pollution is the fact that most forms of pollution do not respect national boundaries. The first major international conference on environmental issues was held in Stockholm, Sweden, in 1972 and was sponsored by the United Nations (UN). This meeting, at which the United States took a leading role, was controversial because many developing countries were fearful that a focus on environmental protection was a means for the developed world to keep the undeveloped world in an economically subservient position. The most important outcome of the conference was the creation of the United Nations Environmental Program (UNEP).

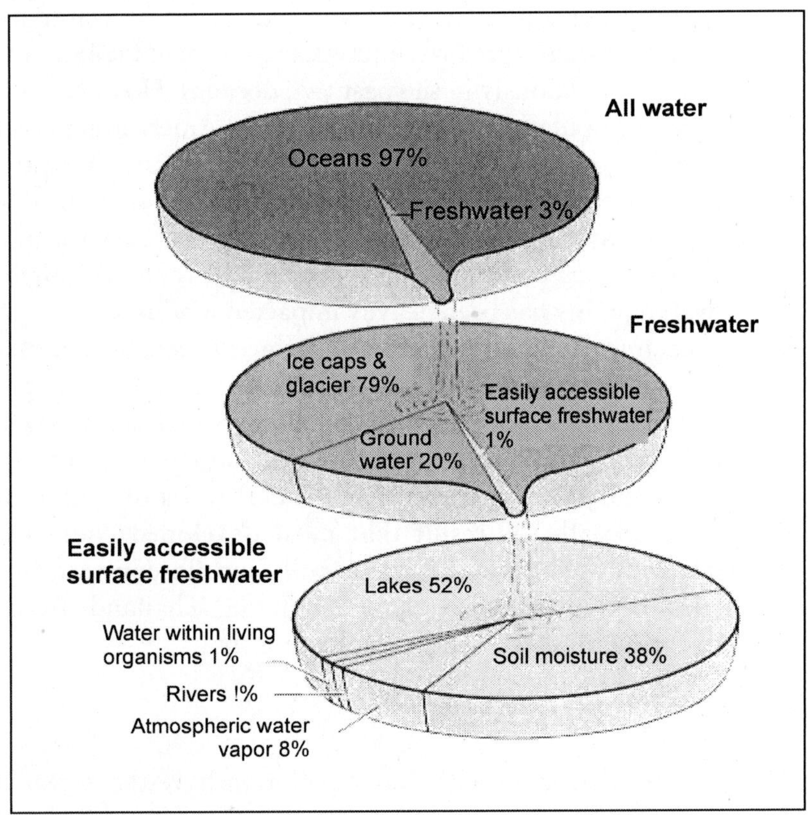

Fig. 8.4: Distribution of World's water.

UNEP was designed to be 'the environmental conscience of the United Nations,' and, in an attempt to allay fears of the developing world, it became the first UN agency to be headquartered in a developing country, with offices in Nairobi, Kenya. In addition to attempting to achieve scientific consensus about major environmental issues, a major focus for UNEP has been the study of ways to encourage sustainable development increasing standards of living without destroying the environment. At the time of UNEP's creation in 1972, only 11 countries had environmental agencies. Ten years later that number had grown to 106, of which 70 were in developing countries.

Water Quality

Water quality is closely linked to water use and to the state of economic development. In industrialized countries, bacterial contamination of surface water caused serious health problems in major cities throughout the mid 1800's. By the turn of the century, cities in Europe and North America began building sewer networks to route domestic wastes downstream of water intakes. Development of these sewage networks and waste treatment facilities in urban areas has expanded tremendously in the past two decades. However, the rapid growth of the urban population (especially in Latin America and Asia) has outpaced the ability of governments to expand sewage and water infrastructure. While waterborne diseases have been eliminated in the developed world, outbreaks of cholera and other similar diseases still occur with alarming frequency in the developing countries. Since World War II and the birth of the 'chemical age', water quality has been heavily impacted worldwide by industrial and agricultural chemicals. Eutrophication of surface waters from human and agricultural wastes and nitrification of groundwater from agricultural practices has greatly affected large parts of the world. Acidification of surface waters by air pollution is a recent phenomenon and threatens aquatic life in many area of the world. In developed countries, these general types of pollution have occurred sequentially with the result that most developed countries have successfully dealt with major surface water pollution. In contrast, however, newly industrialized countries such as China, India, Thailand, Brazil, and Mexico are now facing all these issues simultaneously.

Conclusion

Clearly, the problems associated with water pollution have the capabilities to disrupt life on our planet to a great extent. Congress has passed laws to try to combat water pollution thus acknowledging the fact that water pollution is, indeed, a serious issue. But the government alone cannot solve the entire problem. It is ultimately up to us, to be informed, responsible and involved when it comes to the problems we face with our water. We must become familiar with our local water resources and learn about ways for disposing harmful household wastes so they don't end up in sewage treatment plants that can't handle them or landfills not designed to receive hazardous materials. In our yards, we must determine whether additional nutrients are needed before fertilizers are applied, and look for alternatives where fertilizers might run off into surface waters. We have to preserve existing trees and plant new trees and shrubs to help prevent soil erosion and promote infiltration of water into the soil. Around our houses, we must keep litter, pet waste, leaves, and grass clippings out of gutters and storm drains. These are just a few of the many ways in which we, as humans, have the ability to combat water pollution. As we head into the 21st century, awareness and education will most assuredly continue to be the two most important ways to prevent water pollution. If these measures are not taken and water pollution continues, life on earth will suffer severely.

Global environmental collapse is not inevitable. But the developed world must work with the developing world to ensure that new industrialized economies do not add to the world's environmental problems. Politicians must think of sustainable development rather than economic expansion. Conservation strategies have to become more widely accepted, and people must learn that energy use can be dramatically diminished without sacrificing comfort. In short, with the technology that currently exists, the years of global environmental mistreatment can begin to be reversed.

8.5 Wastewater

Wastewater is any water that has been adversely affected in quality by anthropogenic influence. It comprises liquid waste discharged by domestic residences, commercial properties, industry, and/or agriculture and can encompass a wide range of potential contaminants and concentrations. In the most common usage, it refers to the municipal wastewater that contains a broad spectrum of contaminants resulting from the mixing of wastewaters from different sources.

Sewage is correctly the subset of wastewater that is contaminated with faeces or urine, but is often used to mean any waste water. "Sewage" includes domestic, municipal, or industrial liquid waste products disposed of, usually via a pipe or sewer or similar structure, sometimes in a cesspool emptier.

The physical infrastructure, including pipes, pumps, screens, channels etc. used to convey sewage from its origin to the point of eventual treatment or disposal is termed sewerage.

Wastewater Origin

- Human waste, usually from lavatories: (fæces, used toilet paper, wipes, urine, other bodily fluids) also known as black water
- Cesspit leakage
- Septic tank discharge
- Sewage treatment plant discharge

- Washing water (personal, clothes, floors, dishes, etc.) also known as greywater or sullage
- Rainfall collected on roofs, yards, hard-standings, etc. (traces of oils and fuel but generally clean)
- Groundwater infiltrated into sewerage.
- Surplus manufactured liquids from domestic sources (drinks, cooking oil, pesticides, lubricating oil, paint, cleaning liquids, etc.)
- Urban rainfall run-off from roads, car-parks, roofs, side-walks or pavements (contains oils, animal faeces, litter, fuel residues, rubber residues, metals from vehicle exhausts etc)
- Seawater ingress (salt, micro-biota, high volumes)
- Direct ingress of river water (micro-biota, high volumes)
- Direct ingress of man-made liquids (illegal disposal of pesticides, used oils, etc.)
- Highway drainage (oil, de-icing agents, rubber residues)
- Storm drains (almost anything including cars, shopping trolleys, trees, cattle etc.)
- Black water—surface water contaminated by sewage
- Industrial waste:

  - Industrial site drainage (silt, sand, alkali, oil, chemical)
  - Industrial cooling waters (biocides, heat, slimes, silt)
  - Industrial process waters
  - Organic—bio-degradable—includes waste from abattoirs and creameries and ice-cream manufacture.
  - Organic—non bio-degradable or difficult to treat—for example Pharmaceutical or Pesticide manufacturing
  - Inorganic—for example from the metalworking industry
  - extreme pH—from acid/alkali manufacturing, metal plating
  - Toxic—*e.g.* from metal plating, cyanide production, pesticide manufacturing
  - Solids and Emulsions—*e.g.* Paper manufacturing, food stuffs, lubricating and hydraulic oil manufacture
  - Agricultural drainage—direct and diffuse

Wastewater Constituents

The composition of wastewater varies widely. This is a partial list of what it may contain:

- Water (> 95%) which is often added during flushing to carry the waste down a drain.
- Pathogens such as bacteria, viruses, prions and parasitic worms.
- Non-pathogenic bacteria (> 100,000/ml for sewage).
- Organic particles such as faeces, hairs, food, vomit, paper fibers, plant material, humus, etc.
- Soluble organic material such as urea, fruit sugars, soluble proteins, drugs, pharmaceuticals, etc.
- Inorganic particles such as sand, grit, metal particles, ceramics, etc.
- Soluble inorganic material such as ammonia, road-salt, sea-salt, cyanide, hydrogen sulfide, thiocyanates, thiosulfates, etc.
- Animals such as protozoa, insects, arthropods, small fish, etc.
- Macro-solids such as sanitary napkins, nappies/diapers, condoms, needles, children's toys, dead pets, body parts, etc.
- Gases such as hydrogen sulfide, carbon dioxide, methane, etc.
- Emulsions such as paints, adhesives, mayonnaise, hair colorants, emulsified oils, etc.
- Toxins such as pesticides, poisons, herbicides, etc.

Wastewater Quality Indicators

Any oxidizable material present in a natural waterway or in an industrial wastewater will be oxidized both by biochemical (bacterial) or chemical processes. The result is that the oxygen content of the water will be decreased. Basically, the reaction for biochemical oxidation may be written as:

$$\text{Oxidizable material} + \text{bacteria} + \text{nutrient} + O_2 \rightarrow CO_2 + H_2O$$
$$+ \text{oxidized inorganics such as } NO_3 \text{ or } SO_4$$

Oxygen consumption by reducing chemicals such as sulfides and nitrites is typified as follows:

$$S^- + 2\,O_2 \rightarrow SO_4^-$$
$$NO_2^- + \tfrac{1}{2}\,O_2 \rightarrow NO_3^-$$

Since all natural waterways contain bacteria and nutrient, almost any waste compounds introduced into such waterways will initiate biochemical reactions (such as shown above). Those biochemical reactions create what is measured in the laboratory as the Biochemical oxygen demand (BOD).

Oxidizable chemicals (such as reducing chemicals) introduced into a natural water will similarly initiate chemical reactions (such as shown above).

Those chemical reactions create what is measured in the laboratory as the Chemical oxygen demand (COD).

Both the BOD and COD tests are a measure of the relative oxygen-depletion effect of a waste contaminant. Both have been widely adopted as a measure of pollution effect. The BOD test measures the oxygen demand of biodegradable pollutants whereas the COD test measures the oxygen demand of biogradable pollutants plus the oxygen demand of non-biodegradable oxidizable pollutants.

The so-called 5-day BOD measures the amount of oxygen consumed by biochemical oxidation of waste contaminants in a 5-day period. The total amount of oxygen consumed when the biochemical reaction is allowed to proceed to completion is called the Ultimate BOD. The Ultimate BOD is too time consuming, so the 5-day BOD has almost universally been adopted as a measure of relative pollution effect.

There are also many different COD tests. Perhaps, the most common is the 4-hour COD.

It should be emphasized that there is no generalized correlation between the 5-day BOD and the Ultimate BOD. Likewise, there is no generalized correlation between BOD and COD. It is possible to develop such correlations for a specific waste contaminant in a specific wastewater stream... but such correlations cannot be generalized for use with any other waste contaminants or wastewater streams.

The laboratory test procedures for the determining the above oxygen demands are detailed in the following sections of the "Standard Methods For the Examination Of Water and Wastewater" available at www.standardmethods.org

- 5-day BOD and Ultimate BOD: Sections 5210B and 5210C
- COD: Section 5220

Sewage Disposal

In some urban areas, sewage is carried separately in sanitary sewers and runoff from streets is carried in storm drains. Access to either of these is typically through a manhole. During high precipitation periods a sanitary sewer overflow can occur, causing potential public health and ecological damage.

Sewage may drain directly into major watersheds with minimal or no treatment. When untreated, sewage can have serious impacts on the quality of an environment and on the health of people. Pathogens can cause a variety of illnesses. Some chemicals pose risks even at very low concentrations and can remain a threat for long periods of time because of bioaccumulation in animal or human tissue.

Treatment

There are numerous processes that can be used to clean up waste waters depending on the type and extent of contamination. Most wastewater is treated in industrial-scale wastewater treatment plants (WWTPs) which may include physical, chemical and biological treatment processes. However, the use of septic tanks and other On-Site Sewage Facilities (OSSF) is widespread in rural areas, serving up to one quarter of the homes in the U.S. The most important aerobic treatment system is the activated sludge process, based on the maintenance and recirculation of a complex biomass composed by micro-organisms able to absorb and adsorb the organic matter carried in the wastewater. Anaerobic processes are widely applied in the treatment of industrial wastewaters and biological sludge. Some wastewater may be highly treated and reused as reclaimed water. For some waste waters ecological approaches using reed bed systems such as constructed wetlands may be appropriate. Modern systems include tertiary treatment by micro filtration or synthetic membranes. After membrane filtration, the treated wastewater is indistinguishable from waters of natural origin of drinking quality. Nitrates can be removed from wastewater by microbial denitrification, for which a small amount of methanol is typically added to provide the bacteria with a source of carbon. Ozone Waste Water Treatment is also growing in popularity, and requires the use of an ozone generator, which decontaminates the water as Ozone bubbles percolate through the tank.

Disposal of wastewaters from an industrial plant is a difficult and costly problem. Most petroleum refineries, chemical and petrochemical plants have onsite facilities to treat their wastewaters so that the pollutant concentrations in the treated wastewater comply with the local and/or national regulations regarding disposal of wastewaters into community treatment plants or into rivers, lakes or oceans.

Reuse

Treated wastewater can be reused as drinking water (Singapore), in industry (cooling towers), in artificial recharge of aquifers, in agriculture (70% of Israel's irrigated agriculture is based on highly purified wastewater) and in the rehabilitation of natural ecosystems (Florida's Everglades).

Woods Hole Oceanographic Institution and Harbor Branch Oceanographic Institution use wastewater for breeding algae. The wastewater from domestic and industrial sources contain rich organic compounds, which accelerate the growth of algae. This algae can be used to produce algal fuels

8.6 Sewage Treatment

Sewage treatment, or *domestic wastewater treatment*, is the process of removing contaminants from wastewater, both runoff (effluents) and domestic. It includes physical, chemical and biological processes to remove physical, chemical and biological contaminants. Its objective is to produce a waste stream (or treated effluent) and a solid waste or sludge suitable for discharge or reuse back into the environment. This material is often inadvertently contaminated with many toxic organic and inorganic compounds.

Sewage is created by residences, institutions, hospitals and commercial and industrial establishments. It can be treated close to where it is created (in septic tanks, biofilters or aerobic treatment systems), or collected and transported via a network of pipes and pump stations to a municipal treatment plant (see sewerage and pipes and infrastructure). Sewage collection and treatment is typically subject to local, state and federal regulations and standards. Industrial sources of wastewater often require specialized treatment processes.

The sewage treatment involves three stages, called *primary*, *secondary* and *tertiary treatment*. First, the solids are separated from the wastewater stream. Then dissolved biological matter is progressively converted into a solid mass by using indigenous, water-borne microorganisms. Finally, the biological solids are neutralized then disposed of or re-used, and the treated water may be disinfected chemically or physically (for example by lagoons and micro-filtration). The final effluent can be discharged into a stream, river, bay, lagoon or wetland, or it can be used for the irrigation of a golf course, green way or park. If it is sufficiently clean, it can also be used for groundwater recharge.

Description

Raw influent (sewage) includes household waste liquid from toilets, baths, showers, kitchens, sinks, and so forth that is disposed of via sewers. In many areas, sewage also includes liquid waste from industry and commerce. The draining of household waste into greywater and blackwater is becoming more common in the developed world, with greywater being permitted to be used for watering plants or recycled for flushing toilets. A lot of sewage also includes some surface water from roofs or hard-standing areas. Municipal wastewater therefore includes residential, commercial, and industrial liquid waste discharges, and may include stormwater runoff. Sewage systems capable of handling stormwater are known as combined systems or combined sewers. Such systems are usually avoided since they complicate and thereby reduce the efficiency of sewage treatment plants owing to their seasonality. The variability in flow also leads to often larger than necessary, and subsequently more expensive, treatment facilities. In addition, heavy storms that contribute more flows than the treatment plant can handle may overwhelm the sewage treatment system, causing a spill or overflow (called a combined sewer overflow, or CSO, in the United States). It is preferable to have a separate storm drain system for stormwater in areas that are developed with sewer systems.

As rainfall runs over the surface of roofs and the ground, it may pick up various contaminants including soil particles and other sediment, heavy metals, organic compounds, animal waste, and oil and grease. Some jurisdictions require stormwater to receive some level of treatment before being discharged directly into waterways. Examples of treatment processes used for stormwater include sedimentation basins, wetlands, buried concrete vaults with various kinds of filters, and vortex separators (to remove coarse solids).

The site where the raw wastewater is processed before it is discharged back to the environment is called a wastewater treatment plant (WWTP). The order and types of mechanical, chemical and biological systems that comprise the wastewater treatment plant are typically the same for most developed countries:

- Mechanical Treatment
  - Influx (Influent)
  - Removal of large objects
  - Removal of sand and grit
  - Pre-precipitation
- Biological Treatment
  - Oxidation bed (oxidizing bed) or aeration system
  - Post precipitation
- *Chemical Treatment* (this step is usually combined with settling and other processes to remove solids, such as filtration. The combination is referred to in the U.S. as physical-chemical treatment.

Treatment Stages

Primary Treatment

Primary treatment removes the materials that can be easily collected from the raw wastewater and disposed of. The typical materials that are removed during primary treatment include fats, oils, and greases (also referred to as FOG), sand, gravels and rocks (also referred to as grit), larger settleable solids and floating materials (such as rags and flushed feminine hygiene products). This step is done entirely with machinery.

Removal of Large Objects from Influent Sewage

In primary treatment, the influent sewage water is strained to remove all large objects that are deposited in the sewer system, such as rags, sticks, tampons, cans, fruit, etc. This is most commonly done with a manual or automated mechanically raked screen. The raking action of a mechanical bar screen is typically paced according to the accumulation on the bar screens and/or flow rate. The bar screen is used because large solids can damage or clog the equipment used later in the sewage treatment plant. The solids are collected in a dumpster and later disposed in a landfill.

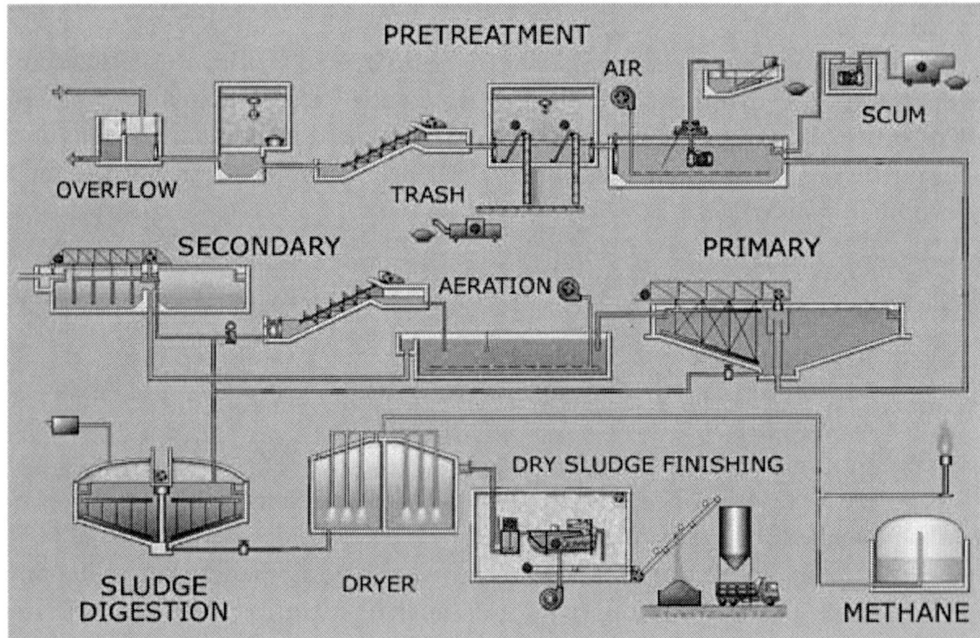

Fig. 8.5: Typical Process Flow Diagram.

Sand and Grit Removal

Primary treatment also typically includes a sand or grit channel or chamber where the velocity of the incoming wastewater is carefully controlled to allow sand grit and stones to settle, while keeping the majority of the suspended organic material in the water column. This equipment is called a detritor or sand catcher. Sand, grit, and stones need to be removed early in the process to avoid damage to pumps and other equipment in the remaining treatment stages. Sometimes there is a sand washer (grit classifier) followed by a conveyor that transports the sand to a container for disposal. The contents from the sand catcher may be fed into the incinerator in a sludge processing plant, but in many cases, the sand and grit is sent to a landfill.

Sedimentation

Many plants have a sedimentation stage where the sewage is allowed to pass slowly through large tanks, commonly called "primary clarifiers" or "primary sedimentation tanks". The tanks are large enough that sludge can settle and floating material such as grease and oils can rise to the surface and be skimmed off. The main purpose of the primary clarification stage is to produce both a generally homogeneous liquid capable of being treated biologically and a sludge that can be separately treated or processed. Primary settling tanks are usually equipped with mechanically driven scrapers that continually drive the collected sludge towards a hopper in the base of the tank from where it can be pumped to further sludge treatment stages.

Secondary Treatment

Secondary treatment is designed to substantially degrade the biological content

Fig. 8.6: Primary sedimentation tank at a rural treatment plant.

of the sewage such as are derived from human waste, food waste, soaps and detergent. The majority of municipal and industrial plants treat the settled sewage liquor using aerobic biological processes. For this to be effective, the biota require both oxygen and a substrate on which to live. There are number of ways in which this is done. In all these methods, the bacteria and protozoa consume biodegradable soluble organic contaminants (*e.g.* sugars, fats, organic short-chain carbon molecules, etc.) and bind much of the less soluble fractions into floc. Secondary treatment systems are classified as *fixed film* or suspended growth. Fixed-film treatment process including trickling filter and rotating biological contactors where the biomass grows on media and the sewage passes over its surface. In *suspended growth systems*—such as activated sludge—the biomass is well mixed with the sewage and can be operated in a smaller space than fixed-film systems that treat the same amount of water. However, fixed-film systems are more able to cope with drastic changes in the amount of biological material and can provide higher removal rates for organic material and suspended solids than suspended growth systems.

Roughing filters are intended to treat particularly strong or variable organic loads, typically industrial, to allow them to then be treated by conventional secondary treatment processes. Characteristics include typically tall, circular filters filled with open synthetic filter media to which wastewater is applied at a relatively high rate. They are designed to allow high hydraulic loading and a high flow-through of air. On larger installations, air is forced through the media using blowers. The resultant wastewater is usually within the normal range for conventional treatment processes.

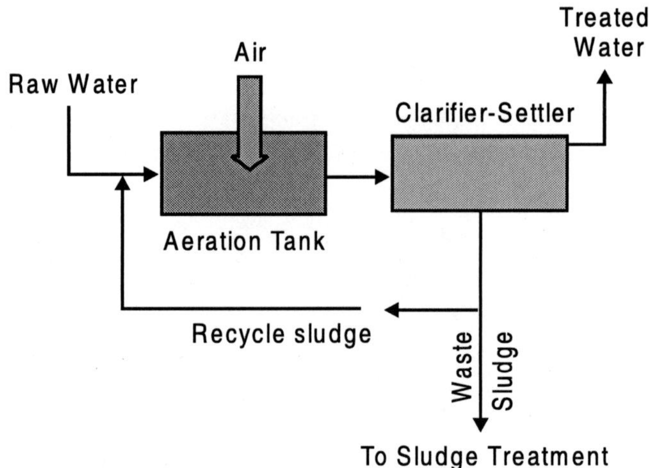

Fig. 8.7: A generalized schematic diagram of an activated sludge process

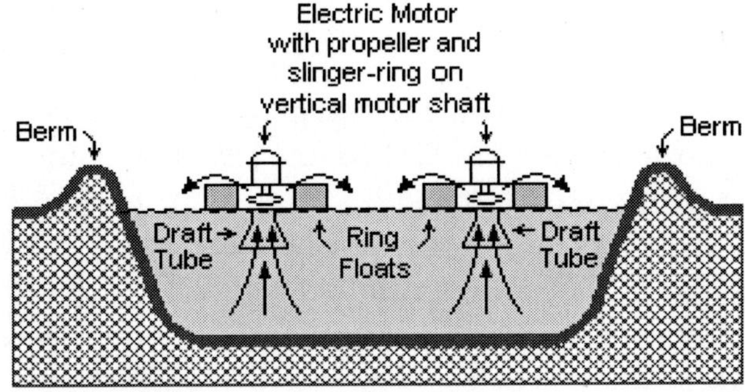

Note: The ring floats are tethered to posts on the berms.

Fig. 8.8: A Typical Surface-Aerated Basin (using motor-driven floating aerators).

Activated Sludge

In general, activated sludge plants encompass a variety of mechanisms and processes that use dissolved oxygen to promote the growth of biological floc that substantially removes organic material.

The process traps particulate material and can, under ideal conditions, convert ammonia to nitrite and nitrate and ultimately to nitrogen gas.

Surface-Aerated Basins

Most biological oxidation processes for treating industrial wastewaters have in common the use of oxygen (or air) and microbial action. Surface-aerated basins achieve 80 to 90 per cent removal of Biochemical Oxygen Demand with retention times of 1 to 10 days. The basins may range in depth from 1.5 to 5.0 metres and use motor-driven aerators floating on the surface of the wastewater.

In an aerated basin system, the aerators provide two functions: they transfer air into the basins required by the biological oxidation reactions, and they provide the mixing required for dispersing the air and for contacting the reactants (that is, oxygen, wastewater and microbes). Typically, the floating surface aerators are rated to deliver the amount of air equivalent to 1.8 to 2.7 kg O_2/kW·h. However, they do not provide as good mixing as is normally achieved in activated sludge systems and therefore aerated basins do not achieve the same performance level as activated sludge units.

Biological oxidation processes are sensitive to temperature and, between 0 °C and 40 °C, the rate of biological reactions increase with temperature. Most surface aerated vessels operate at between 4°C and 32°C.

Fluidized Bed Reactors

The carbon absorption following biological treatment is particularly effective in reducing both the BOD and COD to low levels. A fluidized bed reactor is a combination of the most common stirred tank packed bed, continuous flow reactors. It is very important to chemical engineering because of its excellent heat and mass transfer characteristics. In a fluidized bed reactor, the substrate is passed upward through the immobilized enzyme bed at a high velocity to lift the particles. However the velocity must not be so high that the enzymes are swept away from the reactor entirely. This causes low mixing; these type of reactors are highly suitable for the exothermic reactions. It is most often applied in immobilized enzyme catalysis.

Trickling filter bed using plastic media

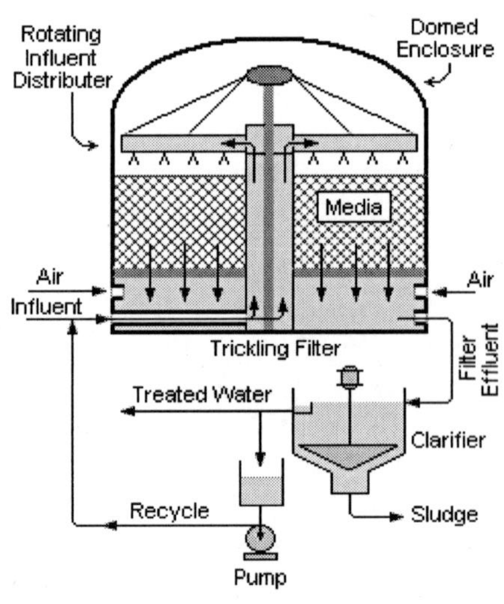

Schematic diagram of a complete trickle filter process in waste treatment plants.

Filter beds (oxidising beds) In older plants and plants receiving more variable loads, trickling filter beds are used where the settled sewage liquor is spread onto the surface of a deep bed made up of coke (carbonised coal), limestone chips or specially fabricated plastic media. Such media must have high surface areas to support the biofilms that form. The liquor is distributed through perforated rotating arms radiating from a central pivot. The distributed liquor trickles through this bed and is collected in drains at the base. These drains also provide a source of air which percolates up through the bed, keeping it aerobic. Biological films of bacteria, protozoa and fungi form on the media's surfaces and eat or otherwise reduce the organic content.

This biofilm is grazed by insect larvae and worms which help maintain an optimal thickness. Overloading of beds increases the thickness of the film leading to clogging of the filter media and ponding on the surface.

Biological Aerated Filters

Biological Aerated (or Anoxic) Filter (BAF) or Biofilters combine filtration with biological carbon reduction, nitrification or denitrification. BAF usually includes a reactor filled with a filter media. The media is either in suspension or supported by a gravel layer at the foot of the filter. The dual purpose of this media is to support highly active biomass that is attached to it and to filter suspended solids. Carbon reduction and ammonia conversion occurs in aerobic mode and sometime achieved in a single reactor while nitrate conversion occurs in anoxic mode. BAF is operated either in upflow or downflow configuration depending on design specified by manufacturer.

Fig. 8.9: Secondary Sedimentation tank at a rural treatment plant.

Membrane Bioreactors

Membrane bioreactors (MBR) combines activated sludge treatment with a membrane liquid-solid separation process. The membrane component uses low pressure microfiltration or ultra filtration membranes and eliminates the need for clarification and tertiary filtration. The membranes are typically immersed in the aeration tank (however, some applications utilize a separate membrane tank). One of the key benefits of a membrane bioreactor system is that it effectively overcomes the limitations associated with poor settling of sludge in conventional activated sludge (CAS) processes. The technology permits bioreactor operation with considerably higher mixed liquor suspended solids (MLSS) concentration than CAS systems, which are limited by sludge settling. The process is typically operated at MLSS in the range of 8,000-12,000 mg/L, while CAS are operated in the range of 2,000-3,000 mg/L. The elevated biomass concentration in the membrane bioreactor process allows for very effective removal of both soluble and particulate biodegradable materials at higher loading rates. Thus increased Sludge Retention Times (SRTs)—usually exceeding 15 days—ensure complete nitrification even in extremely cold weather.

The cost of building and operating a MBR is usually higher than conventional wastewater treatment, however, as the technology has become increasingly popular and has gained wider acceptance throughout the industry, the life-cycle costs have been steadily decreasing. As well, in developed urban areas where the footprint of the treatment plant is considered a limiting factor MBR facilities can be considered a desirable option.

Secondary Sedimentation

The final step in the secondary treatment stage is to settle out the biological floc or filter material and produce sewage water containing very low levels of organic material and suspended matter.

Rotating biological contactors (RBCs) are mechanical secondary treatment systems, which are robust and capable of withstanding surges in organic load. RBCs were first installed in Germany in 1960 and have since been developed and refined into a reliable operating unit. The rotating disks support the growth of bacteria and micro-organisms present in the sewage, which

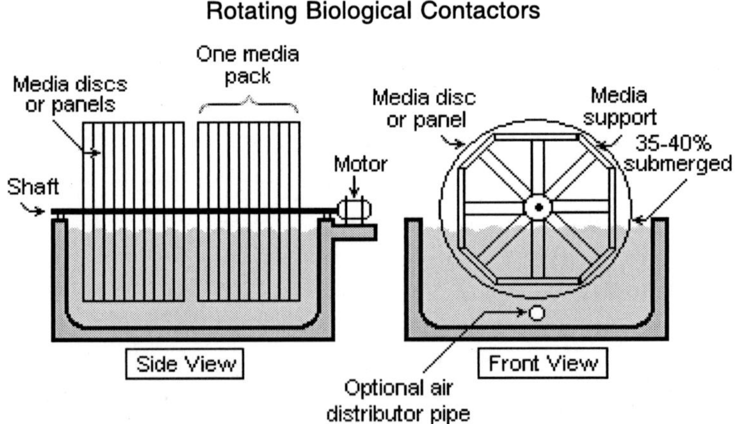

Fig. 8.10: Schematic diagram of a typical rotating biological contactor (RBC). The treated effluent clarifier/settler is not included in the diagram.

breakdown and stabilise organic pollutants. To be successful, micro-organisms need both oxygen to live and food to grow. Oxygen is obtained from the atmosphere as the disks rotate. As the micro-organisms grow, they build up on the media until they are sloughed off due to shear forces provided by the rotating discs in the sewage. Effluent from the RBC is then passed through final clarifiers where the micro-organisms in suspension settle as a sludge. The sludge is withdrawn from the clarifier for further treatment.

Tertiary Treatment

Tertiary treatment provides a final stage to raise the effluent quality before it is discharged to the receiving environment (sea, river, lake, ground, etc.). More than one tertiary treatment process may be used at any treatment plant. If disinfection is practiced, it is always the final process. It is also called "effluent polishing".

Filtration

Sand filtration removes much of the residual suspended matter. Filtration over activated carbon removes residual toxins.

Lagooning

Lagooning provides settlement and further biological improvement through storage in large man-made ponds or lagoons. These lagoons are highly aerobic and colonization by native macrophytes, especially reeds, is often encouraged. Small filter feeding invertebrates such as Daphnia and species of Rotifera greatly assist in treatment by removing fine particulates.

Constructed Wetlands

Constructed wetlands include engineered reedbeds and a range of similar methodologies, all of which provide a high degree of aerobic biological improvement and can often be used instead of secondary treatment for small communities, also see phytoremediation. One example is a small reedbed used to clean the drainage from the elephants' enclosure at Chester Zoo in England.

Nutrient Removal

Wastewater may contain high levels of the nutrients nitrogen and phosphorus.

Excessive release to the environment can lead to a build up of nutrients, called eutrophication, which can in turn encourage the overgrowth of weeds, algae, and cyanobacteria (blue-green algae). This may cause an algal bloom, a rapid growth in the population of algae. The algae numbers are unsustainable and eventually most of them die. The decomposition of the algae by bacteria uses up so much of oxygen in the water that most or all of the animals die, which creates more organic matter for the bacteria to decompose. In addition to causing deoxygenation, some algal species produce toxins that contaminate drinking water supplies. Different treatment processes are required to remove nitrogen and phosphorus.

Nitrogen Removal

The removal of nitrogen is effected through the biological oxidation of nitrogen from ammonia (nitrification) to nitrate, followed by denitrification, the reduction of nitrate to nitrogen gas. Nitrogen gas is released to the atmosphere and thus removed from the water.

Nitrification itself is a two-step aerobic process, each step facilitated by a different type of bacteria. The oxidation of ammonia (NH_3) to nitrite (NO_2^-) is most often facilitated by *Nitrosomonas* spp. (nitroso referring to the formation of a nitroso functional group). Nitrite oxidation to nitrate (NO_3^-), though traditionally believed to be facilitated by *Nitrobacter* spp. (nitro referring the formation of a nitro functional group), is now known to be facilitated in the environment almost exclusively by *Nitrospira* spp.

Denitrification requires anoxic conditions to encourage the appropriate biological communities to form. It is facilitated by a wide diversity of bacteria. Sand filters, lagooning and reed beds can all be used to reduce nitrogen, but the activated sludge process (if designed well) can do the job the most easily. Since denitrification is the reduction of nitrate to dinitrogen gas, an electron donor is needed. This can be, depending on the wastewater, organic matter (from faeces), sulfide, or an added donor like methanol.

Sometimes the conversion of toxic ammonia to nitrate alone is referred to as tertiary treatment.

Phosphorus Removal

Phosphorus can be removed biologically in a process called enhanced biological phosphorus removal. In this process, specific bacteria, called polyphosphate accumulating organisms, are selectively enriched and accumulate large quantities of phosphorus within their cells (up to 20% of their mass). When the biomass enriched in these bacteria is separated from the treated water, these biosolids have a high fertilizer value.

Phosphorus removal can also be achieved by chemical precipitation, usually with salts of iron (*e.g.* ferric chloride) or aluminum (*e.g.* alum). The resulting chemical sludge is difficult to handle and the added chemicals can be expensive. Despite this, chemical phosphorus removal requires significantly smaller equipment footprint than biological removal, is easier to operate and can be more reliable in areas that have wastewater compositions that make biological phosphorus removal difficult.

Disinfection

The purpose of disinfection in the treatment of wastewater is to substantially reduce the number of microorganisms in the water to be discharged back into the environment. The effectiveness of disinfection depends on the quality of the water being treated (*e.g.*, cloudiness, pH, etc.), the type of disinfection being used, the disinfectant dosage (concentration and time), and other environmental variables. Cloudy water will be treated less successfully since solid matter can shield organisms, especially from ultraviolet light or if contact times are low. Generally, short contact times, low doses and high flows all militate against effective disinfection. Common methods of disinfection include ozone, chlorine, or ultraviolet light. Chloramine, which is used for drinking water, is not used in wastewater treatment because of its persistence.

Chlorination remains the most common form of wastewater disinfection in North America due to its low cost and long-term history of effectiveness. One disadvantage is that chlorination of residual organic material can generate chlorinated-organic compounds that may be carcinogenic or harmful to the environment. Residual chlorine or chloramines may also be capable of chlorinating organic material in the natural aquatic environment. Further, because residual chlorine is toxic to aquatic species, the treated effluent must also be chemically dechlorinated, adding to the complexity and cost of treatment.

Ultraviolet (UV) light can be used instead of chlorine, iodine, or other chemicals. Because no chemicals are used, the treated water has no adverse effect on organisms that later consume it, as may be the case with other methods. UV radiation causes damage to the genetic structure of bacteria, viruses, and other pathogens, making them incapable of reproduction. The key disadvantages of UV disinfection are the need for frequent lamp maintenance and replacement and the need for a highly treated effluent to ensure that the target microorganisms are not shielded from the UV radiation (*i.e.*, any solids present in the treated effluent may protect microorganisms from the UV light). In the United Kingdom, light is becoming the most common means of disinfection because of the concerns about the impacts of chlorine in chlorinating residual organics in the wastewater and in chlorinating organics in the receiving water. Edmonton, Alberta, Canada also uses UV light for its water treatment.

Ozone O_3 is generated by passing oxygen O_2 through a high voltage potential resulting in a third oxygen atom becoming attached and forming O_3. Ozone is very unstable and reactive and oxidizes most organic material it comes in contact with, thereby destroying many pathogenic microorganisms. Ozone is considered to be safer than chlorine because, unlike chlorine which has to be stored on site (highly poisonous in the event of an accidental release), ozone is generated onsite as needed. Ozonation also produces fewer disinfection by-products than chlorination. A disadvantage of ozone disinfection is the high cost of the ozone generation equipment and the requirements for special operators.

Package Plants and Batch Reactors

In order to use less space, treat difficult waste, deal with intermittent flow or achieve higher environmental standards, a number of designs of hybrid treatment plants have been produced. Such plants often combine all or at least two stages of the three main treatment stages into one combined stage. In the UK, where a large number of sewage treatment plants serve small populations, package plants are a viable alternative to building discrete structures for each process stage.

One type of system that combines secondary treatment and settlement is the sequencing batch reactor (SBR). Typically, activated sludge is mixed with raw incoming sewage and mixed and aerated. The resultant mixture is then allowed to settle producing a high quality effluent. The settled sludge is run off and re-aerated before a proportion is returned to the head of the works. SBR plants are now being deployed in many parts of the world including North Liberty, Iowa, and Llanasa, North Wales.

The disadvantage of such processes is that precise control of timing, mixing and aeration is required. This precision is usually achieved by computer controls linked to many sensors in the plant. Such a complex, fragile system is unsuited to places where such controls may be unreliable, or poorly maintained, or where the power supply may be intermittent.

Package plants may be referred to as *high charged* or *low charged*. This refers to the way the biological load is processed. In high charged systems, the biological stage is presented with a high organic load and the combined floc and organic material is then oxygenated for a few hours before being charged again with a new load. In the low charged system the biological stage contains a low organic load and is combined with floculate for a relatively long time.

Sludge Treatment and Disposal

The sludges accumulated in a wastewater treatment process must be treated and disposed of in a safe and effective manner. The purpose of digestion is to reduce the amount of organic matter and the number of disease-causing microorganisms present in the solids. The most common treatment options include anaerobic digestion, aerobic digestion, and composting.

The choice of a wastewater solid treatment method depends on the amount of solids generated and other site-specific conditions. However, in general, composting is most often applied to smaller-scale applications followed by aerobic digestion and then lastly anaerobic digestion for the larger-scale municipal applications.

Anaerobic Digestion

Anaerobic digestion is a bacterial process that is carried out in the absence of oxygen. The process can either be *thermophilic* digestion, in which sludge is fermented in tanks at a temperature of 55°C, or *mesophilic*, at a temperature of around 36°C. Though allowing shorter retention time (and thus smaller tanks), thermophilic digestion is more expensive in terms of energy consumption for heating the sludge.

One major feature of anaerobic digestion is the production of biogas, which can be used in generators for electricity production and/or in boilers for heating purposes.

Composting

Composting is also an aerobic process that involves mixing the sludge with sources of carbon such as sawdust, straw or wood chips. In the presence of oxygen, bacteria digest both the wastewater solids and the added carbon source and, in doing so, produce a large amount of heat.

Thermal Depolymerization

Thermal depolymerization uses hydrous pyrolysis to convert reduced complex organics to oil.

Sludge Disposal

When a liquid sludge is produced, further treatment may be required to make it suitable for final disposal. Typically, sludges are thickened (dewatered) to reduce the volumes transported off-site for disposal. There is no process which completely eliminates the need to dispose of biosolids. There is, however, an additional step some cities are taking to superheat the wastewater sludge and convert it into small pelletized granules that are high in nitrogen and other organic materials. This product is then sold to local farmers and turf farms as a soil amendment or fertilizer, reducing the amount of space required to dispose of sludge in landfills

Treatment in the Receiving Environment

Many processes in a wastewater treatment plant are designed to mimic the natural treatment processes that occur in the environment, whether that environment is a natural water body or the ground. If not overloaded, bacteria in the environment will consume organic contaminants, although this will reduce the levels of oxygen in the water and may significantly change the overall ecology of the receiving water. Native bacterial populations feed on the organic contaminants, and the numbers of disease-causing microorganisms are reduced by natural environmental conditions such as predation exposure to ultraviolet radiation, for example. Consequently, in cases where the receiving environment provides a high level of dilution, a high degree of wastewater treatment may not be required. However, recent evidence has demonstrated that very low levels of certain contaminants in wastewater, including hormones (from animal husbandry and residue from human hormonal contraception methods) and synthetic materials such as phthalates that mimic hormones in their action, can have an unpredictable adverse impact on the natural biota and potentially on humans if the water is re-used for drinking water. In the US and EU, uncontrolled discharges of wastewater to the environment are not permitted under law, and strict water quality requirements are to be met. A significant threat in the coming decades will be the increasing uncontrolled discharges of wastewater within rapidly developing countries.

Sewage Treatment in Developing Countries

There are few reliable figures on the share of the wastewater collected in sewers that is being treated in the world. In many developing countries the bulk of domestic and industrial wastewater is discharged without any treatment or after primary treatment only. In Latin America about 15 per cent of collected wastewater passes through treatment plants (with varying levels of actual treatment). In Venezuela, a below average country in South America with respect to wastewater treatment, 97 percent of the country's sewage is discharged raw into the environment. In a relatively developed Middle Eastern country such as Iran, Tehran's majority of population has totally untreated sewage injected to the city's groundwater. Most of sub-Saharan Africa is without wastewater treatment.

Water utilities in developing countries are chronically underfunded because of low water tariffs, the inexistence of sanitation tariffs in many cases, low billing efficiency (*i.e.* many users that are billed do not pay) and poor operational efficiency (*i.e.* there are overly high levels of staff, there are high physical losses, and many users have illegal connections and are thus not being billed). In addition, wastewater treatment typically is the process within the utility that receives the least attention, partly because enforcement of environmental standards is poor. As a result of all these factors, operation and maintenance of many wastewater treatment plants is poor. This is evidenced by the frequent breakdown of equipment, shutdown of electrically operated equipment due to power outages or to reduce costs, and sedimentation due to lack of sludge removal. Developing countries as diverse as Egypt, Algeria, China or Colombia have invested substantial sums in wastewater treatment without achieving a significant impact in terms of environmental improvement. Even if wastewater treatment plants are properly operating, it can be argued that the environmental impact is limited in cases where the assimilative capacity of the receiving waters (ocean with strong currents or large rivers) is high, as it is often the case.

Benefits of wastewater treatment compared to benefits of sewage collection in developing countries

Waterborne diseases that are prevalent in developing countries, such as typhus and cholera, are caused primarily by poor hygiene practices and the absence of improved household sanitation facilities. The public health impact of the discharge of untreated wastewater is comparatively much lower. Hygiene promotion, on-site sanitation and low-cost sanitation thus are likely to have a much greater impact on public health than wastewater treatment.

9

Environmental Nanotechnology

9.1 Nanotechnology

Nanotechnology refers to a field of applied science and technology whose theme is the control of matter on the atomic and molecular scale, generally 100 nanometers or smaller, and the fabrication of devices or materials that lie within that size range.

Overview

Nanotechnology is a highly multidisciplinary field, drawing from a number of fields such as applied physics, materials science, interface and colloid science, device physics, supramolecular chemistry (which refers to the area of chemistry that focuses on the noncovalent bonding interactions of molecules), self-replicating machines and robotics, chemical engineering, mechanical engineering, biological engineering, and electrical engineering. Grouping of the sciences under the umbrella of "nanotechnology" has been questioned on the basis that there is little actual boundary-crossing between the sciences that operate on the nano-scale. Instrumentation is the only area of technology common to all disciplines; on the contrary, for example pharmaceutical and semiconductor industries do not "talk with each other". Corporations that call their products "nanotechnology" typically market them only to a certain industrial cluster.

Two main approaches are used in nanotechnology. In the "bottom-up" approach, materials and devices are built from molecular components which assemble themselves chemically by principles of molecular recognition. In the "top-down" approach, nano-objects are constructed from larger entities without atomic-level control. The impetus for nanotechnology comes from a renewed interest in Interface and Colloid Science, coupled with a new generation of analytical tools such as the atomic force microscope (AFM), and the scanning tunneling microscope (STM). Combined with refined processes such as electron beam lithography and molecular beam epitaxy, these instruments allow the deliberate manipulation of nanostructures, and lead to the observation of novel phenomena.

Examples of nanotechnology are the manufacture of polymers based on molecular structure, and the design of computer chip layouts based on surface science. Despite the promise of nanotechnologies such as quantum dots and nanotubes, real commercial applications have mainly used the advantages of colloidal nanoparticles in bulk form, such as suntan lotion, cosmetics, protective coatings, drug delivery, and stain resistant clothing.

Origins

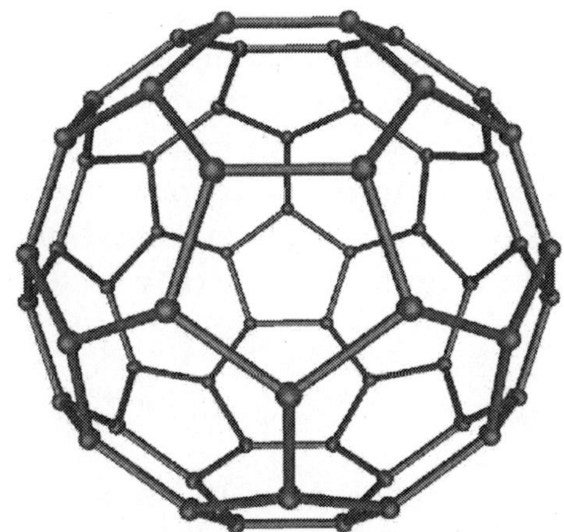

Fig. 9.1: Buckminsterfullerene C_{60}, also known as the buckyball, is the simplest of the carbon structures known as fullerenes. Members of the fullerene family are a major subject of research falling under the nanotechnology umbrella.

The first use of the concepts in 'nano-technology' (but predating use of that name) was in "There's Plenty of Room at the Bottom," a talk given by physicist Richard Feynman at an American Physical Society meeting at Caltech on December 29, 1959. Feynman described a process by which the ability to manipulate individual atoms and molecules might be developed, using one set of precise tools to build and operate another proportionally smaller set, so on down to the needed scale. In the course of this, he noted, scaling issues would arise from the changing magnitude of various physical phenomena: gravity would become less important, surface tension and Van der Waals attraction would become more important, etc. This basic idea appears plausible, and exponential assembly enhances it with parallelism to produce a useful quantity of end products. The term "nanotechnology" was defined by Tokyo Science University Professor Norio Taniguchi in a 1974 paper as follows: "'Nano-technology' mainly consists of the processing of, separation, consolidation, and deformation of materials by one atom or by one molecule." In the 1980s the basic idea of this definition was explored in much more depth by Dr. K. Eric Drexler, who promoted the technological significance of nano-scale phenomena and devices through speeches and the books Engines of Creation: The Coming Era of Nanotechnology (1986) and *Nanosystems: Molecular Machinery, Manufacturing, and Computation*, and so the term acquired its current sense. Engines of Creation: The Coming Era of Nanotechnology is considered the first book on the topic of nanotechnology. Nanotechnology and nanoscience got started in the early 1980s with two major developments; the birth of cluster science and the invention of the scanning tunneling microscope (STM). This development led to the discovery of fullerenes in 1986 and carbon nanotubes a few years later. In another development, the synthesis and properties of semiconductor nanocrystals was studied; This led to a fast increasing number of metal oxide nanoparticles of quantum dots. The atomic force microscope was invented six years after the STM was invented. In 2000, the United States National Nanotechnology Initiative was founded to coordinate Federal nanotechnology research and development.

Fundamental Concepts

One nanometer (nm) is one billionth, or 10^{-9} of a meter. To put that scale in context, the comparative size of a nanometer to a meter is the same as that of a marble to the size of the earth. Or another way of putting it: a nanometer is the amount a man's beard grows in the time it takes him to raise the razor to his face.

Typical carbon-carbon bond lengths, or the spacing between these atoms in a molecule, are in the range 0.12-0.15 nm, and a DNA double-helix has a diameter around 2 nm. On the other hand, the smallest cellular lifeforms, the bacteria of the genus Mycoplasma, are around 200 nm in length.

Larger to Smaller: A Materials Perspective

A number of physical phenomena become pronounced as the size of the system decreases. These include statistical mechanical effects, as well as quantum mechanical effects, for example the "quantum size effect" where the electronic properties of solids are altered with great reductions in particle size. This effect does not come into play by going from macro to micro dimensions. However, it becomes dominant when the nanometer size range is reached. Additionally, a number of physical (mechanical, electrical, optical, etc.) properties change when compared to macroscopic systems. One example is the increase in surface area to volume ratio altering mechanical, thermal and catalytic properties of materials. Novel *mechanical* properties of nanosystems are of interest in the nanomechanics research. The catalytic activity of nanomaterials also opens potential risks in their interaction with biomaterials.

Fig. 9.2: Image of reconstruction on a clean Au(100) surface, as visualized using scanning tunneling microscopy. The positions of the individual atoms composing the surface are visible.

Materials reduced to the nanoscale can show different properties compared to what they exhibit on a macroscale, enabling unique applications. For instance, opaque substances become transparent (copper); inert materials become catalysts (platinum); stable materials turn combustible (aluminum); solids turn into liquids at room temperature (gold); insulators become conductors (silicon). A material such as gold, which is chemically inert at normal scales, can serve as a potent chemical catalyst at nanoscales. Much of the fascination with nanotechnology stems from these quantum and surface phenomena that matter exhibits at the nanoscale.

Simple to Complex: A Molecular Perspective

Modern synthetic chemistry has reached the point where it is possible to prepare small molecules to almost any structure. These methods are used today to produce a wide variety of useful chemicals such as pharmaceuticals or commercial polymers. This ability raises the question of extending this kind of control to the next-larger level, seeking methods to assemble these single molecules into supramolecular assemblies consisting of many molecules arranged in a well defined manner.

These approaches utilize the concepts of molecular self-assembly and/or supramolecular chemistry to automatically arrange themselves into some useful conformation through a bottom-up approach. The concept of molecular recognition is especially important: molecules can be designed so that a specific conformation or arrangement is favored due to non-covalent intermolecular forces. The Watson-Crick basepairing rules are a direct result of this, as is the specificity of an enzyme being targeted to a single substrate, or the specific folding of the protein itself. Thus, two or more components can be designed to be complementary and mutually attractive so that they make a more complex and useful whole.

Such bottom-up approaches should be able to produce devices in parallel and much cheaper than top-down methods, but could potentially be overwhelmed as the size and complexity of the desired assembly increases. Most useful structures require complex and thermodynamically unlikely arrangements of atoms. Nevertheless, there are many examples of self-assembly based on molecular recognition in biology, most notably Watson-Crick basepairing and enzyme-substrate interactions. The challenge for nanotechnology is whether these principles can be used to engineer novel constructs in addition to natural ones.

Molecular Nanotechnology: A Long-Term View

Molecular nanotechnology, sometimes called molecular manufacturing, is a term given to the concept of engineered nanosystems (nanoscale machines) operating on the molecular scale. It is especially associated with the concept of a molecular assembler, a machine that can produce a desired structure or device atom-by-atom using the principles of mechanosynthesis. Manufacturing in the context of productive nanosystems is not related to, and should be clearly distinguished from, the conventional technologies used to manufacture nanomaterials such as carbon nanotubes and nanoparticles.

When the term "nanotechnology" was independently coined and popularized by Eric Drexler (who at the time was unaware of an earlier usage by Norio Taniguchi) it referred to a future manufacturing technology based on molecular machine systems. The premise was that molecular-scale biological analogies of traditional machine components demonstrated molecular machines were possible: by the countless examples found in biology, it is known that sophisticated, stochastically optimised biological machines can be produced.

It is hoped that developments in nanotechnology will make possible their construction by some other means, perhaps using biomimetic principles. However, Drexler and other researchers have proposed that advanced nanotechnology, although perhaps initially implemented by biomimetic means, ultimately could be based on mechanical engineering principles, namely, a manufacturing technology based on the mechanical functionality of these components (such as gears, bearings, motors, and structural members) that would enable programmable, positional assembly to atomic specification (PNAS-1981). The physics and engineering performance of exemplar designs were analyzed in Drexler's book *Nanosystems*.

But Drexler's analysis is very qualitative and does not address very pressing issues, such as the "fat fingers" and "Sticky fingers" problems. In general it is very difficult to assemble devices on the atomic scale, as all one has to position atoms are other atoms of comparable size and stickyness. Another view, put forth by Carlo Montemagno, is that future nanosystems will be hybrids of silicon technology and biological molecular machines. Yet another view, put forward by the late Richard Smalley, is that mechanosynthesis is impossible due to the difficulties in mechanically manipulating individual molecules.

This led to an exchange of letters in the ACS publication Chemical & Engineering News in 2003. Though biology clearly demonstrates that molecular machine systems are possible, non-biological molecular machines are today only in their infancy. Leaders in research on non-biological molecular machines are Dr. Alex Zettl and his colleagues at Lawrence Berkeley Laboratories and UC Berkeley. They have constructed at least three distinct molecular devices whose motion is controlled from the desktop with changing voltage: a nanotube nanomotor, a molecular actuator, and a nanoelectromechanical relaxation oscillator.

An experiment indicating that positional molecular assembly is possible was performed by Ho and Lee at Cornell University in 1999. They used a scanning tunneling microscope to move an individual carbon monoxide molecule (CO) to an individual iron atom (Fe) sitting on a flat silver crystal, and chemically bound the CO to the Fe by applying a voltage.

Current Research

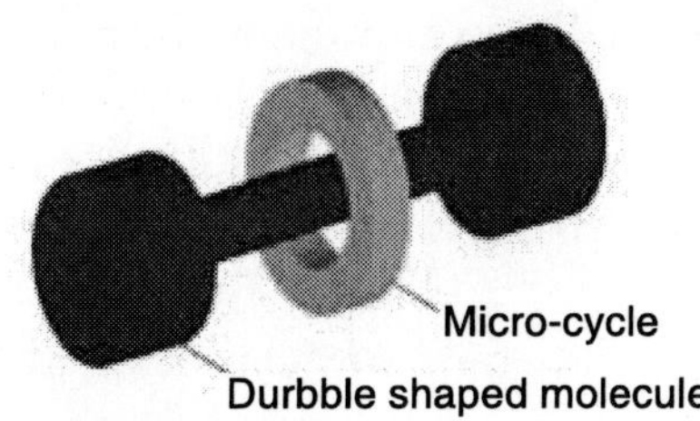

Graphical representation of a rotaxane, useful as a molecular switch.

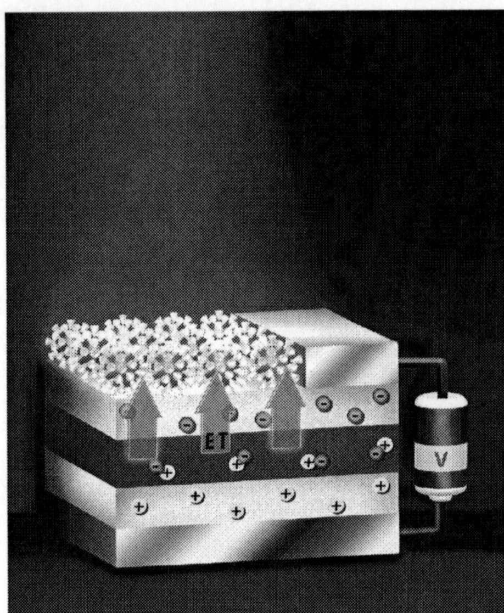

Fig. 9.3: This device transfers energy from nano-thin layers of quantum wells to nanocrystals above them, causing the nanocrystals to emit visible light.

Nanomaterials

This includes subfields which develop or study materials having unique properties arising from their nanoscale dimensions.

- *Interface and Colloid Science* has given rise to many materials which may be useful in nanotechnology, such as carbon nanotubes and other fullerenes, and various nanoparticles and nanorods.
- can also be used for *bulk applications*; most present commercial applications of nanotechnology are of this flavor.
- Progress has been made in using these materials for medical applications; see *Nanomedicine*.

Bottom-up Approaches

These seek to arrange smaller components into more complex assemblies.

- *DNA nanotechnology* utilizes the specificity of Watson-Crick basepairing to construct well-defined structures out of DNA and other nucleic acids.
- Approaches from the field of "classical" chemical synthesis also aim at designing molecules with well-defined shape (*e.g.* bis-peptides)
- More generally, *molecular self-assembly* seeks to use concepts of

supramolecular chemistry, and molecular recognition in particular, to cause single-molecule components to automatically arrange themselves into some useful conformation.

Top-down Approaches

These seek to create smaller devices by using larger ones to direct their assembly.

- Many technologies descended from conventional *solid-state silicon methods* for fabricating microprocessors are now capable of creating features smaller than 100 nm, falling under the definition of nanotechnology. Giant magnetoresistance-based hard drives already on the market fit this description, as do atomic layer deposition (ALD) techniques. Peter Grünberg and Albert Fert received the Nobel Prize in Physics for their discovery of Giant magnetoresistance and contributions to the field of spintronics in 2007.
- Solid-state techniques can also be used to create devices known as *nanoelectromechanical systems* or NEMS, which are related to microelectromechanical systems or MEMS.
- Atomic force microscope tips can be used as a nanoscale "write head" to deposit a chemical upon a surface in a desired pattern in a process called *dip pen nanolithography*. This fits into the larger subfield of nanolithography.

Functional Approaches

These seek to develop components of a desired functionality without regard to how they might be assembled.

- *Molecular electronics* seeks to develop molecules with useful electronic properties. These could then be used as single-molecule components in a nanoelectronic device. For an example see rotaxane.
- Synthetic chemical methods can also be used to create *synthetic molecular motors*, such as in a so-called nanocar.

Speculative

These subfields seek to anticipate what inventions nanotechnology might yield, or attempt to propose an agenda along which inquiry might progress. These often take a big-picture view of nanotechnology, with more emphasis on its societal implications than the details of how such inventions could actually be created.

- *Molecular nanotechnology* is a proposed approach which involves manipulating single molecules in finely controlled, deterministic ways. This is more theoretical than the other subfields and is beyond current capabilities.
- *Nanorobotics* centers on self-sufficient machines of some functionality operating at the nanoscale. There are hopes for applying nanorobots in medicine, but it may not be easy to do such a thing because of several drawbacks of such devices. Nevertheless, progress on innovative materials and methodologies has been demonstrated with some patents granted about new nanomanufacturing devices for future commercial applications, which also progressively helps in the development towards nanorobots with the use of embedded nanobioelectronics concept.
- *Programmable matter* based on artificial atoms seeks to design materials whose properties can be easily and reversibly externally controlled.
- Due to the popularity and media exposure of the term nanotechnology, the words *picotechnology* and *femtotechnology* have been coined in analogy to it, although these are only used rarely and informally.

Tools and Techniques

The first observations and size measurements of nano-particles were made during the first decade of the 20th century. They are mostly associated with the name of Zsigmondy who made detailed studies of gold sols and other nanomaterials with sizes down to 10 nm and less. He published a book in 1914. He used ultramicroscope that employs a *dark field* method for seeing particles with sizes much less than light wavelength.

There are traditional techniques developed during 20th century in Interface and Colloid Science for characterizing nanomaterials. These are widely used for *first generation* passive nanomaterials specified in the next section.

These methods include several different techniques for characterizing particle size distribution. This characterization is imperative because many materials that are expected to be nano-sized are actually aggregated in solutions. Some of methods are based on light scattering. Other apply ultrasound, such as ultrasound attenuation spectroscopy for testing concentrated nano-dispersions and microemulsions.

There is also a group of traditional techniques for characterizing surface charge or zeta potential of nano-particles in solutions. These information is required for proper system stabilzation, preventing its aggregation or flocculation. These methods include microelectrophoresis, electrophoretic light scattering and electroacoustics. The last one, for instance colloid vibration current method is suitable for characterizing concentrated systems.

Next group of nanotechnological techniques include those used for fabrication of nanowires, those used in semiconductor fabrication such as deep ultraviolet lithography, electron beam lithography, focused ion beam machining, nanoimprint lithography, atomic layer deposition, and molecular vapor deposition, and further including molecular self-assembly techniques such as those employing di-block copolymers. However, all of these techniques preceded the nanotech era, and are extensions in the development of scientific advancements rather than techniques which were devised with the sole purpose of creating nanotechnology and which were results of nanotechnology research.

There are several important modern developments. The atomic force microscope (AFM) and the Scanning Tunneling Microscope (STM) are two early versions of scanning probes that launched nanotechnology. There are other types of scanning probe microscopy, all flowing from the ideas of the scanning confocal microscope developed by Marvin Minsky in 1961 and the scanning acoustic microscope (SAM) developed by Calvin Quate and coworkers in the 1970s, that made it possible to see structures at the nanoscale. The tip of a scanning probe can also be used to manipulate nanostructures (a process called positional assembly). Feature-oriented scanning-positioning methodology suggested by Rostislav Lapshin appears to be a promising way to implement these nanomanipulations in automatic mode. However, this is still a slow process because of low scanning velocity of the microscope. Various techniques of nanolithography such as dip pen nanolithography, electron beam lithography or nanoimprint lithography were also developed. Lithography is a top-down fabrication technique where a bulk material is reduced in size to nanoscale pattern.

The top-down approach anticipates nanodevices that must be built piece by piece in stages, much as manufactured items are made. Scanning probe microscopy is an important technique both for characterization and synthesis of nanomaterials. and scanning tunneling microscopes can be used to look at surfaces and to move atoms around. By designing different tips for these microscopes, they can be used for carving out structures on surfaces and to help guide self-assembling structures. By using, for example, feature-oriented scanning-positioning approach, atoms can be moved around on a surface with scanning probe microscopy techniques. At present, it is expensive and time-consuming for mass production but very suitable for laboratory experimentation.

In contrast, bottom-up techniques build or grow larger structures atom by atom or molecule by molecule. These techniques include chemical synthesis, self-assembly and positional assembly. Another variation of the bottom-up approach is molecular beam epitaxy or MBE. Researchers at Bell Telephone Laboratories like John R. Arthur. Alfred Y. Cho, and Art C. Gossard developed and implemented MBE as a research tool in the late 1960s and 1970s. Samples made by MBE were key to the discovery of the fractional quantum Hall effect for which the 1998 Nobel Prize in Physics was awarded. MBE allows scientists to lay down atomically-precise layers of atoms and, in the process, build up complex structures. Important for research on semiconductors, MBE is also widely used to make samples and devices for the newly emerging field of spintronics.

Newer techniques such as Dual Polarisation Interferometry are enabling scientists to measure quantitatively the molecular interactions that take place at the nano-scale. Borna Mehr Fan Ahad Rafat Talebi

Applications

As of April 24, 2008 The Project on Emerging Nanotechnologies claims that over 609 nanotech products exist, with new ones hitting the market at a pace of 3-4 per week. The project lists all of the products in a database. Most applications are limited to the use of "first generation" passive nanomaterials which includes titanium dioxide in sunscreen, cosmetics and some food products; Carbon allotropes used to produce gecko tape; silver in food packaging, clothing, disinfectants and household appliances; zinc oxide in sunscreens and cosmetics, surface coatings, paints and outdoor furniture varnishes; and cerium oxide as a fuel catalyst.

The National Science Foundation (a major source of funding for nanotechnology in the United States) funded researcher David Berube to study the field of nanotechnology. His findings are published in the monograph "Nano-Hype: The Truth Behind the Nanotechnology Buzz". This published study (with a foreword by Mihail Roco, Senior Advisor for Nanotechnology at the National Science Foundation) concludes that much of what is sold as "nanotechnology" is in fact a recasting of straightforward materials science, which is leading to a "nanotech industry built solely on selling nanotubes, nanowires, and the like" which will "end up with a few suppliers selling low margin products in huge volumes." Further applications which require actual manipulation or arrangement of nanoscale components await further research. Though technologies branded with the term 'nano' are sometimes little related to and fall far short of the most ambitious and transformative technological goals of the sort in molecular manufacturing proposals, the term still connotes such ideas. Thus there may be a danger that a "nano bubble" will form, or is forming already, from the use of the term by scientists and entrepreneurs to garner funding, regardless of interest in the transformative possibilities of more ambitious and far-sighted work.

Implications

Due to the far-ranging claims that have been made about potential applications of nanotechnology, a number of serious concerns have been raised about what effects these will have on our society if realized, and what action if any is appropriate to mitigate these risks.

One area of concern is the effect that industrial-scale manufacturing and use of nanomaterials would have on human health and the environment, as suggested by nanotoxicology research. Groups such as the Center for Responsible Nanotechnology have advocated that nanotechnology should be specially regulated by governments for these reasons. Others counter that overregulation would stifle scientific research and the development of innovations which could greatly benefit mankind.

Other experts, including director of the Woodrow Wilson Center's Project on Emerging Nanotechnologies David Rejeski, have testified that successful commercialization depends on adequate oversight, risk research strategy, and public engagement. More recently local municipalities have passed (Berkeley, CA) or are considering (Cambridge, MA)—ordinances requiring nanomaterial manufacturers to disclose the known risks of their products.

The National Institute for Occupational Safety and Health is conducting research on how nanoparticles interact with the body's systems and how workers might be exposed to nano-sized particles in the manufacturing or industrial use of nanomaterials. NIOSH offers interim guidelines for working with nanomaterials consistent with the best scientific knowledge.

Longer-term concerns center on the implications that new technologies will have for society at large, and whether these could possibly lead to either a post scarcity economy, or alternatively exacerbate the wealth gap between developed and developing nations. The effects of nanotechnology on the society as a whole, on human health and the environment, on trade, on security, on food systems and even on the definition of "human", have not been characterized or politicized.

Health and Environmental Concerns

Some of the recently developed nanoparticle products may have unintended consequences. Researchers have discovered that silver nanoparticles used in socks to reduce foot odor are being released in the wash with possible negative consequences. Silver nanoparticles, which are bacteriostatic, may then destroy beneficial bacteria which are important for breaking down organic matter in waste treatment plants or farms.

A study at the University of Rochester found that when rats breathed in nanoparticles, the particles settled in the brain and lungs, which lead to significant increases in biomarkers for inflammation and stress response.

A major study published more recently in Nature nanotechnology suggests some forms of carbon nanotubes— a poster child for the "nanotechnology revolution"— could be as harmful as asbestos if inhaled in sufficient quantities. Anthony Seaton of the Institute of Occupational Medicine in Edinburgh, Scotland, who contributed to the article on carbon nanotubes said "We know that some of them probably have the potential to cause mesothelioma. So those sorts of materials need to be handled very carefully." In the absence of specific nano-regulation forthcoming from governments, Paull and Lyons (2008) have called for an exclusion of engineered nanoparticles from organic food.

Regulation

Calls for tighter regulation of nanotechnology have occurred alongside a growing debate related to the human health and safety risks associated with nanotechnology. Further, there is significant debate about who is responsible for the regulation of nanotechnology. While some non-nanotechnology specific regulatory agencies currently cover some products and processes (to varying degrees)— by "bolting on" nanotechnology to existing regulations— there are clear gaps in these regimes.

Stakeholders concerned by the lack of a regulatory framework to assess and control risks associated with the release of nanoparticles and nanotubes have drawn parallels with bovine spongiform encephalopathy ('mad cow's disease), thalidomide, genetically modified food), nuclear energy, reproductive technologies, biotechnology, and asbestosis. The Woodrow Wilson Centre's Project on Emerging Technologies conclude that there is insufficient funding for human health and safety research, and as a result there is currently limited understanding of the human health and safety risks associated with nanotechnology.

The Royal Society report identified a risk of nanoparticles or nanotubes being released during disposal, destruction and recycling, and recommended that "manufacturers of products that fall under extended producer responsibility regimes such as end-of-life regulations publish procedures outlining how these materials will be managed to minimize possible human and environmental exposure" (p. xiii). Reflecting the challenges for ensuring responsible life cycle regulation, the Institute for Food and Agricultural Standards has proposed standards for nanotechnology research and development should be integrated across consumer, worker and environmental standards. They also propose that NGOs and other citizen groups play a meaningful role in the development of these standards.

9.2 Nanomaterials

Nanomaterials are materials with morphological features smaller than a micron in at least one dimension. Despite the fact that there is no consensus upon the minimum or maximum size of nanomaterials, some authors restricting their size from 1 to 100 nm, a logical definition would situate the nanoscale between microscale (1 micron) and atomic/molecular scale (about 0.2 nanometers).

Fundamental Concepts

A unique aspect of nanotechnology is the vastly increased ratio of surface area to volume present in many nanoscale materials which opens new quantum mechanical effects, for example the "quantum size effect" where the electronic properties of solids are altered with great reductions in particle size. This effect does not come into play by going from macro to micro dimensions. However, it becomes dominant when the nanometer size range is reached. Additionally, a number of physical properties change when compared to macroscopic systems. Novel mechanical properties of nanomaterials is the subject of nanomechanics research. Their catalytic activity reveals novel properties in the interaction with biomaterials.

Nanotechnology can be thought of as extensions of traditional disciplines towards the explicit consideration of these properties. Additionally, traditional disciplines can be re-interpreted as specific applications of nanotechnology. This dynamic reciprocation of ideas and concepts contributes to the modern understanding of the field. Broadly speaking, nanotechnology is the synthesis and application of ideas from science and engineering towards the understanding and production of novel materials and devices. These products

Materials reduced to the nanoscale can suddenly show very different properties compared to what they exhibit on a macroscale, enabling unique applications. For instance, opaque substances become transparent (copper); inert materials become catalysts (platinum); stable materials turn combustible (aluminum); solids turn into liquids at room temperature (gold); insulators become conductors (silicon). Materials such as gold, which is chemically inert at normal scales, can serve as a potent chemical catalyst at nanoscales. Much of the fascination with nanotechnology stems from these unique quantum and surface phenomena that matter exhibits at the nanoscale.

Nanosize powder particles (a few nanometres in diameter, also called nanoparticles) are potentially important in ceramics, powder metallurgy, the achievement of uniform nanoporosity and similar applications. The strong tendency of small particles to form clumps ("agglomerates") is a serious technological problem that impedes such applications. However, a few dispersants such as ammonium citrate (aqueous) and imidazoline or oleyl alcohol (nonaqueous) are promising additives for deagglomeration.

Size Concerns

Another concern is that the volume of an object decreases as the third power of its linear dimensions, but the surface area only decreases as its second power. This somewhat subtle and unavoidable principle has huge ramifications. For example the power of a drill (or any other machine) is proportional to the volume, while the friction of the drill's bearings and gears is proportional to their surface area. For a normal-sized drill, the power of the device is enough to handily overcome any friction. However, scaling its length down by a factor of 1000, for example, decreases its power by 1000^3 (a factor of a billion) while reducing the friction by only 1000^2 (a factor of "only" a million). Proportionally it has 1000 times less power per unit friction than the original drill. If the original friction-to-power ratio was, say, 1 per cent, that implies the smaller drill will have 10 times as much friction as power. The drill is useless.

This is why, while super-miniature electronic integrated circuits can be made to function, the same technology cannot be used to make functional mechanical devices in miniature: the friction overtakes the available power at such small scales. So while you may see microphotographs of delicately etched silicon gears, such devices are curiosities with limited real world applications, for example in moving mirrors and shutters. Surface tension increases in the same way, causing very small objects to tend to stick together. This could possibly make any kind of "micro factory" impractical: even if robotic arms and hands could be scaled down, anything they pick up will tend to be impossible to put down. The above being said, molecular evolution has resulted in working cilia, flagella, muscle fibers, and rotary motors in aqueous environments, all on the nanoscale. These machines, however, exploit the increase of the frictional forces found at the micro or nanoscale. Unlike an oar, paddle or propeller the mechanics of which are dominated by normal frictional forces (the frictional forces perpendicular to the surface) for propulsion, cilia, etc., develop motion resulting from the exaggerated drag or laminar forces (frictional forces parallel to the surface) present at micro and nano dimensions. To develop meaningful "machines" at the nanoscale, the relevant forces need to be considered. We are faced with the development and design of relevant machines rather than the simple reproductions of macroscopic ones.

All these scaling issues have to be kept in mind while evaluating any kind of nanotechnology.

Materials used in Nanotechnology

Materials referred to as "nanomaterials" generally fall into two categories: fullerenes, and inorganic nanoparticles.

Fullerenes

The fullerenes are a class of allotropes of carbon which conceptually are graphene sheets rolled into tubes or spheres. These include the carbon nanotubes which are of interest due to both their mechanical strength and their electrical properties.

For the past decade, the chemical and physical properties of fullerenes have been a hot topic in the field of research and development, and are likely to continue to be for a long time. In April 2003, fullerenes were under study for potential medicinal use: binding specific antibiotics to the structure to target resistant bacteria and even target certain cancer cells such as melanoma. The October 2005 issue of Chemistry and Biology contains an article describing the use of fullerenes as light-activated antimicrobial agents. In the field of nanotechnology, heat resistance and superconductivity are some of the more heavily studied properties.

A common method used to produce fullerenes is to send a large current between two nearby graphite electrodes in an inert atmosphere. The resulting carbon plasma arc between the electrodes cools into sooty residue from which many fullerenes can be isolated.

There are many calculations that have been done using ab-initio Quantum Methods applied to fullerenes. By DFT and TDDFT methods one can obtain IR, Raman and UV spectra. Results of such calculations can be compared with experimental results.

Nanoparticles

Nanoparticles or nanocrystals made of metals, semiconductors, or oxides are of interest for their mechanical, electrical, magnetic, optical, chemical and other properties. Nanoparticles have been used as quantum dots and as chemical catalysts.

Nanoparticles are of great scientific interest as they are effectively a bridge between bulk materials and atomic or molecular structures. A bulk material should have constant physical properties regardless of its size, but at the nano-scale this is often not the case. Size-dependent properties are observed such as quantum confinement in semiconductor particles, surface plasmon resonance in some metal particles and superparamagnetism in magnetic materials.

Nanoparticles exhibit a number of special properties relative to bulk material. For example, the bending of bulk copper (wire, ribbon, etc.) occurs with movement of copper atoms/clusters at about the 50 nm scale. Copper nanoparticles smaller than 50 nm are considered super hard materials that do not exhibit the same malleability and ductility as bulk copper. The change in properties is not always desirable. Ferroelectric materials smaller than 10 nm can switch their magnetisation direction using room temperature thermal energy, thus making them useless for memory storage. Suspensions of nanoparticles are possible because the interaction of the particle surface with the solvent is strong enough to overcome differences in density, which usually result in a material either sinking or floating in a liquid. Nanoparticles often have unexpected visible properties because they are small enough to confine their electrons and produce quantum effects. For example gold nanoparticles appear deep red to black in solution.

Nanoparticles have a very high surface area to volume ratio. This provides a tremendous driving force for diffusion, especially at elevated temperatures. Sintering can take place at lower temperatures, over shorter time scales than for larger particles. This theoretically does not affect the density of the final product, though flow difficulties and the tendency of nanoparticles to agglomerate complicates matters. The surface effects of nanoparticles also reduces the incipient melting temperature.

Safety of Manufactured Nanomaterials

Nanomaterials behave differently than other similarly-sized particles. It is therefore necessary to develop specialized approaches to testing and monitoring their effects on human health and on the environment. The OECD Chemicals Committee has established the Working Party on Manufactured Nanomaterials to address this issue and to study the practices of OECD member countries in regards to nanomaterial safety.

While nanomaterials and nanotechnologies are expected to yield numerous health and health care advances, such as more targeted methods of delivering drugs, new cancer therapies, and methods of early detection of diseases, they also may have unwanted effects. Increased toxicity is the main concern associated with manufactured nanoparticles.

When materials are made into nanoparticles, their reactivity increases. These more reactive particles can enter the body through the skin, lungs, or digestive tract, and may cause inflammation and damage to the lungs as well as other organs. However, the particles must be absorbed in sufficient quantities in order to pose health risks.

9.3 Nanochemistry

Nanochemistry is the use of synthetic chemistry to make nanoscale building blocks of desired shape, size, composition and surface structure, charge and functionality with an optional target to control self-assembly of these building blocks at various scale-lengths.

International interest in nanoscience research has flourished in recent years, as it becomes an integral part in the development of future technologies. The diverse, interdisciplinary nature of nanoscience means effective communication between disciplines is pivotal in the successful utilization of the science.

Nanochemistry: A Chemical Approach to Nanomaterials is the first textbook for teaching nanochemistry and adopts an interdisciplinary and comprehensive approach to the subject. It presents a basic chemical strategy for making nanomaterials and describes some of the principles of materials self-assembly over 'all' scales. It demonstrates how nanometre and micrometre scale building blocks (with a wide range of shapes, compositions and surface functionalities) can be coerced through chemistry to organize spontaneously into unprecedented structures, which can serve as tailored functional materials. Suggestions of new ways to tackle research problems and speculations on how to think about assembling the future of nanotechnology are given.

Primarily designed for teaching, this book will appeal to graduate and advanced undergraduate students. It is well illustrated with graphical representations of the structure and form of nanomaterials and contains problem sets as well as other pedagogical features such as further reading, case studies and a comprehensive bibliography.

9.4 Nanotechnology and Environment

The Energy & Environmental Systems Institute has been instrumental in the formation of three research entities addressing issues concerning potential effects of emerging nanotechnologies on environmental systems:

- International Consortium for Environment & Nanotechnology Research.
- Center for Biological & Environmental Nanotechnology.
- Smalley Institute for Nanoscale Science & Technology.

Nanotechnologies hold great promise for creating new means of detecting pollutants, cleaning polluted waste streams, recovering materials before they become wastes, and expanding available resources. Like all emerging technologies with great promise, the nanotechnology and nanochemistry industries will present new challenges in ensuring that environmental risks are properly managed.

EESI has played a pioneering role in identifying key research issues surrounding the implementation of environmental nanotechnologies and anticipating possible environmental impacts associated with nanomaterials. Indeed, the research agenda for examining environmental implications of nanotechnologies was publicly articulated for the first time at an international symposium organized by EESI in December 2001, with a follow-up conference in December 2005.

Stakeholder Relationships Initiative

A wide range of organizations generally known as non-governmental organizations (NGOs) play critical roles in the sustainable development process. This research initiative explores how NGOs interact with other stakeholders—business, regulatory institutions, the community—when new technologies with potential impact on sustainable development appear. The development of Nanotechnology is likely to bring about a revolution that could transform every aspect of work and life in fields of activity as varied as biotechnology, pharmaceuticals, information and energy storage, water treatment or air pollution. It constitutes such a breakthrough from all known technology that it is difficult to even assess its impact on sustainability, and this creates public concern. Based on the study of public reaction to the introduction on the market of recent new technological discoveries, the Stakeholder Relationships program seeks to identify how and why behavior models vary with different markets and different cultures when a new technology becomes available.

Center Vision

CBEN explores the interface between dry nanomaterials and aqueous environments that range from pure water to biological and environmental systems.

Transforming Nanotechnology into a Tool to Solve Real-World Problems

CBEN's mission is to discover and develop nanomaterials that enable new medical and environmental technologies.

The mission is accomplished by the following:

- Fundamental examination of the 'wet/dry' interface between nanomaterials, complex aqueous systems, and ultimately our environment (Theme 1).
- Engineering research that focuses on multifunctional nanoparticles that solve problems in environmental and biological engineering (Themes 2, 3).
- Educational programs that develop teachers, students, and citizens who are well informed and enthusiastic about nanotechnology.
- Innovative knowledge transfer that recognize the importance of communicating nanotechnology research to the media, policymakers, and the general public.

This mission is inspired by the observation that because of their small size and unique properties, nanomaterials interact with and control biological systems in entirely new ways. Our research exploits these novel capabilities to develop innovative biomedical and environmental technologies. To ensure that our technologies flourish, our outreach addresses broader issues such as technology transfer, public acceptance, and workforce training.

The Center for Biological and Environmental Nanotechnology fosters the development of this field through an integrated set of programs that aim to address the scientific, technological, environmental, human resource, commercialization, and societal barriers that hinder the transition from nanoscience to nanotechnology.

Research

The Center's research focuses on investigating and developing nanoscience at the "wet/dry" interface. Water, the most abundant solvent present on Earth, is of unique importance as the medium of life. The Center's research activities explore this interface between nanomaterials and aqueous systems at multiple length scales, including interactions with solvents, biomolecules, cells, whole-organisms, and the environment. These explorations form the basis for understanding the natural interactions that nanomaterials will experience outside the laboratory, and also serves as foundational knowledge for designing biomolecular/nanomaterial interactions, solving bioengineering problems with nanoscale materials, and constructing nanoscale materials useful in solving environmental engineering problems.

Education

Given the goal of transforming nanoscience into a strong, vital discipline, the Center must draw new talent into the field. Educational outreach efforts develop programs to identify, recruit, and train the nanoscience workforce of the future. As a centerpiece program, 9th grade teachers in the minority-rich Houston school district are being trained to engage in the more successful but challenging discovery-based teaching style. The Center provides content lectures and tutoring to these educators and offers a meaningful experience in research laboratories. These teachers also identify students to participate in a Center science academy. New curriculum and textbook development, a summer Research Experience for Undergraduates program, and research in Center-funded laboratories extend the Center's educational outreach activities to the undergraduate and graduate levels.

Industrial Connections

In addition to a more traditional industrial affiliates program, the Center embraces the increasing importance of small and startup companies in high technology development by partnering with Rice's Jones Graduate School of Management in an entrepreneurial education program. This provides Center members with the skills needed for the more active interactions such organizations demand, and through associated activities brings scientists, students and business experts together to ensure the formation of successful startups based on Center research. Our interactions with industry have led us to understand that among the greatest barriers to successful commercialization of nanotechnology are concerns over safety, environmental impact, and public education. In addition to our scientific research in these areas, CBEN is addressing these issues more comprehensively through ICON, the International Council on Nanotechnology, a broad-based coalition including representatives of industrial, governmental, academic, and public concerns.

HOUSTON—Whether new nanotechnologies that can help clean up the environment might also harm it will be addressed during a workshop at Rice University Dec. 10. Titled "Nanotechnology and Environment: An Examination of the Potential Benefits and Perils of an Emerging Technology," the free workshop is open to the public.

"Emerging technologies present new opportunities for improving the human condition, but they also have the potential for unforeseen negative environmental consequences," said Mark Wiesner, director of Rice's Energy and Environmental Systems Institute (EESI), which is co-sponsoring the workshop with the Office for Science and Technology of the French Embassy USA.

Wiesner cites Freon as an example. This chemical was hailed as an important advance in refrigeration because it was nontoxic and nonflammable. It replaced highly toxic compounds that had caused numerous accidental deaths from their use in home refrigerators. But decades later, scientists discovered that chlorofluorocarbons such as Freon endangered Earth's ozone layer.

"Nanotechnologies hold great promise for creating new means of detecting pollutants, cleaning polluted waste streams, recovering materials before they become wastes and expanding available resources," Wiesner said. "But the nanotechnology industry is just now emerging, so we need to question whether it presents new environmental challenges so that the products of nanochemistry do not become dangerous environmental pollutants."

Rice is hosting the workshop in affiliation with its new Center for Biological and Environmental Nanotechnology, one of six major Nanoscale Science and Engineering Centers recently announced by the National Science Foundation and the first to focus on applications of nanoscience to biology and the environment.

Nanotechnologies involve materials that are one-billionth of a meter. Research at the Rice center will focus on the use of nanomaterials in water-based systems, ranging in size from biomolecules and cells to whole organisms and the surrounding environment. New nanostructured membranes are being developed for potable water treatment, treatment of hazardous materials and environmental analysis. Such membranes might be used to improve water quality while providing a higher level of security to water-treatment systems.

"We're bringing together researchers from Rice and leading French research institutions in the area of nanotechnology and environment with nanochemistry and environmental researchers to discuss how nanotechnologies might be used to protect our environment and the potential dangers they pose," Wiesner said.

Among the scheduled presenters are Richard Smalley, recipient of the 1996 Nobel Prize for chemistry and director of the Center for Nanoscale Science and Technology at Rice, and Neal Lane, a university professor at Rice who served as President Bill Clinto's science and technology adviser.

Topics to be addressed in presentations at the workshop include:

- carbon nanotubes;
- mineral nanoparticles;
- nano-engineering chemical sensors for environmental applications;
- nanostructured membranes; environmental quality on the "Nano-Coast;"
- transport of nanoparticles in the environment;
- potential for facilitated transport of contaminants by nanomaterials; potential for bio-uptake and bio-accumulation of nanoparticles; and
- implications of nanotechnology for environmental policy and society.

The workshop will be held from 9 a.m. to 4:45 p.m. in Anne and Charles Duncan Hall, McMurtry Auditorium, 6100 Main Street. A reception and poster session will be held from 5 to 6 p.m. Preregistration is not required.

Nanotechnology is an emerging and promising field of research, loosely defined as the study of functional structures with dimensions in the 1-1000 nanometer range. Certainly, many organic chemists have designed and fabricated such structures for decades via chemical synthesis. During the last decade, however, developments in the areas of surface microscopy, silicon fabrication, biochemistry, physical chemistry, and computational engineering have converged to provide remarkable capabilities for understanding, fabricating and manipulating structures at the atomic level.

Research in nanoscience is exploding, both because of the intellectual allure of constructing matter and molecules one atom at a time, and because the new technical capabilities permit creation of materials and devices with significant societal impact. The rapid evolution of this new science and the opportunities for its application promise that nanotechnology will become one of the dominant technologies of the 21st century. Nanotechnology represents a central direction for the future of chemistry that is increasingly interdisciplinary and ecumenical in application.

As practiced at Rice University, there are three distinct nanotechnologies:

- "Wet" nanotechnology, which is the study of biological systems that exist primarily in a water environment. The functional nanometer-scale structures of interest here are genetic material, membranes, enzymes and other cellular components. The success of this nanotechnology is amply demonstrated by the existence of living organisms whose form, function, and evolution are governed by the interactions of nanometer-scale structures.
- "Dry" nanotechnology, which derives from surface science and physical chemistry, focuses on fabrication of structures in carbon (for example, fullerenes and nanotubes), silicon, and other inorganic materials. Unlike the "wet" technology, "dry" techniques admit use of metals and semiconductors. The active conduction electrons of these materials make them too reactive to operate in a "wet" environment, but these same electrons provide the physical properties that make "dry" nanostructures promising as electronic, magnetic, and optical devices. Another objective is to develop "dry" structures that possess some of the same attributes of the self-assembly that the wet ones exhibit.
- Computational nanotechnology, which permits the modeling and simulation of complex nanometer-scale structures. The predictive and analytical power of computation is critical to success in nanotechnology: nature required several hundred million years to evolve a functional "wet" nanotechnology; the insight provided by computation should allow us to reduce the development time of a working "dry" nanotechnology to a few decades, and it will have a major impact on the "wet" side as well.

These three nanotechnologies are highly interdependent. The major advances in each have often come from application of techniques or adaptation of information from one or both of the others. This critical symbiosis can flourish at Rice, where we already have a world-class status in computation, in the biosciences, and in the physical sciences and engineering disciplines that underlie "dry."

10
Hazardous Waste

The term *hazardous waste* comprises all toxic chemicals, radioactive materials, and biologic or infectious waste. These materials threaten workers through occupational exposure and the general public in their homes, communities, and general environment. Exposure to these materials can occur near the site of generation, along the path of its transportation, and near their ultimate disposal sites. Most hazardous waste results from industrial processes that yield unwanted byproducts, defective products, and spilled materials. The generation and disposal of hazardous wastes is controlled through a variety of international and national regulations.

10.1 Introduction

Dr. Hazelwood is the Environmental Affairs Project Manager for IT Corporation. This California based corporation is a full line hazardous waste management company which operates and designs treatment facilities, disposal sites and analytical laboratories at various locations around the United States. IT Corp. operates a fleet of 150 vacuum trucks and also works with companies to help them decide how to handle their waste. It performs site investigations, responds to emergency calls about spills and does heavy industrial cleaning. Founded in 1926, IT Corporation went public in December 1983 and is rapidly expanding nationwide.)

Although conservators are well aware of the dangers involved in working with chemicals on a daily basis and articles have been written suggesting methods for proper storage, most conservators don't know how to go about safely disposing of these chemicals after having used them. Some of these materials are highly toxic and many are incompatible when mixed together. This article should help conservators address the problem of waste disposal. Most recommendations are made to maintain compliance with legislation and regulations in California, which are among the most stringent requirements in the country. After some initial research, the authors' impression is that most states are just trying to identify and cope with the vast problems inherent to the disposal of waste materials. Mechanisms for disposal, particularly of the small quantities produced by conservators, are for the most part only beginning to be put into place. However, the agencies we contacted were most willing to give advice and direction.

The Problem With Chemical Waste

The moment you open and use a can of solvent you are a waste generator. Conservation laboratories may only produce 10 to 15 gallons of waste each year and private conservators only one quart, still the improper disposal of even small quantities may cause unforeseen problems. Chemicals dumped in the back yard will filter down to the water table. It might take years, but they will eventually pollute that water below. Some chemicals washed down the drain produce flammable vapors which can collect in stand pipes and explode.

The bottom line is that California does not exempt small generators and if you do not dispose of waste properly, you are actually breaking the law and may be liable to stiff fines for non-compliance.

Current Legislation

Under federal regulations anyone generating less than 1,000 kilograms of waste chemicals per month is considered a small generator and may dispose of waste in common garbage which is taken to a municipal land fill or a waste facility. It is more likely, however, that the laws in your state are more restrictive. To find out, call the local health department, sewage disposal system, or a waste management company. Kansas, Missouri, New Hampshire, New Jersey, South Carolina and Vermont currently exempt only 100 kilograms per month. Minnesota has proposed rules to exempt 100 kilograms per month. California, Rhode Island and Louisiana are unusual because they have no small generator exemption and any quantity of waste must be handled according to the hazardous waste system.

Disposal Methods in the Conservation Studio

In-house disposal methods should be viewed with caution. Of the following alternatives, only a few are acceptable.

- Most chemicals may not be put into the sewage system. Untreated bleaches (oxidizers) can react with organic material in the sewers.

Sodium chlorate, for example, when mixed with automobile brake fluid will burst into flames in 30 seconds. Solvents, heavy metals, poisons and strong acids and bases can damage a sewer system.

- To quote Dr. Hazelwood, "Dilution is not the Solution to Pollution." The law says that chemicals must be handled as though they retain their original strength.
- Some classes of chemicals may be neutralized and then disposed of down the drain with large amounts of water. Only persons familiar with the chemistry of neutralization reaction should attempt to neutralize their waste. (See following paragraphs on Acids and Alkalies, and Bleaches)
- Most chemical waste may not be disposed of in common garbage. Oxidizers can react with organic waste in the garbage truck and spontaneously combust.
- Disposing of waste solvents, paints, varnishes or other chemicals in the back yard or in an empty lot is against federal law. If caught, severe fines or a jail sentence may be imposed.
- Burning waste solvents is illegal. Some chemical compounds (most notably chlorinated hydrocarbons) form very persistent intermediate products when incinerated. These can compound air pollution problems and can be toxic.
- It is legal to allow small amounts of waste solvent to evaporate. After all, solvent evaporates whenever a varnish film is allowed to dry. When evaporating small amounts of waste, use a fume hood or equivalent. Be mindful of fire and health hazards. (See following paragraphs on Solvents.)
- Some large generators of waste are able to recycle their waste materials. This solution does not apply to the small quantities of mixed waste produced by conservators.
- Waste which cannot be disposed of through an in-house method should be collected and containerized for removal from the lab. (See Collecting and Cataloguing Waste for Removal.)

The materials used by most conservators can be broadly grouped into chemical classes. These classifications are: solvents (including paints and varnishes), detergents, acids and alkalies, bleaches and ethyl ether. As a general rule these classes should not be mixed together in a waste container. The possibility of chemical reaction between incompatible materials is a genuine fire and safety hazard.

Solvents

Solvents, as a class, present a known fire and health hazard and accordingly also present disposal and storage problems. Included in this category are: paint, varnish and polymer residues, as well as true solvents like toluene and naptha. (Diethyl ether is categorized with ether.) Solvent waste should be collected in glass bottles for future removal from the lab. Glass is inert and unlike metal, will not rust through if water is mixed in with the waste. (*See* Collecting and Cataloguing Waste for Removal.)

Small amounts of waste solvent may be allowed to evaporate in a fume hood or equivalent.

It must be emphasized that dumping even water soluble solvents down the drain is not an acceptable practice. Flammable vapors can collect in traps and stand pipes creating a fire hazard.

Detergents

Only detergents can be safely and legally disposed of down the drain without prior treatment. With the exception of triethanolamine, sewage plants are designed to accommodate this waste. Triethanolamine should be disposed of as the waste solvents are. Do not collect this material in a metal can if it has been mixed with water.

Acids and Alkalies

Acids and Alkalies may be disposed of in the sewer system under certain conditions. If the acid or base does not contain dissolved heavy metals, it may be neutralized and then washed down the drain with plenty of water. If neutralization is possible, protective gear such as rubber gloves, an apron and a full face shield should be worn during the process. Acids can be neutralized with sodium bicarbonate (baking soda) or sodium carbonate (soda ash). Alkali can be neutralized with acetic acid (vinegar), or even better, photographic indicating stop bath (add stop until the red color is produced.) While not recommended, if other chemicals are used for neutralization reaction, an indicator like methyl red should be used.

Please note that waste containing dissolved heavy metals should not be disposed of in the sewer system under any conditions. Metals such as copper, zinc, lead, cadmium and mercury are toxic and can kill the bacteria which is introduced at the treatment plant to work on the sewage. This reaction requires the plant to initiate a new treatment cycle.

If neutralization is not possible, the waste must be containerized for removal. (See Collecting and Cataloguing Waste for Removal)

Bleaches

The very mild and dilute bleaches used in conservation treatments should be neutralized before disposal in the sewer system. Many of the bleaches self-neutralize with time. Sodium perborate, hydrogen peroxide, sodium borohydride and chlorine dioxide generating baths, each in dilute aqueous solution, should stand for an hour in order to avoid complications in the sewer system and then be discarded. Chloramine T may be diluted and washed down the drain, as it is often used in water treatment plants. Other bleaches must be neutralized before disposal.

Ether

Ether, diethyl ether and ethyl ether all refer to the same material. Petroleum ether (pet ether) is not the same chemical and is handled like the solvents. Ether is terribly dangerous because it is highly flammable and a terrific explosion hazard. Its vapor is heavier than air and creeps along the floor. If it finds an open flame, the fire can flash back to the container. No one should ever smoke or have an open flame near an open can of ether.

Ether reacts with air to form shock sensitive explosive peroxides. For this reason it should always be kept in a metal container which will inhibit the formation of peroxides. Ether must never be stored in glass jars. Bottles of ether contaminated with peroxides have been known to explode from unscrewing the lid. Be aware that although metal inhibits the formation of peroxides, it does not remove existing peroxides and new peroxides still may form.

As a general rule, ether stored without refrigeration should not be used longer than three months after it is opened. Close attention should be paid to the expiration date on the can. Very old cans of ether often must be disposed of by bomb squads rather than disposal agencies.

Suspect ether can be tested for peroxides with an acidified solution of potassium iodide. A small amount of ether would be poured into a test tube containing a small amount of the test reagent. If a red color appears upon shaking the vial, peroxides have formed. Be extremely careful and contact an expert at once! Also note that if contaminated ether in any type of container is allowed to evaporate, the peroxides will have concentrated in the container, increasing a very real explosion hazard.

To dispose of fresh ether, we recommend letting it evaporate in a safe fume hood (or equivalent.) For large amounts, contact a disposal firm at once. For the reasons stated above, it is not wise to save ether for lengths of time while waiting for a scheduled removal of other waste.

As a footnote, you may be interested to know that law enforcement agencies train dogs to follow the scent of ether because it is used in the manufacture of PCP. The scent of ether often indicates an illicit drug operation to the authorities, so be polite to stray dogs who come to your door.

PCBs

Another disposal problem nearly unique to the conservator, is the handling of small amounts of PCBs (polychlorinated biphenyls), Arochlor mounting medium. It is absolutely illegal to throw Arochlor into the garbage. Even materials contaminated with PCB, a well documented carcinogen, must never be thrown into the garbage. All materials including contaminated tissues, microscope slides, swabs and so on, should be segregated and disposed of by a registered contractor.

Arochlor is categorized in California as an extremely hazardous material and therefore waste may not be disposed of in this state by anyone other than a registered waste hauler.

Dry Waste

Disposal of solid or dry waste is difficult to discuss in general terms, but as a rule solid or dry materials should be kept in that state and not mixed in with liquid waste for disposal. It may be possible to neutralize small amounts of dry waste before disposal. Seek advice from a professional on particular disposal methods especially for toxic and reactive materials.

Chemical Incompatibilities

Some chemicals are not compatible with others. Chemical reactions are fairly common in waste collection containers. Incompatible materials may burst into flames immediately or hours after mixing; emit noxious or toxic gases; or simply bubble and fizz out of the container making a mess.

As a rule, do not mix or store the following chemical classes together:

- ❖ Acids and alkalies
- ❖ Bleaches
- ❖ Oxidizing agents

- Reducing agents
- Solvents and flammables

Learn about which materials are incompatible when mixed together and which are not. Numerous lists and tables of incompatible chemical types have been compiled. Often chemical and supply catalogues contain such listings (i.e. Conservation Materials or MCB Reagents Catalog.)

The following are specific incompatibilities:

- Ammonia with hypochlorite bleach;
- Nitric acid with acetic acid;
- Nitric acid with sulphuric acid;
- 1-Butanol with strong mineral acids;
- n-Butylamine with copper and copper alloys;
- n-n-Dimethyl formamide with halogenated hydrocarbons;
- Ethyl acetate with strong alkalies;
- Ethylene Dichloride with oxidizing materials;
- Ethylene glycol with sulphuric acid;
- MEK peroxide (hardener for polyester casting resin) with anything flammable; and
- 1,1,1 Trichloroethane with caustic soda and caustic potash.

Do not use these chemicals at all:

- benzene (benzine, the petroleum thinner, is safe - remember, benzine is fine, benzene is mean);
- carbon tetrachloride;
- chloroform;
- pyridine;
- phenol (carbolic acid);
- perchloric acid;
- hydrofluoric acid (If hydrofluoric acid must be used, be aware that it is lethal and its vapors can pass through your bones. It will eat through glass and dissolve silicon. It is an absolute must to wear protective clothing. This acid is considered an extremely hazardous material.)

How To Find Help

Assistance should be available in your area for advice on how to inventory, catalogue, collect, containerize and dispose of hazardous waste. Try looking first in the telephone book for the State Department of Health Services or equivalent, the local fire department, a waste management company or waste hauler. Waste management companies and haulers must be registered with the state and with the Environmental Protection Agency (EPA.) There may be a consultant fee for services rendered.

You will not require the complete services of a large waste management company like the IT Corporation, however IT might be willing to offer some guidance as a community service. It is also possible that you will locate a company which is set to handle small quantities of waste. If a solution cannot be found through the obvious channels, then find out which other nearby organizations generate small quantities of varied waste. A university laboratory or a hospital may have problems similar to yours. Ask them how they have set up a collection/disposal system.

Collecting and Cataloguing for Removal

First take an inventory of the chemicals in the lab. Think about upcoming projects and the type and amount of waste they might produce. These initial steps will help the lab schedule and budget for future waste removal. Waste should be collected in a volume which is cost effective to remove.

It is important to maintain a list of all the materials in a waste container and to indicate approximate amounts. A check list attached to the container might be a simple way to keep track. This information will be needed in order to complete the forms which document waste removal and disposal.

Ask the waste management company or waste hauler you are working with about the type of container they would like you to use and how it should be labeled. In most cases, reusing the glass bottles in which the chemical was purchased will be a convenient solution. This approach postpones the problem of disposing of empty bottles. The original labels should be obliterated or removed and the bottles clearly labeled as waste. Open mouth glass bottles also make suitable waste containers. Glass or plastic is preferable to metal because small amounts of water in the waste will not cause rusting. An advantage to glass or plastic is that it is transparent and you can see what is inside. Teflon or polyethylene containers are good and are less apt to break than glass. Waste bottles should be kept closed unless they are under constant ventilation in a fume hood. As noted earlier, ether should be kept in a metal container.

To minimize the risk of breakage or leaka ge, waste containers may be kept in cardboard boxes or even better, a wooden crate. The cardboard boxes that solvents are shipped in are quite good for this purpose. For maximum safety these boxes can be lined with polyethylene and filled in with vermiculite or cat litter around the bottles. This fill is an absorbent material which becomes a containable slush if waste leaks or is spilled. Vermiculite is preferable to cat litter because it is inorganic, non-reactive and nonflammable. It is sold at hardware or plant stores. Even though waste has been containerized, incompatible materials must not be put in the same box. As a final precaution, place a small tray under each bottle. This waste is now ready for transport to an intermediate receiver who will pack or prepare the waste in some way for final disposal.

If you have agreed with your waste hauler to have the waste ready for direct transport to a disposal site, get professional advice on preparation of waste materials. You may be requested to use a DOT (Department of Transportation) steel drum. There are several types, so find out which one is appropriate to your needs.

The DOT drum should be filled with vermiculite before adding chemicals which are compatible with one another. Because disposal sites no longer are willing to accept liquid waste, it must be solidified. When mixed with vermiculite, the waste is made into a slurry. Vermiculite acts like a sponge and can hold a limited amount of liquid before it begins to lose excess. A rule of thumb is that 5 parts vermiculite are used for every 1 part liquid. According to the instruction you have received, the waste may be poured in freely and mixed with some kind of paddle which should be left in the drum, or bottles of waste may be placed inside using the loose vermiculite as cushioning. The latter is certainly a neater procedure. If the waste is poured in as free liquid, you are then confronted with the problem of what to do with the bottles. In states which do not exempt small generators, empty containers which once held hazardous materials, must be disposed of as hazardous waste. If your state regulations do not supersede federal law, then a container holding less than 1" of liquid may be disposed of in common garbage. In any case, a properly filled 55 gallon drum will hold about 10 gallons of waste materials. Keep a list of the contents and be aware that the weight and capacity of the drum may exclude you from legally transporting the drum privately.

Removal

Now that it is legal in California for the public to transport small quantities of hazardous waste (*see* Current Legislation), the question is where to take it. There are only a handful of disposal sites in all of California and this state has more than most. Further, the sites which are geared to accept small quantities of varied waste are even fewer. It is probable then, that you will arrange either to leave the waste at an intermediate station or to have a registered waste hauler pick it up. In most cases there is some cost involved.

Contact the local Health Services Department, fire station, waste haulers, waste management companies, or other small waste generators to ask how disposal is handled in your community. If you work at an organization which operates under the auspices of a government agency, that agency should be able to give some guidance or even include the lab (now that they know that it is a waste generator) in its waste disposal system. It is too expensive and cumbersome for a large company like the IT Corporation to send out a truck for your two bottles, but they will probably accept a delivery from you by prior arrangement if the containers are properly labeled and if it is understood how the manifest forms which record the disposal will be filled out. A fee for disposal would be determined on an individual basis. Containerized Chemical Disposal, Inc. operating out of Monrovia, CA is a waste management company specializing in the needs of the small generator. They offer a complete service which includes advice on cataloguing and collecting waste. They pick up the waste on scheduled "milk runs" and pack the waste for final disposal themselves. Charges are set according to quantity and type of waste, but an average cost may be $100-$125.

The Environmental Health Coalition, an environmental advocacy group in San Diego has initiated a pilot project with city and county grant money. This group has organized a program to pick up household hazardous waste free of charge. Currently collection is limited to household waste and pick ups are only made within certain zip codes. The Coalition would like to extend these services to small business, but first must determine their needs. At the end of the pilot project, recommendations will be given to San Diego on how to organize a continuing program. A similar program exists in Sacramento under the auspices of The Golden Empire Health System Agency. The Los Angeles County Department of Health Services has a "Small Generators Program."

Disposal of waste means filling out a hazardous waste manifest which indicates what kind of waste is in the container, who produced the waste, the name of the transporter and how the waste was disposed of. By signing these forms the generator, the transporter and the disposal site all share some responsibility for the waste, however the government always views the waste as being yours. The forms are sent to the Department of Health Services where a copy is kept on file and a copy is returned to the generator describing how the waste was treated. There is an enormous amount of paperwork involved.

The Conservation Center at the Los Angeles County Museum of Art is currently designing a disposal plan for the department's hazardous waste. The *WAAC newsletter* is interested in publishing any follow up information on this topic. So please write with any questions, comments or findings of your own to either **Chris Stavroudis** at LACMA or to **Caroline Black** at 1644 N. Courtney Ave, LA, CA 90046. **Dr. Hazelwood** has offered to answer questions if you call or write to him at IT Corporation, 23456 Hawthorns Blvd., Suite 220, Torrance, CA 90505, (213) 378-9933.

10.2 Toxic Waste

Toxic waste is waste material, often in chemical form, that can cause death or injury to living creatures. It usually is the product of industry or commerce, but comes also from residential use, agriculture, the military, medical facilities, radioactive sources, and light industry, such as dry cleaning establishments. As with many pollution problems, toxic waste began to be a significant issue during the industrial revolution. The term is often used interchangeably with "hazardous waste," or discarded material that can pose a long-term risk to health or environment. Toxins can be released into air, water, or land.

10.3 Reference Procedure

Purpose

The purpose of this program is to ensure that Marshall University (MU) is in compliance with federal, state, and local regulations pertaining to the handling and disposal of hazardous wastes.

It is the policy of Marshall University (MU) to protect all faculty, staff, students, and visitors from any hazardous exposure to hazardous waste. No employee shall engage in or be required to perform any task involving hazardous waste unless they have been properly trained and are thoroughly familiar with proper waste handling and emergency procedures relevant to their responsibilities during normal facilities operations and emergencies.

Acting as the representative of the Marshall University President, the Environmental Safety and Health Office has prime responsibility for overseeing the program. The Environmental Health and Safety Office will act as program manager.

Scope

The purpose of the Hazardous Waste Management Program is to ensure that proper handling and legal disposal of hazardous wastes is conducted at all Marshall University facilities.

All requirements of the Marshall University Hazardous Waste Management Program will apply to the following:

1. Any liquid, semi-solid, solid or gaseous substance defined as hazardous waste.
2. Waste which consists of or contains a hazardous material.
3. A waste mixture formed by mixing any waste or substance with a hazardous waste.
4. A hazardous sludge, residue, concentrate, or ash originating from a hazardous waste.
5. Hazardous material disposed of to land, accidentally discharged onto land or accidentally spilled onto land.
6. Radioactive materials that emit ionizing radiation.

The hazardous chemical waste disposal program at Marshall University is able to remove and dispose of all the above mentioned types of hazardous chemicals with the EXCEPTION of radioactive waste. The Office of Radiation Safety is responsible for the control, use, and disposal of all radioactive material on Marshall University campuses.

Definitions

Acutely Hazardous Waste—Those specific wastes identified in 40 CFR 261.3 (e). Container - Any Portable device, in which a material is stored, transported, treated, disposed of, or otherwise handled. Conditionally Exempt Small Quantity Generator – Generates no more than 100 kg of hazardous waste, 1 kg of acutely hazardous waste, or 100 kg of contaminated waste from an acutely hazardous waste spill in a month. Accumulates no more than 1,000 kg of hazardous waste at any time.

Disposal—The discharge, deposit, injection, dumping, spilling, leaking, or placing of any solid waste or hazardous waste into or on any land or water so that such solid waste or hazardous waste may enter the environment or be emitted into the air or discharged into any waters.

EPA Identification Number—Number assigned by the Environmental Protection Agency to each generator; transporter; and processing, storage or disposal facility.

Generator—Any person by site, whose act or process producer hazardous waste identified or listed in 40 CFR 261.

Handling—The transportation from one place to another, loading, unloading, pumping or packaging of waste.

Hazardous Waste Manifest—A hazardous waste manifest form is to accompany shipments of hazardous waste as defined by 49 CFR

Hazardous Waste—Hazardous Waste as defined in 40 CFR 261 subpart C & D or Toxicity Characteristic Leaching Procedure (TCLP) as defined in Part 261.

Incompatible Waste—A hazardous waste which is unsuitable for (1) placement in a particular device or facility because it may cause corrosion or decay of containment materials or (2) commingling with another waste or material under uncontrolled conditions because commingling might produce heat or pressure, fire or explosion, violent reaction, toxic dust, mist, fumes, or gases.

Satellite Collection Station—A hazardous waste collection station at or near any point of generation where wastes initially accumulate under the control of the operator of the process generating the waste. Satellite collection stations must comply with the requirements specified under 40 CFR 264.34 (c). Small Quantity Generator (SQG)—A generator who generates between 100 and 1,000 kg of hazardous waste and no more than 1 kg of acutely hazardous waste in one month. Solid Waste—Solid waste as defined in 40 CFR 261.2. Treatment—Any method, technique, or process, including, neutralization, designed to change the physical, chemical, or biological character or composition of any hazardous waste so as to neutralize such waste, or so as to recover energy or material resources from the waste, or so as to render such waste non-hazardous, or less hazardous; safer to transport, store, or dispose of; or amenable for recover, amenable for storage, or reduced in volume. Waste Stream—A waste material generated either one time or routinely at a single generating facility with physical characteristics and chemical composition that do not vary significantly from shipment to shipment.

Responsibilities

1. Environmental Safety and Health Office:

   1.1. Develop, implement and maintain the Hazardous Waste Management Program in compliance with federal, state, and local requirements applicable to the "Small Quantity Generator".

   1.2. Assist departments in complying with the program by providing them with hazardous waste consultation.

1.3. Identification of hazardous waste generators on campus.
1 4. Hazardous waste segregation.
1 5. Hazardous waste pick up.
1.6. Assist departments with the redistribution of usable materials.
1.7. Periodically audits facility for Hazardous Waste Management Compliance.
1.8. Recordkeeping.
1.9. Annual Reporting.
1.10. Biennial Reporting.

The Environmental Safety and Health Office is located in room 209 of the Sorrell Maintenance Building, phone 696-2993. In an emergency call 696-6680 or Marshall University Police Department at 696-4357. The Radiation Safety Office can be reached by calling 969-2751

2. Marshall University Safety Committee: Provide support for the Hazardous Waste Management Program in conjunction with individual building unit safety committees.
3. Generator:

   3.1. Develop procedures to ensure effective compliance with the Hazardous Waste Program.
   3.2. Provide Environmental Safety and Health Office with notification prior to implementing changes that reduce or increase waste streams.
   3.3. Ensure that all appropriate personnel strictly adhere to the Hazardous Waste Management Program.
   3.4. Ensure that employees working with hazardous waste are fully trained and aware of the hazards of exposure and comply with procedures for control.
   3.5. Comply with Federal mandates that require hazardous waste minimization.

4. *Employee*: Although no single set of safety procedures can guarantee accident free employment or place of employment; and, because of the number of potential hazards that may exist or be created in the work environment, employees must first use common sense and good judgment at all times. Each employee assigned to work with hazardous waste shall comply with hazardous waste procedures, whether written or oral, while performing assigned duties.

Standard Operating Procedures

Regulatory Authority: West Virginia Department of Environmental Protection Series XV—Chapter 20, Article 5E of the code of West Virginia-(47 CSR 35); 40 CFR Part 261—265; 29 CFR 1019.120—Hazardous Waste Operations; Marshall University—Spill Policy Hazardous Waste Standard Operating Procedure: To implement the Marshall University Environmental Health and Safety's procedures for properly managing and disposing of hazardous waste in accordance with Environmental Health and Safety's Hazardous Materials Management Program. Exclusions from the Hazardous Waste Standard Operation Procedure include work involving hazardous and radioactive mixed waste management; and biohazard and mixed hazardous waste management.

Procedure

(a) General

1. The Environmental Safety and Health Office shall conduct all assigned hazardous waste management and disposal activities in accordance with all applicable Federal, State regulation, local laws and Marshall University Policies.
2. The Environmental Safety and Health Office shall maintain a list of hazardous waste generators for Marshall University.
3. All Environmental Safety and Health's hazardous material staff shall be trained in accordance with the requirements specified in 40 CFR 265.16.

(b) Generators

The Environmental Safety and Health Office shall insure that: Generators of hazardous waste shall adhere to these procedures. The initial determination of hazardous waste is with the generators. Each Department/ Chair/Chemical Hygiene Officer/Generator is responsible for submission of the Hazardous Waste Form to be forwarded to Environmental Safety and Health Office for proper disposal. Generators are responsible for properly managing hazardous materials until collection for pickup.

1. Use of Containers: Storage of the hazardous chemicals at the generation point may be in their original containers until disposal. These containers must remain intact and not permit any leaks or spills. Only compatible materials shall be placed in the storage containers.

2. If the original container is no longer intact, the unused chemical can be transferred to another container and properly labeled. Caution must be taken by the generators when transferring flammable materials insuring that proper bonding and/or grounding techniques are followed where applicable.
3. Generators must label the container with the following: "Chemical Name and Percentage" as stated on the Safety and Health's Hazardous Waste Disposal Form.
4. Hazardous waste generators are responsible for properly containing, storing and segregating the waste at the point of generation. If there is any question regarding whether a waste or chemical is hazardous, it shall be treated as hazardous until a determination is made by the Environmental Safety and Health Office.
5. The Environmental Safety and Health Office will provide assistance in identifying the appropriate container and proper storage requirements.
6. Containers must be suitable for onsite transportation and interim storage at the designated building/storage areas chosen by the Dean/ Chair/ Department/Chemical Hygiene Officer.
7. Containers that hold hazardous chemicals must be closed during storage except when adding or removing hazardous chemicals.
8. All other potentially hazardous materials (contaminated clothing, rags, etc.) must be placed in a plastic bag and securely closed. The plastic bag will be considered containing hazardous waste and handled accordingly.

(c) Waste Determination for Routine Waste

The final waste determination is made by the Environmental Safety and Health Office based upon routine or non-routine waste characteristics or waste type. The Hazardous Waste Form is reviewed for the following: Recycling, containerization, chemical compatibility, staging and/or storage, and hazardous waste disposal.

(d) Hazardous Waste Storage Areas

Check with the Environmental Safety and Health Office for the approved hazardous chemical storage areas. The personnel directly responsible for the facility must complete classroom or on-the-job training which will allow them to perform their duties in a way that ensures the facility's compliance with the environmental and safety regulations and with the "Chemical Hygiene Plan" for your designated project or operation.

Hazardous Waste storage facilities will be equipped with the following: A telephone capable of summoning emergency assistance, portable fire extinguishers or comparable methods of fire control such as fire suppression system, and spill control equipment. The "Site of Generation" is defined as the laboratory (room) generating the chemical waste. If the chemical waste is removed and consolidated, it is not considered a "Site of Generation." The waste must then be placed in hazardous waste facility where the hazardous waste vendor will dispose of the waste during the next pick-up. The Environmental Safety and Health Office will insure that the Dean/Director/Chair/Chemical Hygiene Officer or the Building Manager does the following:

1. Place the hazardous chemicals within storage areas. Each storage area is then segregated by compatible waste.
2. Only place compatible chemicals near each other within this storage area.
3. A log of chemicals (including amounts) will be maintained at all times within this storage facility.
4. The facility will be maintained and operated to minimize the possibility of a fire, explosion or any release of hazardous chemicals into the air, water or soil.
5. A copy of the Marshall University Spill Policy will be posted in a visible location at the outside of the storage facility.
6. Inspections of the hazardous waste storage facility will be done *weekly* to assure that the above requirements are met. The inspector must sign and date the inspection form.
7. There will be a record of weekly facility inspections maintained for review. Environmental Safety and Health Office will keep a copy of the inspections on file.
8. The Environmental Safety and Health Office will insure that any hazardous chemicals picked up for disposal by the hazardous waste vendor matches the "Hazardous Waste Chemical Disposal Form."

(e) Hazardous Waste Removal

The Environmental Safety and Health Office shall insure the following:

1. Schedule hazardous waste removal with the approved vendor. This will be on a set schedule not to exceed five (5) month intervals.

2. Schedule specialized waste removal with the approved vendor if needed to remove poisons, shock sensitive materials, cylinders or other regulated waste not listed during the regularly scheduled waste removals.
3. The Environmental Safety and Health Office will ensure that someone is on-site during the approved waste vendor's removal days to insure packaging, manifesting and removal is in accordance with the Federal guidelines.
4. The Environmental Safety and Health Office shall sign and retain a copy of the manifest immediately before the hazardous waste is transported from Marshall University owned properties and keep on file indefinitely.
5. The Environmental Safety and Health Office shall retain a copy of the hazardous waste manifest upon arrival at the approved TSDF and keep manifest copy on file indefinitely.
6. The Environmental Safety and Health Office shall retain a copy of the ultimate disposal form from the incinerator and keep disposal copy on file indefinitely.

10.4 *Hazardous Waste Disposal*

Government Defines "Hazardous" Waste

Our definitions of "hazardous" waste are not all inclusive. However, if the waste your company produces falls in any of the categories detailed below, and you are not managing it according to government "hazardous" waste guidelines, please contact Allegheny Environmental for a more detailed explanation. Some of the guidelines and definitions below may sound obvious, unclear, or even illogical...but...disagreement will not excuse a company for non-compliance.

A material can only be a hazardous waste if, it is first,...a *waste*. This can be a technical matter to define but generally a waste is any solid, liquid, or contained gaseous material that you no longer use, and either recycle, throw away, or store until you have enough to treat or dispose of. A waste may remain in storage, go out into the dumpster, be poured down a drain, may be sent for re-use or recycling at another location, or...may be set aside for *hazardous waste disposal*, by a company such as Allegheny Environmental Services which provides *permitted* transportation and treatment/disposal.

Characteristics That Make a Waste "Hazardous"

First, a couple of "exceptions." The below characteristics and criteria may not apply if the waste is listed as an exception in Federal Code of Regulations 40-261.4(b).

The "characteristic" hazardous wastes below all carry a "D" code listing.

1. *Ignitable Waste*: If the waste is a liquid with a flash point less than 140-degrees F. it is in this category. If it is not a liquid but is capable under standard conditions of temperature and pressure of *causing* fire. If it is a *gas or oxidizer* as defined by the Department of Transportation. These wastes will carry a waste code of D001.
   Examples are: Many paint solvents and mineral spirits, paint waste, solvents, alcohol, degreasers, fine carbon particles, flammable gas cylinders.
2. *Corrosive Waste*: Among the criteria is a measured pH equal to or less than 2, or greater than or equal to 12.5 in liquid solution...or able to corrode SAE 1020 steel at a rate greater than 0.25 inches per year...or a solid that, if water is added, will cause the solution to show a pH in the above defined acid or alkaline ranges. These wastes will carry a waste code of D002
   Examples may include: Acids, corrosive cleaners, many photochemicals, strippers, rust removers, batteries, drain cleaners, bases, alkaline liquids.
3. *Reactive Waste*. This category takes in a wide field of materials. Generally waste that will cause any of the following:

   - React violently with water
   - Contains cyanides or sulfides
   - Is unstable
   - Is capable of explosion
   - Forms explosive mixtures with water or generates toxic gases when mixed with water.

   These wastes may carry a waste code of D003.
   Examples: sulfide containing wastes.
4. *Toxic Materials*: Includes thousands of materials including hundreds on government lists. Includes materials that do not pass so-called TCLP tests for toxicity. Poisons. These wastes will carry waste codes of D004 - D043.

Examples: Lead containing materials, rodent poisons, mercury and heavy metals.

The above criteria is just the beginning...

5. *"Listed" Wastes*: The government has published lists *("listed wastes")* which are defined as hazardous just because they are the result of certain industrial operations or because they contain certain chemicals. These lists are know as F, K, P, U. and often go by the title for example as a *F-listed waste.* These lists take in hundreds, even thousands of materials. They are available in the Federal Code of Regulations or you may call Allegheny Environmental Services to find out if your waste is on one of these lists.

   Examples: F-Listed: Hazardous Waste from Non-Specific Sources. Benzene, Carbon tetrachloride, K-Listed: Hazardous Waste from Specific Sources. e.g. Ink sludges containing heavy metals P-Listed: Discarded Commercial Products & Residues On specification and off-specification wastes which are commercial products or manufacturing chemical intermediate which are acutely hazardous wastes based on toxicity and reactivity. Aldicarb and many pesticides, U-Listed: Discarded Commercial Products & Residues On-specification and off-specification wastes which are commercial chemical products or manufacturing chemical intermediates based on toxicity. Creosote, Lindane, and many pesticides.

   And...material contaminated with any of the above material...may then become a "hazardous waste" A common example is soil where gasoline has spilled. Government regulations are so configured so that "dilution is not the solution."

Disposal Regulations and Procedures

Generator Responsibility. The government holds the generator of the waste responsible for its proper disposal and treatment. Since lack of knowledge is no excuse, it is always best to check to see if what you have to dispose of is defined as "hazardous." The company that "gets away" with disposing of a "hazardous" waste in the dumpster or down the sewer stands liable for BIG fines and even criminal enforcement action. Allegheny Environmental has a brief bulletin listing a variety of enforcement actions in a wide range of businesses for various waste violations. The bulletin is free for the asking.

If a waste is hazardous it must be treated and managed as such. That covers storage and labeling, transportation, treatment and disposal. And, there is paperwork. Most companies rely on an expert firm in hazardous waste to deal with their waste management and this is one of the services Allegheny Environmental provides at low cost. Some of the items that must be given attention when you have a "hazardous waste."

- ❖ Proper storage, hazardous labels, record-keeping, employee safety and right-to-know.
- ❖ Legal transportation off-site including manifesting, labeling, DOT and other government approved and permitted transporters must be used, and record keeping.
- ❖ Treatment, incineration, and disposal *only* at EPA permitted facilities.
- ❖ Testing if necessary to confirm a waste is non-hazardous. The burden of proof is on the company that generates the waste.

Examples of Hazardous and "Special" Waste Disposed of by Allegheny Environmental Services, Inc.

Wastes From Metal Work

- ❖ Degreasers and strippers
- ❖ Oils and Coolants
- ❖ Sludges and slag
- ❖ Metal and coating dust
- ❖ Plating waste—all types
- ❖ Coatings

Laboratory Wastes and Chemicals

- ❖ Out-of-date and expired chemicals
- ❖ Poisons, corrosives and reactives
- ❖ Old samples and unknowns
- ❖ Mercury, thermometers, and mercury compounds
- ❖ Contaminated glassware and containers
- ❖ Low-level radioactives
- ❖ Gas cylinders; all kinds of gases
- ❖ Peroxides and all oxidizers
- ❖ Ether and other peroxide forming compounds
- ❖ Organics and Inorganics
- ❖ Explosives
- ❖ Spill residues
- ❖ Scintillation vials
- ❖ Dyes and colors

- Pyphoric materials
- Contaminated filters
- Biological specimens in formaldin solutions
- Confidential research products and byproducts
- Over 60,000 chemicals and reagents

Printing and Photo Waste

- Inks, paints, and dyes
- Blanket wash and solvents
- Filters and screens
- All photochemicals
- Plating and engraving waste

Transportation & Motor Waste

- Oils and greases
- Pit sludge and wastes
- Oil-water mixes and coolants
- Used absorbent
- Spill residues
- Contaminated gasoline

Assorted "Housecleaning" Wastes

- Cleaners
- Paints, thinners, and paint removers
- Spray bottles and aerosols
- Old light ballasts and light bulbs of all types
- Lye, acids, and corrosives
- Duplicating fluid
- Bleach
- Floor treatments, water treatment chemicals, and boiler treatment chemicals
- Office machine toners
- Soaps, detergents, and alkalines
- Surplus and out-of-date product

Wastes in All Sizes, From Any Source

- In drums, pails, bottles, and buckets

- In cubic yard boxes
- In rolloffs and trucks
- In tanks and holding ponds
- Waste without containers
- "Unknown" waste
- Dredging waste
- Adhesives, glues, and binders
- Paint and paint sludge
- Organic solvents
- "Freon" and CFCs (for recycle/recovery)
- Chlorinated solvents
- Tars
- Contaminated wood
- Agrichemicals
- Industrial by-product waste
- Grinding waste and abrasives
- Fireworks
- Fire extinguisher waste
- Filter Press cake and waste from waste processing units
- Construction waste
- Store Close-Out and Unsold merchandise waste
- Batteries of all types
- PCB and Transformer/Capacitor waste
- MACE and tear gas
- Asbestos
- Chemical Plant and Distributor Surplus and Spoils waste
- Dry cleaning waste
- Flammable solids and and liquids
- Pesticides, Herbicides, Fungicides, Insecticides
- Ash and Incineration waste
- Heavy metal waste from any source
- Contaminated soils and remediation waste
- Sludge from waste treatment systems

Hazardous Waste Disposal Through Allegheny Environmental Services,Inc.

Allegheny provides disposal of all hazardous waste indicated above (everything except medical waste) in two service areas.

In addition to bulk waste which the customer has contained in drums, cubic yard boxes, tanks, or pallets, Allegheny provides a turnkey *Lab Pack*

Service to package smaller chemical containers (bottles, pails, vials, in volumes 5-gallons and under). With this service, Allegheny chemists come to your site, package all chemicals according to hazard class making identification of unknown items in the process. All regulatory paperwork is prepared including container DOT transportation labels and markings.

Our *Primary Service Area* consists of Upstate New York and Central and Western Pennsylvania. In this area turnkey, fully permitted and insured pickup and disposal of hazardous and special wastes is available on a regular route schedule. This is usually monthly.

Our *Secondary Service Area* consists of Eastern Pennsylvania and Northern Ohio. The same pickup and disposal service is available on a quarterly route schedule.

Our *Special Service Area* is anywhere in the Eastern United States where waste for disposal consists of large quantities (truck load and more). Custom schedules are arranged for pickup.

All packaging, transportation, and treatment/incineration is by fully permitted and insured processes. Everything is in compliance with federal and state regulations. Our field employees all have the 40-hour OSHA-prescribed Hazardous Materials Training.

10.5 Hazardous Waste Management

The Sierra Club recognizes that hazardous wastes present immediate and long-term problems that threaten the life-sustaining capabilities of the Earth. Our goal is to manage existing wastes in the safest way and ultimately to lead society to the wisest use of all materials so that hazardous wastes are not produced. This will protect worker and community health, conserve resources and improve environmental quality.

Each state or region should inventory its waste management needs, set goals for waste minimization, and avoid permitting excess capability. Waste streams should be kept separate, transported as little as possible, and treated with the best technology. Some secure, retrievable storage may be necessary while safer, better treatment technologies are developed and implemented. The best technology should be used to protect the health and safety of all workers.

1. Technologies

Elimination or reduction of wastes from industrial processes should have the highest priority in a waste management program. Process substitution, waste reduction, reuse, and conversion to nonhazardous substances should be encouraged. Closed systems of manufacture, use, and disposal should be developed. Industrial processes producing wastes that cannot be reused, converted to nonhazardous substances, or safely incinerated should be phased out. If materials and energy recovery are not possible, the best technologies should be used to reduce the level of hazard as much as possible. Waste exchanges should be encouraged by regulatory programs.

Incineration should be considered after all other acceptable options have been evaluated and found to be impracticable for man-aging a particular waste. Regulations must be in place to require the lowest levels of emissions and effluents that are technically feasible. Incinerator design, permitting, and operation must include sufficient safeguards to protect human health and the environment. If the lowest feasible levels of emissions do not fully protect human health and the environment, the facility should not be allowed to operate unless incineration is the least hazardous option for management of an existing problem waste. Because much is still not known about the health and environmental effects of emissions, including the products of incomplete combustion, research should be pursued to further characterize and reduce emissions.

(a) The permitting, siting, and operation of hazardous waste incinerators, including mobile units, must include a public education and participation process that makes all data associated with design, waste streams, transportation, storage, and emissions conveniently available to the public.
(b) All permits should be written to provide for changes or revocation at any time, and should be re-evaluated at least every two years. A mobile incinerator must have a permit that is specific for each site where it operates.
(c) Incinerator siting should minimize risks to the public from emissions and transportation, at the same time avoiding such environmentally sensitive areas as critical wildlife habitat, parks, and wilderness areas. A facility should include a buffer zone and operate inside a containment building where fugitive emissions are controlled.
(d) Trial burns must be conducted on each individual incinerator using waste streams as close as possible in character to those to be burned in normal operations.
(e) All hazardous waste incinerators should be equipped with emission control systems for acid gas, toxics, and particulates. Monitoring should address all potentially toxic emissions, including organics and metals. Monitoring should be based on frequent or continuous

sampling whenever technically feasible. Continuous emissions monitoring should be employed that will fully characterize the operating conditions of the incinerator and should be connected to automatic interlock systems that will not allow continued operation of the incinerator if design values cannot be attained. Ashes, slag, and aqueous wastes should be presumed hazardous and managed appropriately. Management of incinerator residues should be provided for in the design and permitting process.

(f) Monitoring programs should include the collection of on-site and off-site baseline data prior to facility operation so that contamination from the facility can be distinguished from ambient or background pollution levels when operation begins. Established quality control procedures should be used. All monitoring data should be conveniently available to the public and the press from facility operators and regulatory authorities.

(g) Regulatory controls should require operator certification and an operations and maintenance plan. Inspection to determine that the plan is being followed should be both announced and unannounced and should cover all shifts at facilities. Regulations should provide for the establishment of a citizens' over-sight committee which can designate and supervise its own inspector. This person shall be trained and employed at state or facility owner expense and should inspect the facility on a least a weekly basis, independently of state inspections.

(h) Research on other thermal technologies, using nonhazardous chemicals for initial testing, should be encouraged.

(i) Part of the facility's cost of doing business should include the preparation of environmental and public health risk assessments, permit fees, monitoring, closure, insurance, and post-closure contingency funds. A special tax, such as a gross receipts tax, should be imposed on the facility to compensate the host community that bears the health risks. This compensation should be available for improving community emergency response capabilities, additional monitoring, health testing, health care, transportation safety, and other costs attributable to facility impacts. Marine incineration is not acceptable because human or mechanical error or natural disaster could cause a release of toxics that could not be contained or cleaned up. Compliance with strict safety or emissions requirements would be difficult to assure and safer disposal options exist.

Land disposal is not acceptable for hazardous waste. Even so, it may be necessary to landfill the irreducible residues of treatment of hazardous wastes. For that reason, these facilities should be sited and constructed to minimize the possibility of surface or groundwater contamination. Air, land, and water should be monitored for contamination.

Pits, ponds, and lagoons are not acceptable, except possibly for inorganic wastes of low volatility and low leachability. They should be constructed or retrofitted with double liners, leak detection between the liners, and continuous soil and groundwater monitoring. Underground injection is not acceptable because of the potential for present or future groundwater contamination.

Export of hazardous wastes from the United States to other countries is not acceptable because it poses significant health and environmental threats, is technically unnecessary and unjustified, and is unethical. An exception might be a unique and special circumstance where great public health and environmental benefit can be demonstrated.

2. Siting Processes

Programs to site new hazardous waste facilities should be based on the need for facilities and the most appropriate technologies for dealing with the particular wastes. Agencies administering such programs should work with environmental, citizen, trade, small business, educational, and labor groups to evaluate these needs, to ensure the most effective management of the wastes, and to devise comprehensive mechanisms for local monitoring of the facility's impact.

The process for siting new hazardous waste facilities is the key to their success. It should be open and fair to all interests. Specific provisions should include public education well before any site is chosen, development of siting criteria, establishment of an institution to site facilities, and assessment of need for facilities. Funds for studies and monitoring must come from an earmarked source, the operator or proponent paying for them and the community choosing the consultant. A referee can be jointly chosen in case of a dispute. Upfront financial assurance to cover adequate monitoring, maintenance, closure, and post-closure monitoring of the facility should be required, regardless of the financial conditions of the operator. Citizens should have full access to the data upon which the siting decision is made, including but not limited to the proposed procedures of handling the wastes and the specific wastes to be handled. Collaboration, mitigation, and compensation should be recognized by all parties to the negotiation process as legitimate methods for making a facility acceptable to its host community. Community representatives should have an explicit role in facility oversight.

3. Regulatory Programs

Regulations should be dear, specific, and enforceable. Where possible, incentives should be used to encourage the best management practices. Government responsibilities include over-sight of the hazardous waste management system. Agencies involved must be well managed and capable of coordinating the highly complex system involving different levels of government, the private sector, and the public. Decisions need to be made by people who are accountable to the public. Public participation should include full access to data, consideration of the advice of independent technical experts and other citizens, and explanations made in a timely manner. Regulatory responsibilities include:

(a) Procedures for handling hazardous wastes by generators, recyclers, transporters, treaters, and disposers. This should include protection of workers.
(b) Rules governing facility siting, operation, and closure.
(c) Effective enforcement of laws and permit conditions.
(d) Conscientious inspections to ensure proper operation and accurate reporting.
(e) Authority to allow administrative penalties and collect fines for violations, at both state and federal levels.
(f) Cooperation and assistance to the institution conducting the planning and siting process for future facilities.

All new hazardous waste facilities should be constructed totally or partially above existing grade, physically accessible to inspection personnel. The facility and loading areas should be designed to prevent spills and contain any that may occur, including rainfall in outdoor areas. Cleanup facilities should be adequate to deal with the maximum possible leak or spill. These safeguards and facility monitoring apply to existing, expanding, or new facilities. Regulations should address all generators of hazardous wastes. Every effort should be made to educate small businesses and the general public in reduction and proper handling of small quantities of hazardous wastes such as discarded pesticides and solvents. Municipal or county collection and treatment systems should be developed so these substances do not contaminate municipal landfills, sewage treatment systems, or groundwater. An effective national pretreatment program for industrial waste water is essential to ensure that liquid hazardous wastes do not degrade surface, ground, or marine waters, damage publicly owned treatment works, or create sewage sludge that cannot be used safely as a soil improvement. Aqueous wastes should not be evaporated if hazardous compounds will be vaporized; they must be treated.

4. Liability

Hazardous waste generators, transporters, treaters, and disposers should be held strictly, jointly, and severally liable for any injury which may occur to property, human health and welfare, or the environment. Procedures should be established to hold individuals responsible for their actions. Legislation should be enacted to ease the burden of proving the causal relationship and any injury it may have caused. There should be both civil and animal penalties for violators.

Legislation should be passed to allow citizen suits to enforce hazardous waste laws and regulations. Financial responsibility (insurance or bonding) should be required of all generators, transporters, treaters, and disposers to assure their ability to compensate for any damage.

5. Cleanup

A waste management program should include a clear set of priorities governing the process for cleanup of accidents and abandoned sites. It should have funding so that trained and protected personnel with equipment to carry out the cleanup are available when needed. Sites need to be carefully evaluated for appropriate containment and treatment of the waste. The full cost of cleanup shall be recovered from the party whenever possible.

6. Funding

A fair national mechanism for funding government hazardous waste programs should be devised. Assessments should be based on degree of hazard and progress made in reducing tonnage and toxicity through best management practices. Research must be funded, preferably through such assessments, to increase the amount and quality of data on the effects of hazardous waste on human health and the environment. All states should establish comprehensive mandatory birth defect and cancer registries. Research is needed to assure that the full range of wastes that are toxic, including mixtures, will be regulated. More accurate technologies for site monitoring should be developed.

7. Compensation

Public and private sectors should cooperate with the communities where hazardous waste facilities are located so that acute and chronic health problems can be identified and addressed quickly. The Sierra Club supports notification and full disclosure to those who have been or may be exposed to hazardous wastes in the community or workplace. Where warranted, a program of victim's assistance or compensation should be provided.

Adopted by the Board of Directors, May 5-6, 1984; amended May 2-3, 1987, and March 18-19, 1989

Guidance on Hazardous Waste Incinerators by the Hazardous Materials Committee

[*Note*: This statement should be quoted in full and not excerpted.]

Incinerators should not be used to manage hazardous waste unless it can be demonstrated that there is no other technically feasible method for management of a specific waste. EPA's new incinerator regulations are not yet in place. Adequate regulatory frameworks do not exist at the state and federal level to protect public health and the environment from the hazards of incineration. Technologies other than incineration are more appropriate for managing many hazardous wastes, and a region must give priority to siting and permitting such facilities. The programs that direct manufacturers to produce less waste are neither strong enough nor fully implemented. Disposers must have to document that they have done all they can to avoid, minimize, recycle, or otherwise treat a waste before sending it to an incinerator. Incinerator permits must allow only such residuals to be accepted.

10.6 Hazardous Waste Management Services

Treatment Storage and Disposal Facilities are typically the last link of the hazardous waste management chain.

The Hazardous Waste Management Services Program allows the TSDF to receive electronic manifests, track waste inventory and compile reports for federal and state regulating agencies.

The program also allows TSDFs to manage hazardous waste for client generators. Program modules provide powerful tools for managing waste from generation to final treatment, storage or disposal.

Comprehensive reporting capability and data archive makes waste information available on demand.

- Send & receive electronic waste documents.

- Compile data for annual & biennial reports.
- Manage waste for clients.
- Track waste location & destination.
- Access generator, transporter & TSD information on demand.
- Access data from electronic archive.
- Compile comprehensive waste reports .

10.7 Transport

Transporters can receive waste manifests electronically at the site of pickup or anywhere there is an Internet connection. Print hard-copys for fast reference or retain hard-copy for records.

Digital signatures make the process paperless.

The program displays waste status while in route and updates automatically when the TSDF confirms receipt of the waste and electronic manifest.

The program records and archive all activity and recall data for reporting purposes.

- Receive & send electronic manifests.
- Print manifest forms for fast reference or hard copy.
- Access waste information & emergency response numbers.
- Recall manifest & shipment information from archive.

11

Pharmacology, Hospital Wastes Control and Treatment

11.1 Pharmacology

Pharmacology is the study of how drugs interact with living organisms to produce a change in function. If substances have medicinal properties, they are considered *pharmaceuticals*. The field encompasses drug composition and properties, interactions, toxicology, therapy, and medical applications and antipathogenic capabilities. Pharmacology is not synonymous with pharmacy, which is the name used for a profession. Though in common usage the two terms are confused at times. Pharmacology deals with how drugs interact within biological systems to affect function, while pharmacy is a medical science concerned with the safe and effective use of medicines.

The origins of clinical pharmacology date back to the Middle Ages in Avicenna's *The Canon of Medicine*, Peter of Spain's *Commentary on Isaac*, and John of St Amand's *Commentary on the Antedotary of Nicholas*. Pharmacology as a scientific discipline did not further advance until the mid-

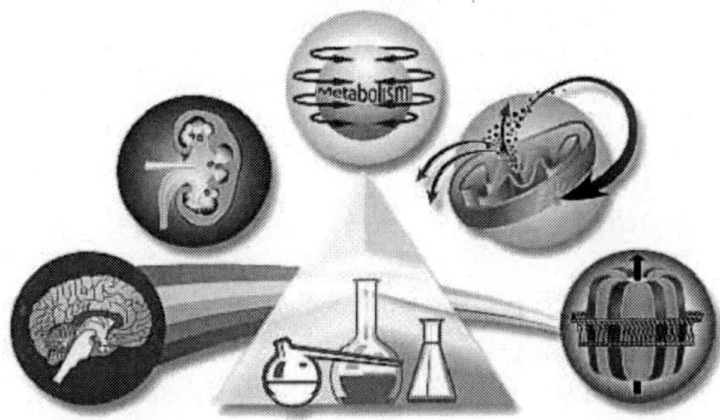

Fig. 11.1: A variety of topics involved with pharmacology, including neuropharmacology, renal pharmacology, human metabolism, intracellular metabolism, and intracellular regulation.

19th century amid the great biomedical resurgence of that period. Before the second half of the nineteenth century, the remarkable potency and specificity of the actions of drugs such as morphine, quinine and digitalis were explained vaguely and with reference to extraordinary chemical powers and affinities to certain organs or tissues. The first pharmacology department was set up by Buchheim in 1847, in recognition of the need to understand how therapeutic drugs and poisons produced their effects.

The word Pharmacology comes from Greek: *pharmakon* meaning drug, *logos*, "knowledge". Early pharmacologists focused on natural substances, mainly plant extracts. Pharmacology developed in the 19th century as a new biomedical science that applied the principles of scientific experimentation to therapeutic contexts.

Divisions

Pharmacology as a chemical science is practiced by pharmacologists. Subdisciplines include

- *Clinical pharmacology*—The medical field of medication effects on humans.
- *Neuro-* and *psychopharmacology* (effects of medication on behavior and nervous system functioning).
- *Pharmacogenetics* (clinical testing of genetic variation that gives rise to differing response to drugs).
- *Pharmacogenomics* (application of genomic technologies to new drug discovery and further characterization of older drugs).
- *Pharmacoepidemiology* (study of effects of drugs in large numbers of people).
- *Toxicology* study of the effects of poisons.
- *Theoretical pharmacology*
- *Posology*—How medicines are dosed?
- *Pharmacognosy*—Deriving medicines from plants.

Scientific Background

The study of chemicals requires intimate knowledge of the biological system affected. With the knowledge of cell biology and biochemistry increasing, the field of pharmacology has also changed substantially. It has become possible, through molecular analysis of receptors, to design chemicals that act on specific cellular signaling or metabolic pathways by affecting sites directly on cell-surface receptors (which modulate and mediate cellular signaling pathways controlling cellular function).

A chemical has, from the pharmacological point-of-view, various properties. Pharmacokinetics describes the effect of the body on the chemical (*e.g.* half-life and volume of distribution), and pharmacodynamics describes the chemical's effect on the body (desired or toxic).

When describing the pharmacokinetic properties of a chemical, pharmacologists are often interested in *ADME*:

- *Absorption*—How is the medication absorbed (through the skin, the intestine, the oral mucosa)?
- *Distribution*—How does it spread through the organism?
- *Metabolism*—Is the medication converted chemically inside the body, and into which substances. Are these active? Could they be toxic?
- *Excretion*—How is the medication eliminated (through the bile, urine, breath, skin)?

Medication is said to have a narrow or wide *therapeutic index* or *therapeutic window*. This describes the ratio of desired effect to toxic effect. A compound with a narrow therapeutic index (close to one) exerts its desired effect at a dose close to its toxic dose. A compound with a wide therapeutic index (greater than five) exerts its desired effect at a dose substantially below its toxic dose. Those with a narrow margin are more difficult to dose and administer, and may require therapeutic drug monitoring (examples are warfarin, some antiepileptics, aminoglycoside antibiotics). Most anti-cancer drugs have a narrow therapeutic margin: toxic side-effects are almost always encountered at doses used to kill tumors.

Medicine Development and Safety Testing

Development of medication is a vital concern to medicine, but also has strong economical and political implications. To protect the consumer and prevent abuse, many governments regulate the manufacture, sale, and administration of medication. In the United States, the main body that regulates pharmaceuticals is the Food and Drug Administration and they enforce standards set by the United States Pharmacopoeia. In the European Union, the main body that regulates pharmaceuticals is the EMEA and they enforce standards set by the European Pharmacopoeia.

If the structure of a medicine is altered slightly, this will slightly alter the medicine's properties. This means when a useful activity has been identified, chemists will make many similar compounds called analogues, to attempt and maximise the beneficial effects. This development phase can take up to 3 years and is expensive.

These new analogues need to be developed. It needs to be determined how safe the medicine is for human consumption, its stability in the human body and the best form for dispensing, like tablet or aerosol. After extensive testing, which can take up to 6 years the new medicine is ready for marketing and selling.

As a result of the long time required to develop analogues and test a new medicine and the fact that of every 5000 potential new medicines typically only one will ever reach the open market, this is an expensive way of doing things, costing millions of dollars. To recoup this outlay pharmaceutical companies may do a number of things: Carefully research the demand for their potential new product before spending an outlay of company funds.

Obtain a patent on the new medicine preventing other companies from producing that medicine for a certain allocation of time.

Drug Legislation and Safety

In the United States, the Food and Drug Administration (FDA) is responsible for creating guidelines for the approval and use of drugs. The FDA requires that all approved drugs fulfill two requirements:

1. The drug must be found to be effective against the disease for which it is seeking approval.
2. The drug must meet safety criteria by being subject to extensive animal and controlled human testing.

Gaining FDA approval usually takes several years to attain. Testing done on animals must be extensive and must include several species to help in the evaluation of both the effectiveness and toxicity of the drug. The dosage of any drug approved for use is intended to fall within a range in which the drug produces a therapeutic effect or desired outcome.

The safety and effectiveness of prescription drugs in the U.S. is regulated by the federal Prescription Drug Marketing Act of 1987.

The Medicines and Healthcare products Regulatory Agency (MHRA) has a similar role in the UK.

Education

The study of pharmacology is offered in many universities worldwide.

Again, pharmacology education programs differ from pharmacy programs.

Students of pharmacology are trained as researchers, studying the effects of substances in order to better understand the mechanisms which might lead to new drug discoveries for example. Whereas as pharmacy student will eventually work in a pharmacy dispensing medications or some other position focused on the patient, pharmacologist will typically work within a laboratory setting.

Some higher educational institutions combine pharmacology and toxicology into a single program as does Michigan State University. Michigan State University offers PhD training in Pharmacology & Toxicology with an optional Environmental Toxicology specialization. They also offer a Professional Science Masters in Integrative Pharmacology.

11.2 Clinical Pharmacology

Clinical Pharmacology is the science of drugs and their clinical use. It is underpinned by the basic science of pharmacology, with added focus on the *application* of pharmacological principles and methods in the real world. It has a broad scope, from the discovery of new target molecules, to the effects of drug usage in whole populations.

Clinical pharmacology connects the gap between medical practice and laboratory science. The main objective is to promote the safety of prescription, maximise the drug effects and minimise the side effects. It is important that there be association with pharmacists skilled in areas of drug information, medication safety and other aspects of pharmacy practice related to clinical pharmacology.

Clinical pharmacologists usually have a rigorous medical and scientific training which enables them to evaluate evidence and produce new data through well designed studies. Clinical pharmacologists must have access to an adequate number of out-patients for clinical care, teaching and education, and research as well be supervised by medical specialists. Their responsibilities to patients include, but are not limited to analyzing adverse drug effects, therapeutics, and toxicology including reproductive toxicology, cardiovascular risks, perioperative drug management and psychopharmacology.

In addition, the application of genetic, biochemical, or virotherapuetical techniques has led to a clear appreciation of the mechanisms involved in drug action.

Branches

- *Pharmacodynamics*—finding out what drugs do to the body and how. This includes not just the cellular and molecular aspects, but also more relevant clinical measurements. For example, not just the biology of salbutamol, a beta2-adrenergic receptor agonist, but the peak flow rate of both healthy volunteers and real patients.

- *Pharmacokinetics*—what happens to the drug while in the body. This involves the body systems for handling the drug, usually divided into the following classifcation:
  - Absorption
  - Distribution
  - Metabolism
  - Elimination
- *Rational Prescribing*—using the right medication, at the right dose, using the right route and frequency of administration for the patient, and stopping the drug approppropriately.
- Adverse Drug Effects
- Toxicology
- Drug interactions
- *Drug development*—usually culminating in some form of clinical trial.

11.3 Hospital Waste Fact Sheet

Hospital Waste Management means the management of waste produced by hospitals using such techniques that will help to check the spread of diseases through it.

The Story So Far

The management of waste poses to be a major problem in most of the countries, especially hospital waste. It is an ongoing problem for many countries. In recent years, medical waste disposal has posed even more difficulties with the appearance of disposable needles, syringes, and other similar items. Pakistan is also facing this problem. Around 250,000 tonnes of medical waste is annually produced from all sorts of health care facilities in the country. This type of waste has a bad affect on the environment by contaminating the land, air and water resources.

According to a report, 15 tonnes of waste is produced daily in Punjab. The rate of generation is 1.8 kilograms per day per bed. The province houses 250 hospitals with a total capacity of 41,000 beds.

Different Types

Hospital wastes are categorised according to their weight, density and constituents. The World Health Organisation (WHO) has classified medical waste into different categories. These are:

- *Infectious*: material-containing pathogens in sufficient concentrations or quantities that, if exposed, can cause diseases. This includes waste from surgery and autopsies on patients with infectious diseases;
- *Sharps*: disposable needles, syringes, saws, blades, broken glasses, nails or any other item that could cause a cut;
- *Pathological*: tissues, organs, body parts, human flesh, fetuses, blood and body fluids;
- *Pharmaceuticals*: drugs and chemicals that are returned from wards, spilled, outdated, contaminated, or are no longer required;
- *Radioactive*: solids, liquids and gaseous waste contaminated with radioactive substances used in diagnosis and treatment of diseases like toxic goiter; and
- *Others*: waste from the offices, kitchens, rooms, including bed linen, utensils, paper, etc.

Guidelines

There are Guidelines for Hospital Waste Management In Pakistan since 1998 prepared by the Environmental Health Unit, of the Ministry of Health, Government of Pakistan, giving detailed information and covering all aspects of safe hospital waste management in the country, including the risk associated with the waste, formation of a waste management team in hospitals, their responsibilities, plan, collection, segregation, transportation, storage, disposal methods, containers, and their color coding, waste minimisation techniques, protective clothing, etc.

A project was implemented in January, 2000 in the biggest hospital in every province by the Ministry of Health in Islamabad, in collaboration with WHO.

Improper Disposal

Hospitals and public health care units are supposed to safeguard the health of the community. However, the waste produced by the medical care centers if disposed off improperly, can pose an even greater threat than the original diseases themselves.

Pakistan is also facing such problems. There are no systematic approaches to medical waste disposal. Hospital wastes are simply mixed with the municipal waste in collecting bins at roadsides and disposed off similarly

Some waste is simply buried without any appropriate measure. The reality is that while all the equipment necessary to ensure the proper management of hospital waste probably exists, the main problem is that the staff fails to prepare and implement an effective disposable policy.

In Lahore, like most of the cities in Pakistan, there are no proper measures taken for the management of hospital waste. The standard practice of hospital waste disposal is dumping it in the M.C.L. container wherever situated.

Disposable syringes and needles are also not disposed off properly. Some patients, who routinely use syringes at home, do not know how to dispose them off properly. They just throw them in a dustbin or other similar places, because they think that these practices are inexpensive, safe, and easy solution to dispose off a potentially dangerous waste item.

How does Hospital Waste Affect Us?

If hospital waste is not managed properly it proves to be harmful to the environment. It not only poses a threat to the employees working in the hospital, but also to the people surrounding that area.

Infectious waste can cause diseases like Hepatitis A & B, AIDS, Typhoid, Boils, etc. A common practice in Pakistan is the reuse of disposable syringes. People pick up used syringes from the hospital waste and sell them. Many drug addicts also reuse the syringes that can cause AIDS and other dangerous and contagious diseases. If a syringe, previously used by an AIDS patient, is reused, it can affect the person using it. So, the hospital staff should dispose off the syringes properly, by cutting the needles of the syringes with the help of a cutter, so that the needle ca not be reused. When waste containing plastics are burnt, Dioxin is produced, which can cause Cancer, birth defects, decreased psychomotor ability, hearing defects, cognitive defects and behavioral alternations in infants. Flies also sit on the uncovered piles of rotting garbage. This promotes mechanical transmissions of fatal diseases like Diarrhea, Dysentery, Typhoid, Hepatitis and Cholera. Under moist conditions, mosquitoes transmit many types of infections, like Malaria and Yellow fever. Similarly, dogs, cats and rats also transmit a variety of diseases, including Plague and Flea born fever, as they mostly live in and around the refuse. A high tendency of contracting intestinal, parasitic and skin diseases is found in workers engaged in collecting refuse.

Solution

Some steps should be taken for the minimisation of hospital waste. Before any clear improvement can be made in medical waste management, consistent and scientifically based definitions must be established as to what is meant by medical waste and its components, and what the goals are. Plans and policies should be laid down for this purpose. Then the waste should be segregated. Imposing segregated practices within hospitals to separate biological and chemical hazardous waste will result in a clean solid waste stream, which can be recycled easily. If proper segregation is achieved through training, clear standards, and tough enforcement, then resources can be turned to the management of the small portion of the waste stream needing special treatment. New emphasis should be put on the reduction of waste, workers' safety should be ensured through education, training and proper personal protective equipment.

Incinerators: A Solution or a Threat?

Incineration has been the treatment method of choice for medical waste for two important reasons. First, incineration has always been thought to be the best method of eliminating any infectious organisms that are present in medical waste. Second, incineration has been economical for hospitals because it substantially reduces the volume to be disposed of in a landfill. Waste disposal costs have historically been based on the volume to be disposed. Both of these assumptions behind medical waste incineration are no longer able to support objective scrutiny. Waste is burnt at very high temperatures, that produce emissions full of acidic gases, heavy metals, toxic organisms and dioxins. There is a lot of ash produced by an incinerator as well.

Incinerators for medical and municipal waste have been linked to severe public health threats and pollution. The combination of intense public opposition to incineration and increasingly strict environmental pollution regulation has forced the closure or cancellation of many incinerators in industrialised countries.

Incinerators are fast becoming an obsolete technology in many developed countries as they are moving towards safer and more economical alternative approaches to medical and municipal waste management. As a result, many incinerator companies are targeting overseas markets where people are not yet aware of the serious health and environmental threats associated with incineration or the many advantages of alternatives. Incinerator companies are now targeting Asia, Africa, and Latin America to sell their toxic technology. Researchers came to the conclusion that Dioxin, as well as mercury and other toxic substances, are emitted when waste is burnt in an incinerator. Dioxin and related chlorinated organic compounds are extremely potent toxic substances that produce a remarkable variety of adverse effects in human and animals at extremely low doses.

Mercury is also bio-accumulative and is toxic to the kidneys and nervous system. Readily converted to its organic form in the environment, mercury interferes with normal brain development.

Techniques to be Used

Various alternative technologies for incineration are available at hospitals in many developed countries. As these techniques are either too complicated or very expensive, they are not being used in Pakistan. Though, these techniques should also be applied here, for proper waste disposal.

Steam Autoclaving

Steam Autoclaving is the most widely used and most efficient alternative medical-waste-treatment technology. Most available autoclaves are designed to handle both biohazard and normal hospital wastes simultaneously. However, they cannot treat pathological animal wastes, chemotherapy wastes, and low level radioactive wastes. These wastes have to be treated separately. Medical waste autoclaves usually jointly operate with a shredder, and a compactor.

In autoclaves, the effects of heat from saturated steam and increased pressure decontaminate medical waste by inactivating and destroying microorganisms.

There are two types of autoclaves, gravity displacement and pre-vacuum. Those designed for medical waste are mostly pre-vacuum.

Chemical Treatment

In chemical treatment systems, an anti-microbial chemical, such as sodium hypochlorite, chlorine dioxide, or per acetic acid, decontaminates the medical waste. Most chemical treatment systems, currently in use, operate at ambient temperature.

Microwave Radiation

In Microwave Radiation, medical waste enters the system by batch or continuous mode, where it is wetted with steam or water and heated by microwave radiation at de-contaminating temperatures.

Other Thermal Systems

Some systems use a combination of infrared radiation and forced hot-air convection to treat the waste. The waste then is compacted, preparing it for landfill. Other systems use gamma radiation to heat the waste to disinfecting temperatures. A portion of the solid residue obtained is recycled, while the remainder is disposed. Several other thermal systems currently under development use steam, oil, electricity or some form of radiation as their source of heat.

Disposal of Pathological Waste

As mentioned above, Pathological waste (body parts, research animals, etc.) cannot be disposed off by autoclaving. For disposal of such waste, either Crematoria (burning of the body) or burial should be performed.

Training

The hospital staff should be trained in such a manner that they help in disposing off the waste properly.

WWF's Position Statement

WWF advocates for safer waste disposal techniques and wishes to educate the broader public about dioxin, mercury and other endocrine disrupting chemicals. WWF is ready to join a campaign that makes explicit links between environmental contamination and public health. WWF also advocates for safe waste disposal methods like waste reduction, then waste segregation, which is crucial to reduce the volume and toxicity of the medical waste stream. Then stress is laid on the assessment and recycling of hospital waste. Waste assessment will give hospital staff a clear idea of how their waste is managed on a daily basis. Recycling reduces pollution from resource extraction and manufacturing products, and pollution associated with incineration, landfills and other waste disposal methods. WWF-P questions the policy followed by the developed world as they export their incinerators to developing countries. Despite being fully aware of the unmanageable pollution problems caused by waste-burners. WWF-P urges the general public to demand a 'Right to Know' basis about the kinds of waste management technologies proposed to be adopted by the Government.

11.4 Methods Adopt to Dispose Off the Hospital Waste

Hospital waste are classified such as: infectious, sharps, pathological, pharmaceuticals, radioactive, & others. are well disposed to prevent transmission of diseases by separating biological and chemical hazardous waste such as infectious: material containing pathogens in sufficient concentrations or quantities that, if exposed, can cause diseases. This includes waste from surgery and autopsies on patients with infectious diseases;

- *Sharps*: disposable needles, syringes, saws, blades, broken glasses, nails or any other item that could cause a cut;
- *Pathological*: tissues, organs, body parts, human flesh, fetuses, blood and body fluids; Pharmaceuticals: drugs and chemicals that are returned from wards, spilled, outdated, contaminated, or are no longer required;
- *Radioactive*: solids, liquids and gaseous waste contaminated with radioactive substances used in diagnosis and treatment of diseases like toxic goiter; and
- *Others*: waste from the offices, kitchens, rooms, including bed linen, utensils, paper, etc. hospitals in develop countries has incinerators it has been the treatment method of choice for medical waste for two important reasons. First, incineration has always been thought to be the best method of eliminating any infectious organisms that are present in medical waste. Second, incineration has been economical for hospitals because it substantially reduces the volume to be disposed of in a landfill. Waste disposal costs have historically been based on the volume to be disposed. Both of these assumptions behind medical waste incineration are no longer able to support objective scrutiny. Waste is burnt at very high temperatures, that produce emissions full of acidic gases, heavy metals, toxic organisms and dioxins. There is a lot of ash produced by an incinerator as well; but one disadvantage about incinerator it causes air pollution & that would be another problem. however, now a days technologies are more high-tech, there are techniques are applied for waste disposal but either too complicated & very expensive....such as—

1. Steam Autoclaving is the most widely used and most efficient alternative medical-waste-treatment technology. Most available autoclaves are designed to handle both biohazard and normal hospital wastes simultaneously. However, they cannot treat pathological animal wastes, chemotherapy wastes, and low level radioactive wastes. These wastes have to be treated separately.
   Medical waste autoclaves usually jointly operate with a shredder, and a compactor (to minimize the waste volume).
   In autoclaves, the effects of heat from saturated steam and

increased pressure decontaminate medical waste by inactivating and destroying microorganisms.

There are two types of autoclaves, gravity displacement and pre-vacuum. Those designed for medical waste are mostly pre-vacuum.

2. Chemical treatment systems, an anti-microbial chemical, such as sodium hypochlorite, chlorine dioxide, or per acetic acid, decontaminates the medical waste. Most chemical treatment systems, currently in use, operate at ambient temperature.
3. Microwave Radiation, medical waste enters the system by batch or continuous mode, where it is wetted with steam or water and heated by microwave radiation at de-contaminating temperatures.
4. Some systems use a combination of infrared radiation and forced hot-air convection to treat the waste. The waste then is compacted, preparing it for landfill. Other systems use gamma radiation to heat the waste to disinfecting temperatures. A portion of the solid residue obtained is recycled, while the remainder is disposed. Several other thermal systems currently under development use steam, oil, electricity or some form of radiation as their source of heat.
5. Pathological waste (body parts, research animals, etc.) cannot be disposed off by autoclaving. For disposal of such waste, either Crematoria (burning of the body) or burial should be performed.

11.5 Affordable Assistance for Hospital Waste and Hazardous Material Management

Welcome to your own consultant for hazardous material and waste management issues! Most hospital hazmat and waste managers wear many hats, but having an intimate knowledge of the regulations for these issues shouldn't have to be one of them. *Hospital Waste Management* serves hospitals in Washington State.

Hospital Waste Management is your own consultant to help you comply with local, Washington state and federal waste and hazmat regulations. We can sort out the answers to your questions about which hazardous materials you have on site and how best to store them, as well as how to most appropriately manage your waste streams.

Does your Environmental Services staff know how to respond to a hazardous material spill? Do you even know what your options are? *Hospital Waste Management* can sort out your options and, if you elect to train an onsite emergency response team, train them according to Washington's Emergency

Response Standard, Chapter 296-824 WAC. Proper training safeguards your staff and helps to avoid costly fines and settlements with Labor & Industries.

Hospital waste comprises all of the following waste types:

- Solid,
- Biohazardous (regulated medical, infectious),
- Dangerous (flammable, corrosive, toxic, or persistent in the environment),
- Radioactive (primarily short-lived isotopes),
- Universal (batteries, fluorescent lamps),
- Electronic (computers, CRTs, medical devices), and
- Special (solid corrosive dangerous waste, including SodaSorb™ and BaraLyme™ from anesthesiology).

Is your facility managing these waste streams properly? Do you have a Hazardous Material & Waste Management Plan in effect that guides your staff in properly managing these waste streams?

A good deal of a hospital's waste goes down the sanitary sewer, which is usually acceptable, but frequently not. Do you know where your garbage compactor drainage is going? Most loading dock drains are piped to stormwater systems and go directly to the nearest pond, creek or river.

We at *Hospital Waste Management* can assess your waste stream management, develop a hazmat inventory for your facility (required by Joint Commission), train in-house staff to respond safely and competently to hazmat spills (and know when to call in better-trained or equipped responders), and give you the confidence that you're facility complies with local, state and federal hazmat and waste management regulations. All at an affordable cost.

If you need to train your hospital employees in hazmat response awareness, I have developed a self-learning module that takes an employee just 15 to 20 minutes to complete and fulfills L & I's FR/AL requirements in WAC 296-824, the *Emergency Response Standard.* This course is based upon Microsoft's popular PowerPoint™ software and can be loaded onto a stand-alone computer or your facility's intranet. If you want to train an in-house hazmat response team, we teach a First Responder/Operations Level course for your staff to safely and completely clean up hospital hazmat spills.

11.6 Terms of Reference

Introduction

Most wastes generated by hospitals and medical clinics are non-hazardous general wastes from hospital organization activities (*i.e.*, including kitchen wastes, office materials, workshop residuals) and patient processing activities in wards which are not handling infectious diseases (*i.e.*, first aid packaging, used but emptied disposable bed liners and diapers, disposable masks, pharmaceutical packaging, etc.). After source segregation of recyclables, disposal is typically by sanitary landfill.

Potentially hazardous wastes from hospitals and clinics which have a pathogenic, chemical, explosive, or radioactive nature are called "medical wastes". Medical wastes include the following:

- Pathological wastes (*i.e.*, body parts, aborted fetus, tissue and body fluids from surgery; and dead infected laboratory animals);
- Infectious waste (*i.e.*, surgical dressings and bandages, infected laboratory beddings, infectious cultures and stocks from laboratories, and all waste from patients in isolation wards handling infectious diseases);
- Sharps (*i.e.*, needles, syringes, used instruments, broken glass);
- Pharmaceutical wastes (*i.e.*, soiled or out-of-date pharmaceutical products);
- Chemical wastes (*i.e.*, spent solvents, disinfectants, pesticides and diagnostic chemicals);
- Aerosols (*i.e.*, aerosol containers or gas canisters which may explode if incinerated or punctured);
- Radioactive wastes (*i.e.*, sealed sources in instruments, and open sources used in vitro diagnosis or nuclear medical therapy); and
- Sludges from any on-site wastewater treatment facilities may be potentially hazardous.

Pathological wastes should be destroyed by incineration under high heat (*i.e.*, over 900°C with an afterburner temperature at over 800°C), although some countries require burial of human pathological wastes at official cemeteries for religious reasons. To reach these temperatures and have adequate afterburning and pollution control typically requires development of a regional medical waste facility. Smaller individual hospital or clinic incinerators may not be able to reach these temperatures and afterburning retention periods. Volatilized metals (such as arsenic, mercury, lead) and dioxins and furans could result from inadequate burning temperatures and retention periods.

Other procedures to consider may include chemical disinfection or sterilization (*i.e.*, irradiation, microwave, autoclave, or hydroclave) followed by secure landfill disposal of residuals. In some cases, following complete disinfection, some wastes may be recycled. For example, recycling by specialized contractors is sometimes arranged after disinfection of thick plastics, such as intravenous bags and tubs, and syringes.

Pharmaceutical wastes require destruction, secure land disposal or return to the manufacturer for destruction through chemical or incineration methods.

Chemical wastes need to be source segregated according to their recycling potential and compatibility; and those which are non-recyclable may require stabilization, neutralization, encapsulation, or incineration.

Hospital wastewater treatment sludges require treatment (*i.e.*, anaerobic digestion, composting, incineration, etc.) which raises temperatures to levels that destroy pathogenic microorganisms.

Radioactive medical therapy and diagnosis in high-income countries are divided into two categories: "open sources" which derive from direct use of the radiochemical substance, and "sealed sources" which involve indirect use of the substance within a sealed apparatus or equipment unit. Only open sources tend to result in radioactive wastes, as sealed sources are returned to the manufacture for recycling when exhausted or no longer required. Radioactive wastes typically include isotopes such as technetium 99, gallium 67, iodine 125, iodine 131, cesium 137, iridium 192, thallium 201, and thallium 204. These wastes are seldom present in low-income and middle-income developing countries, because the hospitals do not have the equipment and technology to generate these wastes. If generated, these wastes should be stored safely until the radioactivity has declined to acceptable levels and then disposed with general refuse to sanitary landfill. The half-lives of commonly used medical radionuclides for therapy, diagnosis, or imaging range from 6 hours to several days. Storage on-site in a secured chamber is typically recommended for a period of 10 half-lives, or for one to two months.

The overall quantity of wastes generated in hospitals varies according to the income level of the country. For developing countries, the data base is limited, but it appears that the following range of quantities is likely:

- ❖ General waste which is not contaminated, and can be handled with general municipal refuse: 1.0 to 2.0 kg/bed/day; and
- ❖ Contaminated medical waste which needs special management, and is considered potentially hazardous: 0.2 to 0.8 kg/bed/day.

Low-income countries would tend to generate medical wastes on the low end of this range, while middle-income countries would tend to generate medical wastes on the upper end of this range. The study area is within a [] income country, based on ranking criteria established by the World Bank and published in its annual development report.

Medical wastes, if not properly managed, pose a risk to the personnel who are handling these wastes, including custodial personnel and waste collectors, as well as to those providing disposal or picking through the wastes for recyclables. There is the danger that syringes will be recovered from transfer depots and disposal sites by waste pickers for recycling (*i.e.*, by drug users). Contaminated containers for collection of medical wastes are not usually dedicated to only one site, but are circulated throughout cities as each skip truck brings an empty container to the hospital or clinic and removes the full one while it covers its daily collection route for general refuse.

Incineration is generally considered the preferred technology for some, if not all, medical wastes. At a minimum, infected tissue, body parts, and laboratory animal carcasses are generally recommended to be incinerated. On-site incinerators operating on a batch basis or regional incinerators operating on a continuous basis are considered appropriate technology. Because of the cost of meeting stringent air pollution control emission standards, many high-income countries are taking steps to steam sterilize, irradiate, chemically disinfect, or gas/vapor sterilize some of the medical wastes.

One hospital incinerator with a capacity of 0.75 tonne/hour, operating on a continuous feed, could cost from $US 0.5 to 1.0 million to implement. Air pollution control systems, if they are added to meet 1995 USA standards, could cost another $ 0.5 to 1.0 million to implement. Incinerators which operate on a batch basis are typically dedicated to one hospital, as their capacity is limited to less than 1 tonne/day. Regional incinerators would typically be designed to operate on a continuous feed basis.

These equipment costs do not include transportation, customs, and setup costs within the study area. Transportation and setup may add about 10 per cent to these costs. If government imports the equipment, especially as it is for waste management purposes, customs may not need to be paid. However, if the private sector is building the facility and needs to import the equipment, customs could add about to these costs. Civil works and land costs which are local costs may add about 30 per cent to these costs.

While the costs/tonne of treatment/destruction are likely to be high (about $100 to $300/tonne depending on the level of pollution control required), the low quantities of medical wastes in developing countries would result in a costs which generally would be less than 1 per cent of the most hospital's operating budget, exclusive of salaries. Therefore, the proper treatment/ destruction facilities are likely to be affordable. Hospitals interviewed in various developing countries have indicated a willingness to pay to cover these costs.

Hospital waste treatment/destruction facilities could be implemented through one or more Design, Build, Own, and Operate (DBOO) or Design, Build, Operate and Transfer (DBOT) concession agreements of 10 to 15 years duration. Or the government could implement the facilities and arrange for service contracts of 2 to 5 years for operation and maintenance. Each hospital would be required to pay tipping fees which fully cover the costs of investment, debt service and operation. As part of the privatization agreement, the company providing the treatment/destruction services could also be awarded the task of also providing collection of the wastes from each hospital and maintaining a manifest system to track the waste from source to ultimate disposal.

Secured sanitary landfill is generally considered the preferred technology for medical wastes which do not require incineration or disinfection, such as packaging materials and general kitchen wastes. Nevertheless, special measures to fence and control access to the area of land filling for medical wastes are essential. No waste picking should be allowed in the secured area. Also, the machinery for compacting refuse should not come in direct contact with the waste. Instead, the waste should be dumped into a trench and a adequate layer of soil dumped over the waste. Only thereafter is it recommendable that the machinery work over the soil covered waste to compact it and grade the surface so that infiltration of rainwater is minimized.

Study Objectives

The feasibility study will assess the technology options for medical waste treatment/destruction. The study will result in recommendations which outline proposed numbers, sizes, and types of medical waste treatment/destruction facilities. Technologies to be considered include incineration, irradiation, chemical disinfection and sterilization.

For purposes of the proposed study on hospital waste management, the following objectives are to be addressed:

- Determine the quantity and character of hazardous medical wastes generated by hospitals and clinics in the study area;
- Evaluate the progress being made in source segregation and develop recommendations for improving the source segregation systems of hospitals and clinics in the study area;
- Estimate the capacity requirements for existing hospital treatment/ destruction facilities for the study area;
- Determine the optimum technology for cost-effective and environmentally

- Based on transport distances and economies of scale, as well as available sites for implementation, determine the number and size of hospital waste treatment/destruction facilities needed;
- Provide a preliminary design, including a typical site layout, and estimate land, capital, operating, and staffing requirements for each of the hospital waste treatment/destruction facilities recommended; and
- Assess the environmental impact issues of implementing each the hospital waste facilities recommended and recommend appropriate mitigative measures to enable the facilities to meet [] environmental requirements.

Scope of Work

Task 1: Waste Quantity and Character

Determine the quantity and character of medical wastes generated in the study area, including pathological, infectious, sharps, pharmaceutical, chemical, aerosol, and radioactive wastes. As part of the effort to make this determination, accomplish the following activities.

Based on the records kept by the solid waste authorities within the study area and the hospitals, determine the volume and weight of medical wastes being collected. If data does not exist, weigh hospital waste loads for a period of at least 4 days.

Visually describe the composition (on a percent wet weight basis) of medical wastes to be managed by treatment/-destruction facilities, such as the contaminated paper products, plastics, fabrics, wood, rubber, cloth, pharmaceutical, tissue, body and bedding materials.

Estimate the calorific heating value of combined mix of medical wastes, on both a dry and wet weight basis, based on the apparent contents of the waste, from the above visual observations. This will involve examining wastes at least large hospitals and examining medical wastes being discharged at disposal sites.

Sample, on an accepted random sampling basis, at least 4 loads of medical waste arriving at disposal sites. Conduct laboratory analyses of the calorific values and moisture contents of the samples. Report results in terms of wet "as received" lower heating value (in kcal/kg and BTU/pound), dry higher heating value (in kcal/kg and BTU/pound), and moisture content (percent on a wet weight basis).

Assess whether there are liquid wastes from the hospitals which could be burned in conjunction with the solid medical wastes and might add to the solvents, strippers, thinners, phenols, resins, and emulsions. Assess whether the addition of these liquid wastes would compromise the air emissions from the proposed treatment/destruction facility. Estimate the quantities of the liquid wastes which could be burned with the solid medical wastes.

Task 2: Source Segregation Systems

Visit at least hospitals to review their systems of medical waste segregation, storage, and disposal. Estimate the percentage of wastes which are being segregated, out of the total being generated. Inspect the storage facilities and estimate the pre-collection volume of medical waste being generated and segregated.

For each of the hospitals visited, estimate the volume/bed/day of refuse. If there are wide variations among the hospitals visited, determine whether the variance is related to compliance with the source segregation system. Estimate the total quantity of medical waste which would be generated in the study area if all hospitals were fully implementing adequate source segregation. Provide an estimated breakdown in terms of the quantity of medical waste requiring: (i) special storage for radiation decay, (ii) treatment/ destruction in a medical waste facility, and (iii) amenable to recovery and recycling.

Task 3: Project Waste Quantities and Characteristics

Based on the economic level of the study area, economic growth projections, population growth projections, trends in hospital waste generation and source segregation, project the quantity and characteristics of medical wastes which are expected to be generated over the next 20 years.

Task 4: Regulatory Requirements

Determine all pollution control standards to be met by a medical waste treatment/destruction facility. Particularly determine the air emission standards which are currently required law and which would be likely to be required in the next 10 years. Assess the corresponding air pollution control requirements for particulates removal, flue gas scrubbing, and dioxin removal. Assess the costs versus the pollution control differences between dry versus wet scrubbing systems.

Outline the environmental permitting, building permitting, and other permitting requirements and procedures which treatment/destruction facilities for medical wastes would need to address. Also outline any public participation or public hearing requirements and procedures. For each requirement, list the lead agency to be contacted. Assess the typical time demands for proposed facilities to obtain permits and address environmental impact assessment and public participation requirements.

Task 5: Treatment/Destruction Options

For the types, quantities and sizes of materials included in the study area's medical wastes, assess alternative technologies and facility sizes for treatment and destruction. The assessment shall compare the alternatives on the basis of capital cost, operating cost, ease of operation, local availability of spare parts, local availability of operational skills, demonstrated reliability, durability, and environmental impacts. The technologies to be considered include: incineration, irradiation, sterilization, and chemical disinfection, and secured landfill. On the basis of this assessment, recommend a process flow for economic and environmentally sound management of medical wastes in the study area.

Task 6: Residuals

For the recommended process flow which would provide treatment/ destruction of study area's medical wastes, assess the quantity and characteristics of residuals. Include assessment of process residuals (such as incinerator ash), as well as pollution control residuals (such as flue gas cleaning sludge, particulates, spent filters, and spent activated carbon).

Task 7: Strategic Location and Sizing

For the recommended treatment/destruction system, assess whether there are significant economies-of-scale to be considered. Also, examine the travel times and distances to drive in the study area from the various centers of medical waste generation to potential locations for treatment/destruction facilities. Also, examine the travel times and distances to drive from the facilities to the location for residuals disposal. Economically analyze whether the study area would be best served by one or more than one treatment/destruction facility. Determine the optimum number and location(s) of facilities.

Task 8: Preliminary Design

Develop a model process flow diagram and site layout for the recommended treatment/destruction facilities. Include treatment processes for wastewater, cooling water, drainage, odor pollution, and air pollution in the model process flow diagram. Include facilities for parking, gate control, weighing loads, administration, worker sanitation and washing/changing, worker cafeteria and training, and truck washing/disinfection in the model site layout. Provide a conceptual floor layout for each of the buildings recommended with the site layout. Assess spatial requirements for the facilities, as a function of their recommended medical waste handling capacities.

Determine the electrical power supply available and the type of fuel (*i.e.*, oil, natural gas) available for operating the facility. Assess the potential for waste-to-energy conversion and which type of energy recovery would be preferred, such as steam, hot water, hot air, thermal liquid, electricity. Outline user requirements, such as steam pressure requirements or hot water requirements.

Task 9: Land and Investment Requirements

Based on the spatial requirements estimated above, determine how much land is required for each of the recommended facilities. Outline the land acquisition issues and constraints which might exist in the study area, including human resettlement issues and constraints. Based on local land values and resettlement costs, estimate the costs of land acquisition.

Assess building requirements, including foundation requirements for the study area. For the model facility designs developed above, and in keeping with local building requirements, develop budgetary estimates for implementation. Include investment costs for site preparation, construction of civil works, stationary equipment, and mobile equipment.

Task 10: Operating Requirements

Determine the cost of consumable supplies and utilities associated with operating the proposed treatment/destruction facilities. For any potential materials recycling and/or energy recovery, estimate the revenue potential based on current market prices.

List the manpower requirements for the proposed treatment/destruction facilities, including managers, planners, administrators, supervisors, operators, guards, and attendants. Estimate the cost of salaries required to operate the facilities.

Based on the cost of consumables, salaries, insurances, registrations, investment depreciation, and debt service, estimate the total annual cost to own, operate and maintain the proposed facilities. Estimate the cost/tonne of waste processed if operating at 70 per cent capacity initially and 90 per cent capacity within 10 years.

Task 11: Environmental Study and Mitigation

Prepare an environmental report which reviews the environmental issues related to the proposed treatment/destruction facilities. These reviews are to be conducted in accordance with local environmental impact assessment guidance, as well as the World's Operational Directive 4.01, "Environmental Assessment". For adverse impacts identified within the reviews, outline mitigative measures which need to be included within the proposed design (including mitigative by wastewater treatment, air pollution control, odor control, etc., processes). Further outline mitigative measures which should be included within the operational procedures. In addition, provide a monitoring program for monitoring throughout implementation and operation activities. If any of the proposed sites for the facilities have inhabitants or tribal nomadic dwellers, address the World Bank's requirements under Operational Directive 4.30, "Involuntary Resettlement" or any other relevant guidance provided by the agencies participating in this project.

Task 12: Implementation Schedule

Develop an anticipated schedule for securing all required permits, including environmental permits, for implementation of the proposed facilities. Include time for public participation, as appropriate.

Develop an implementation schedule for siting, land acquisition, human resettlement, land preparation, construction, training, demonstration, and start-up of the proposed facilities. Include scheduled steps to advertise tenders, evaluate bids, and negotiate contracts.

Task 13: Implementation Strategy

Assess the alternative ways in which the proposed facilities could be implemented and provide adequate discussion of the pros and cons of each alternative to enable decision-making. Include consideration of the following: (a) the local or central government designs, builds, owns and operates the facilities; (b) the local or central government designs, builds and owns the facilities and contracts for operation by government; (c) the local or central government designs, builds and owns the facilities and leases them to the private sector for their operation; (d) the local or central government develops design performance requirements and gives a concession to the private sector to design, build, own, and operate facilities; (e) the local or central government licenses private firms to compete with each other to design, build, own, and operate facilities; or (f) hospitals collectively organize a semi-private enterprise to design, build, own, and operate facilities.

Task 14: Financial Package

Recommend financing arrangements for project implementation. Based on whether the money for implementation is to be borrowed by the hospitals, the government, or invested by the private sector, provide a financial package which would enable final design and procurement activities to begin immediately at the conclusion of the study.

Task 15: Regulatory Framework

Review the existing regulations, strategies, policies, and enforcement practices at the local and central government level concerning the management of medical wastes, and the more general topic of hazardous wastes. Identify limitations and deficiencies in the regulatory framework. Develop specific recommendations on areas which need to be improved within the regulatory framework, so that hospitals, communities, and waste haulers have the appropriate incentives and disincentives to provide proper waste management.

Task 16: Technical Seminar

Provide a seminar to government officials and hospital administrators on the findings of this feasibility study, upon completion of the draft final report. Obtain their review comments during the seminar and address their comments in finalization of the report.

Study Team

The team to conduct the feasibility study will need to have extensive experience in hospital waste management and design of treatment/destruction facilities. The team will need to have practical knowledge of the pros and cons of various hospital waste treatment/destruction options. The team will also need to be familiar with the assessment of technology options under the range of unique skill, management and financial conditions which exist in developing countries. The team will need to be qualified to put together a financial proposal for implementation of the study recommendations. Resumes of the qualifications and experience of the key members of the team will be the key criteria used to evaluate proposals.

Reports

Provide a diagnostic report after completion of Tasks 1 to 4, within 2 months of the commencement date of the contract. Provide an interim report after completion of Tasks 5 to 10, within 4 months of the commencement date of the contract. Provide a final draft report after completion of Tasks 11 to 15, within 6 months of the commencement date of the contract. Conduct the technical seminar required under Task 16, and then issue the final report within 8 months of the commencement date of the contract. Ten copies of each report are to be provided.

11.7 Hospital Waste: An Environmental Hazard and Its Management

Hospital is a place of almighty, a place to serve the patient. Since beginning, the hospitals are known for the treatment of sick persons but we are unaware about the adverse effects of the garbage and filth generated by them on human body and environment. Now it is a well established fact that there are many adverse and harmful effects to the environment including human beings which are caused by the "Hospital waste" generated during the patient care. Hospital waste is a potential health hazard to the health care workers, public and flora and fauna of the area. Hospital acquired infection, transfusion transmitted diseases, rising incidence of Hepatitis B, and HIV, increasing land and water pollution lead to increasing possibility of catching many diseases. Air pollution due to emission of hazardous gases by incinerator such as Furan, Dioxin, Hydrochloric acid etc. have compelled the authorities to think seriously about hospital waste and the diseases transmitted through improper disposal of hospital waste.

A modern hospital is a complex, multidisciplinary system which consumes thousands of items for delivery of medical care and is a part of physical environment. All these products consumed in the hospital leave some unusable leftovers *i.e.* hospital waste. The last century witnessed the rapid mushrooming of hospital in the public and private sector, dictated by the needs of expanding population. The advent and acceptance of "disposable" has made the generation of hospital waste a significant factor in current scenario.

Classification of Hospital Waste

1. *General waste:* Largely composed of domestic or house hold type waste. It is non-hazardous to human beings, *e.g.* kitchen waste, packaging material, paper, wrappers, plastics.
2. *Pathological waste:* Consists of tissue, organ, body part, human foetuses, blood and body fluid. It is hazardous waste.

3. *Infectious waste:* The wastes which contain pathogens in sufficient concentration or quantity that could cause diseases. It is hazardous *e.g.* culture and stocks of infectious agents from laboratories, waste from surgery, waste originating from infectious patients.
4. *Sharps:* Waste materials which could cause the person handling it, a cut or puncture of skin *e.g.* needles, broken glass, saws, nail, blades, scalpels.
5. *Pharmaceutical waste:* This includes pharmaceutical products, drugs, and chemicals that have been returned from wards, have been spilled, are outdated, or contaminated.
6. *Chemical waste:* This comprises discarded solid, liquid and gaseous chemicals *e.g.* cleaning, house keeping, and disinfecting product.
7. *Radioactive waste:* It includes solid, liquid, and gaseous waste that is contaminated with radionucleides generated from in-vitro analysis of body tissues and fluid, in-vivo body organ imaging and tumour localization and therapeutic procedures.

Biomedical Waste

Any solid, fluid and liquid or liquid waste, including it's container and any intermediate product, which is generated during the diagnosis, treatment or immunisation of human being or animals, in research pertaining thereto, or in the production or testing of biological and the animal waste from slaughter houses or any other similar establishment. All biomedical waste are hazardous. In hospital it comprises of 15 per cent of total hospital waste.

Rationale of Hospital Waste Management

Hospital waste management is a part of hospital hygiene and maintenance activities. In fact only 15 per cent of hospital waste *i.e.* "Biomedical waste" is hazardous, not the complete. But when hazardous waste is not segregated at the source of generation and mixed with no hazardous waste, then 100 per cent waste becomes hazardous. The question then arises that what is the need or rationale for spending so much resources in terms of money, man power, material and machine for management of hospital waste? The reasons are:

- ❖ Injuries from sharps leading to infection to all categories of hospital personnel and waste handler.
- ❖ Nosocomial infections in patients from poor infection control practices and poor waste management.
- ❖ Risk of infection outside hospital for waste handlers and scavengers and at time general public living in the vicinity of hospitals.

- Risk associated with hazardous chemicals, drugs to persons handling wastes at all levels.
- "Disposable" being repacked and sold by unscrupulous elements without even being washed.
- Drugs which have been disposed of, being repacked and sold off to unsuspecting buyers.
- Risk of air, water and soil pollution directly due to waste, or due to defective incineration emissions and ash.

Approach for Hospital Waste Management

Based on Bio-medical Waste (Management and Handling) Rules 1998, notified under the Environment Protection Act by the Ministry of Environment and Forest (Government of India).

1. Segregation of Waste

Segregation is the essence of waste management and should be done at the source of generation of Bio-medical waste *e.g.* all patient care activity areas, diagnostic services areas, operation theatres, labour rooms, treatment rooms etc. The responsibility of segregation should be with the generator of biomedical waste *i.e.* doctors, nurses, technicians etc. (medical and paramedical personnel). The biomedical waste should be segregated as per categories mentioned in the rules.

2. Collection of Bio-medical Waste

Collection of bio-medical waste should be done as per Bio-medical waste (Management and Handling) Rules. At ordinary room temperature the collected waste should not be stored for more than 24 hours.

Type of container and colour code for collection of bio-medical waste.

Category	*Waste class*	*Type of container*	*Colour*
1.	Human anatomical waste	Plastic	Yellow
2.	Animal waste	-do-	-do-
3.	Microbiology and Biotechnology waste	-do-	Yellow/Red
4.	Waste sharp	Plastic bag puncture proof containers	Blue/White Translucent
5.	Discarded medicines and Cytotoxic waste	Plastic bags	Black
6.	Solid (biomedical waste)	-do-	Yellow
7.	Solid (plastic)	Plastic bag puncture proof containers	Blue/White Translucent
8.	Incineration waste	Plastic bag	Black
9.	Chemical waste (solid)	-do-	-do-

3. Transportation

Within hospital, waste routes must be designated to avoid the passage of waste through patient care areas. Separate time should be earmarked for transportation of bio-medical waste to reduce chances of it's mixing with general waste. Desiccated wheeled containers, trolleys or carts should be used to transport the waste/plastic bags to the site of storage/treatment.

Trolleys or carts should be thoroughly cleaned and disinfected in the event of any spillage. The wheeled containers should be so designed that the waste can be easily loaded, remains secured during transportation, does not have any sharp edges and is easy to clean and disinfect. Hazardous biomedical waste needing transport to a long distance should be kept in containers and should have proper labels. The transport is done through desiccated vehicles specially constructed for the purpose having fully enclosed body, lined internally with stainless steel or aluminium to provide smooth and impervious surface which can be cleaned. The drivers compartment should be separated from the load compartment with a bulkhead. The load compartment should be provided with roof vents for ventilation.

4. Treatment of Hospital Waste

Treatment of waste is required:

- To disinfect the waste so that it is no longer the source of infection.
- To reduce the volume of the waste.
- Make waste unrecognizable for aesthetic reasons.
- Make recycled items unusable.

4.1 General Waste

The 85 per cent of the waste generated in the hospital belongs to this category. The, safe disposal of this waste is the responsibility of the local authority.

4.2 Bio-medical Waste: 15 Per Cent of Hospital Waste

- *Deep burial:* The waste under category 1 and 2 only can be accorded deep burial and only in cities having less than 5 lakh population.
- *Autoclave and microwave treatment* Standards for the autoclaving and microwaving are also mentioned in the Biomedical waste (Management and Handling) Rules 1998. All equipment installed/ shared should meet these specifications. The waste under category

3,4,6,7 can be treated by these techniques. Standards for the autoclaving are also laid down.

- *Shredding:* The plastic (IV bottles, IV sets, syringes, catheters etc.), sharps (needles, blades, glass etc) should be shredded but only after chemical treatment/microwaving/autoclaving. Needle destroyers can be used for disposal of needles directly without chemical treatment.
- *Secured landfill::* The incinerator ash, discarded medicines, cytotoxic substances and solid chemical waste should be treated by this option.
- *Incineration:* The incinerator should be installed and made operational as per specification under the BMW rules 1998 and a certificate may be taken from CPCB/State Pollution Control Board and emission levels etc should be defined. In case of small hospitals, facilities can be shared. The waste under category 1, 2, 3, 5, 6 can be incinerated depending upon the local policies of the hospital and feasibility. The polythene bags made of chlorinated plastics should not be incinerated.

It may be noted that there are options available for disposal of certain category of waste. The individual hospital can choose the best option depending upon the facilities available and its financial resources. However, it may be noted that depending upon the option chosen, correct colour of the bag needs to be used.

5. Safety Measures

5.1 All the generators of bio-medical waste should adopt universal precautions and appropriate safety measures while doing therapeutic and diagnostic activities and also while handling the bio-medical waste.

5.2 *It should be ensured that*:

- Drivers, collectors and other handlers are aware of the nature and risk of the waste.
- Written instructions, provided regarding the procedures to be adopted in the event of spillage/accidents.
- Protective gears provided and instructions regarding their use are given.
- Workers are protected by vaccination against tetanus and hepatitis B.

6. Training

- Each and every hospital must have well planned awareness and training programme for all category of personnel including administrators (medical, paramedical and administrative).

- All the medical professionals must be made aware of Bio-medical Waste (Management and Handling) Rules 1998.
- To institute awards for safe hospital waste management and universal precaution practices.
- Training should be conducted to all categories of staff in appropriate language/medium and in an acceptable manner.

7. Management and Administration

Heads of each hospital will have to take authorization for generation of waste from appropriate authorities as notified by the concerned State/U.T. Government, well in time and to get it renewed as per time schedule laid down in the rules. Each hospital should constitute a hospital waste management committee, chaired by the head of the Institute and having wide representation from all major departments. This committee should be responsible for making Hospital specific action plan for hospital waste management and its supervision, monitoring and implementation. The annual reports, accident reports, as required under BMW rules should be submitted to the concerned authorities as per BMW rules format.

8. Measures for Waste Minimization

As far as possible, purchase of reusable items made of glass and metal should be encouraged. Select non PVC plastic items. Adopt procedures and policies for proper management of waste generated, the mainstay of which is segregation to reduce the quantity of waste to be treated. Establish effective and sound recycling policy for plastic recycling and get in touch with authorised manufactures.

9. Coordination between Hospital and Outside Agencies

- *Municipal authority*: As quite a large percentage of waste (in India upto 85%), generated in Indian hospitals, belong to general category (non-toxic and non-hazardous), hospital should have constant interaction with municipal authorities so that this category of waste is regularly taken out of the hospital premises for land fill or other treatment.
- *Co-ordination with Pollution Control Boards*: Search for better methods technology, provision of facilities for testing, approval of certain models for hospital use in conformity with standards 'aid down.

12

Health, Safety and Environment

Risk, the official journal of the Risk Assessment & Policy Association, is a refereed, interdisciplinary periodical exploring public and private efforts to manage science and technology for net reduction in the probability, severity and adverse quality of health, safety and environmental impacts of natural and artificial hazards.

12.1 Health, Safety & Environment Policy of BRIT

BRIT production facilities are under specified regulatory control with respect to radiation, environmental and industrial safety. The regulatory requirement of radiation and environment safety is formulated on the basis of well-established international/national radiation protection standards and to protect staff as well as public from the effects of ionizing radiation.

A comprehensive and systematic environmental monitoring is carried out in order to monitor the effluents and flue gas discharges. The laboratory air is filtered through high efficiency particulate filters and discharged through stack. The radioactivity concentrations in these effluents are insignificant and are well within the limits set by AERB. Liquid effluents after proper dilution are discharged into the municipal sewers after monitoring. Basic IAEA/AERB guidelines for discharge control and relevant acts in India are adhered to enforcing the environmental protection.

For ensuring radiation safety, personnel monitoring is carried out with TLD badges and bio-assay/whole body counting. Radiation area monitoring is carried out on a continuous basis for ensuring personnel safety. Adequate radiation shielding is provided in the working plants. Annual medical check up of radiation workers is carried out with follow-up records.

The industrial safety programmes are implemented and built into the system such that the rate of occurrence of industrial accidents is extremely low. Accident Prevention Methodology is implemented with prevention campaign and making the people safety conscious.

12.2 National Policy on Safety, Health and Environment at Work Place

1. Preamble

1.1 The Constitution of India enshrines detailed provisions for the rights of the citizens and other persons and for the principles to be followed by the States in the governance of the country labeled as "Directive Principles of State Policy".

1.2 These Directive Principles provide a) for securing the health and strength of employees, men and women, b) that the tender age of children are not abused c) that citizens are not forced by economic necessity to enter avocations unsuited to their age or strength, d) just and humane conditions of work and maternity relief are provided and e) that the Government shall take steps, by suitable legislation or in any other way, to secure the participation of employee in the management of undertakings, establishments or other organisations engaged in any industry.

1.3 On the basis of these Directive Principles, and international instruments, the Government of India declares its policy, priorities and strategies, purposes through the exercise of its power. The Government is committed to regulate all economic activities among the several states within the country and with foreign nations for management of safety and health risks at workplaces and to provide measures for protection of national assets and for the general welfare to assure, as far as possible, every working man and woman in the nation safe and healthy working condition and to preserve human resources.

1.4 The formulation of policy, priorities and strategies in occupational safety, health and environment at work places, is undertaken by national authorities in consultation with social partners for agreement and involvement for ensuring set goals/objectives.

1.5 Government of India firmly believes that without safe, clean environment and healthful working conditions, social justice and economic growth cannot be achieved.

1.6 The changing job patterns and working relationships, the rise in self employment, greater sub-contracting, outsourcing of work, homework and the increasing number of employees working away from their establishment, pose problem of management of occupational safety and health risks at workplaces. New safety hazards and health risks will be appearing along with the transfer and adoption of new technologies. In addition, many of the well known conventional hazards will continue to be present at the workplace till the risks arising from exposure to these hazards are brought under adequate control.

1.7 Particular attention needs to be paid to the hazardous occupations and of employees in precarious conditions such as migrant employees and various vulnerable groups of employees.

1.8 The increasing use of chemicals, exposure to physical, chemical and biological agents with hazard potential unknown to people; the indiscriminate use of agro-chemicals including pesticides, agricultural machineries and equipment, and their impact on health and safety of exposed population; industries with major accident risks; effects of computer controlled technologies and alarming influence of stress at work in many modern jobs may pose serious safety, health and environmental risks.

1.9 The fundamental purpose of this National Policy on Safety, Health, and Environment at workplace, is not only to eliminate the incidence of work related injuries, diseases, fatalities, disaster and loss of national assets and ensuring achievement of a high level of occupational safety, health and environment performance but also to enhance the well-being of the employee and society, at large.

2. Goals

With a view to improve the occupational safety and health performance year by year, it is essential to—

2.1 providing a statutory framework on OSH in respect of all sectors of economic activities, designing suitable control systems of compliance, enforcement and incentives for better compliance.

2.2 providing administrative and technical support services.

2.3 providing a system of incentives to employers and employees to achieve higher health and safety standards.

2.4 establishing and developing the research and development capability in emerging areas of risk and effective control measures.

2.5 developing a proper interface between the work and the human resource through a system of skill improvement.

2.6 focusing prevention strategies and monitoring performance through improved data collection system on work related injury and disease.

2.7 developing and providing required technical manpower and knowledge in the areas of safety, health and environment at workplaces in different sectors.

3. Objectives

3.1 The policy seeks to bring the national objectives into focus as a step towards improvement in safety, health and environment at workplace performance. The objectives are to achieve:

a) Continuous reduction in the incidence of work related injuries, fatalities, diseases, disaster and loss of national assets.
b) Continuous reduction in the cost of work place injuries and diseases.
c) Extend coverage of work related injuries, fatalities, and diseases for a more comprehensive data base as a means of better performance and monitoring.
d) d) Continuous enhancement of community awareness regarding safety, health and environment at workplace related areas.

4. Action Programme

For the purpose of achieving the above referred objectives and goals, the Government of India draws out the action programme referred hereunder:

4.1. Enforcement

4.1.1 by providing an effective enforcement program;

4.1.2 by effectively enforcing all applicable laws and regulations concerning safety, health and environment at workplaces in all economic activities with such technical variations as may be necessary for which there shall be adequate and qualified inspection services.

4.1.3 By creating a "National Safety, Health and Environment at Workplace Fund" through cess to enable the effective implementation of the policy.

4.1.4 by enabling that employers, employees and others have separate but complementary responsibilities and rights with respect to achieving safe and healthful working conditions;

4.1.5 by amending progressively the existing laws dealing with safety, health and environment in line with the international instruments.

4.1.6 by monitoring the adoption of national standards by regulatory authorities.

4.1.7 by facilitating the sharing of best practices and learning between national and international regulatory authorities.

4.1.8 by developing new enforcement methods including innovative sanctions that encourage and ensure improved workplace performance.

4.1.9 by progressively promulgating a General Enabling Legislation on Safety, Health And Environment At Workplaces.

4.2 National Standards

4.2.1 by appropriately developing standards, codes of practices and manuals on safety, health and environment for uniformity at the national level in all economic activities consistent with international standards and implementation by the stake holders in true spirit.

4.3 Compliance

4.3.1 by encouraging the appropriate government to assume the fullest responsibility for the administration and enforcement of occupational safety, health and environment at workplace laws by providing grants to assist in identifying their needs and responsibilities in the area of safety, health and environment at workplace, to develop plans and programmes in accordance with the provisions of the Acts, and to conduct experimental and demonstration projects in connection therewith;

4.3.2 by calling upon the cooperation of social partners in supervision of application of legislations and regulations relating to safety, health and environment at work place.

4.3.3 by developing guidance on OHS management systems, strengthening voluntary actions and establishing auditing mechanisms which can test and authenticate management systems.

4.3.4 by providing specific measures to prevent catastrophes, and to co-ordinate and make coherent the actions to be taken at different levels, particularly in the industrial zones with high potential risks for employees and the surrounding population are situated;

4.3.5 by recognizing the best safety and health efforts and facilitating others to emulate their examples.

4.3.6 by encouraging to adopt and commit to "Responsible Care" and/ or "Corporate Social Responsibility" to improve safety, health and environment at workplace performance.

4.3.7 by constituting National Accreditation Board to recognize institutions, professionals and services relating to safety, health and environment at workplace for uniformity.

4.4 Awareness

4.4.1 by increasing awareness on safety, health and environment at workplace through appropriate means.

4.4.2 by providing forums for consultations with employers' representatives, employees representatives, and community on matters of national concern relating to safety, health and environment at work place with the overall objective of creating awareness and enhancing national productivity.

4.4.3 by encouraging joint labour-management efforts to preserve, protect and promote national assets and to eliminate injuries and diseases arising out of employment.

4.4.4 by maximizing gains from the substantial investment in awareness campaigns by sharing experience and learning.

4.4.5 by including safety, health and environment at work place in schools, technical, medical, professional and vocational courses.

4.4.6 by securing good liaison arrangements with the International organisations.

4.4.7 by providing medical criteria which will assure in so far as practicable that no employee will suffer diminished health, functional capacity, or life expectancy as a result of his work experience and that in the event of such occupational diseases having been contracted, suitably compensated.

4.4.8 by establishing occupational health services aimed at protection and promotion of employee health and improvement of working conditions and by providing employee in different sectors of economic activities access to these services.

4.4.9 by providing for appropriate reporting procedures with respect to OSH to help achieve the objectives and to accurately describe the nature of the occupational safety and health problem with a view to carry out national project study, surveys to identify problem areas and pragmatic strategies.

4.5 Research and Development

4.5.1 by providing for research in the field of safety, health and environment at workplace, including the social and psychological factors involved, and by developing innovative methods, techniques, and approaches for dealing with safety, health and environment at workplace problems which will help in establishing standards;

4.5.2 by exploring ways to discover latent diseases, establishing causal connections between diseases and work environmental conditions, and conducting other research relating to safety, health and environment at workplace problems.

4.6 Occupational Safety and Health Skills Development

4.6.1 by building upon advances already made through employer and employee initiative for providing safe and healthful working conditions;

4.6.2 by providing for training programs to increase the number and competence of personnel engaged in the field of occupational safety, health and environment at workplace;

4.6.3 by integrating safety, health and environment at workplace, and by arranging training programmes for vocational, professional and enforcement agencies including persons employed.

4.6.4 by providing information and advice, in an appropriate manner, to employers and employee organisations, with a view to eliminating hazards or reducing them as far as practicable;.

4.7 Data Collection

4.7.1 by compiling statistics relating to safety, health and environment at work places, prioritizing key issues for action, conducting national studies/ surveys/projects through governmental and non-governmental organisations and ensuring compliance.

4.7.2 by reinforcing and sharing of national occupational safety, health and environment at work place information amongst different stake holders through a national network system on OSH.

4.7.3 by extending data coverage relevant to work-related injury and disease, including measures of exposure, and occupational groups that are currently excluded, such as self-employed people.

4.7.4 by extending data systems to allow timely reporting and provision of information.

4.7.5 by exploring partnerships to address areas where there is overlap between public health and occupational risk.

4.8 Practical Guidance

4.8.1 by providing practical guidance and encouraging employers and employees in their efforts to reduce the incidence of occupational safety and health risks at their places of employment, and to stimulate employers and employees to institute new programmes and to perfect existing programs for providing safe and healthful working conditions;

4.8.2 by giving effect to the decisions by a Courts of Law or other tribunals involving question of principles relating to the application of safety, health and environment at work.

4.8.3 by developing the means for improved access to information.

4.8.4 by facilitating sharing of practical guidance developed within industry sectors and jurisdictions.

4.9 Incentives

4.9.1 by innovative financial and non-financial incentives.

4.10 Review

4.10.1 An initial review shall be carried out to ascertain the current status of safety, health and environment at workplace in all economic activities.

4.10.2 National Policy and the action programme shall be reviewed at least once in 5 years or earlier if felt necessary to assess relevance of the National Goals and objectives.

5. Summary

5.1 There is a need to develop the co-operation of social partners to meet the challenges ahead in the assessment and control of workplace risks by mobilizing local resources and extending protection to under-served working population and vulnerable groups where social protection is meager.

5.2 We are committed to review the National Policy on Safety, Health and Environment at Workplace and legislation under tripartite consultation; improve enforcement, compilation and analysis of statistics; develop special programmes for hazardous occupations and other specific sectors; set up training mechanisms; create nation-wide awareness; arrange for the mobilization of available resources and expertise.

5.3 The National Policy and programme envisages total commitment and demonstration by all concerned stake holders such as government and social partners. India will certainly and steadily march towards economic prosperity through dedicated and concerted efforts consistent with the requirements of safety, health and environment at work place thereby improving the quality of work-life.

12.3 Health, Safety & Environment: An Overview

Workplace for All

- ❖ Global Unions deplore the fact that over two million women and men die each year from unsustainable froms of work and 160 million more become victims of work-related diseases. They also deplore the conditions fostered by globalisation which induce the replacement of safe and healthy workplaces in one part of the world by more dangerous working environments in others.

- Social dumping inherent in the exporting of technology and work processes, innovation, machines, commodities, and chemicals or chemical products for use in workplaces of recipient countries must be addressed as a matter of priority.
- By adopting social and employment linkages to economic and environment planning for the next decade, the World Summit on Sustainable Development (WSSD) has created new opportunities for trade unions to better introduce poverty, equity and workplace issues within national and local plans to address the world's current destructive production and consumption patterns.
- Trade unions are taking leading roles to better integrate sustainable development with occupational health and safety for workers and to strengthen the bases for promoting forms of Decent Work and the well-being for workers and communities. This is in keeping with the commitment to integrate the three pillars of sustainable development (economic, environmental and social) within public and trade union programmes of work.
- Trade unions are taking leading roles to better integrate sustainable development with occupational health and safety for workers and to strengthen the bases for promoting forms of Decent Work and the well-being for workers and communities. This is in keeping with the commitment to integrate the three pillars of sustainable development (economic, environmental and social) within public and trade union programmes of work.

13

Theoretical Chemistry, Quantum Chemistry and Computational Chemistry

13.1 Theoretical Chemistry

Theoretical chemistry involves the use of physics to explain or predict chemical phenomena. In recent years, it has consisted primarily of quantum chemistry, *i.e.*, the application of quantum mechanics to problems in chemistry. Theoretical chemistry may be broadly divided into electronic structure, dynamics, and statistical mechanics. In the process of solving the problem of predicting chemical reactivities, these may all be invoked to various degrees. Other "miscellaneous" research areas in theoretical chemistry include the mathematical characterization of bulk chemistry in various phases (*e.g.* the study of chemical kinetics) and the study of the applicability of more recent math developments to the basic areas of study (*e.g.* for instance the possible application of principles of topology to the study of electronic structure.) The latter area of theoretical chemistry is sometimes referred to as mathematical chemistry.

Much of this may be categorized as computational chemistry, although computational chemistry usually refers to the application of theoretical chemistry in an applied setting, usually with some approximation scheme such as certain types of post Hartree-Fock, Density Functional Theory, semiempirical methods (like for instance PM3) or force field methods. Some chemical theorists apply statistical mechanics to provide a bridge between the microscopic phenomena of the quantum world and the macroscopic bulk properties of systems.

Theoretical attacks on chemical problems go back to the earliest days, but until the formulation of the Schrödinger equation by the Austrian physicist Erwin Schrödinger, the techniques available were rather crude and speculative. Currently, much more sophisticated theoretical approaches, based on Quantum Field Theory and Nonequilibrium Green Function Theory are in vogue.

Branches of Theorietical Chemistry

- *Quantum chemistry*: The application of quantum mechanics to chemistry
- *Computational chemistry*: The application of computer codes to chemistry
- *Molecular modelling*: Methods for modelling molecular structures without necessarily referring to quantum mechanics. Examples are molecular docking, protein-protein docking, drug design, combinatorial chemistry.
- *Molecular dynamics*: Application of classical mechanics for simulating the movement of the nuclei of an assembly of atoms and molecules.
- *Molecular mechanics*: Modelling of the intra- and inter-molecular interaction potential energy surfaces via a sum of interaction forces.
- *Mathematical chemistry*: Discussion and prediction of the molecular structure using mathematical methods without necessarily referring to quantum mechanics.
- *Theoretical chemical kinetics*: Theoretical study of the dynamical systems associated to reactive chemicals and their corresponding differential equations.

Closely Related Disciplines

Historically, the major field of application of theoretical chemistry has been in the following fields of research:

- *Atomic physics*: The discipline dealing with electrons and atomic nuclei.
- *Molecular physics*: The discipline of the electrons surrounding the molecular nuclei and of movement of the nuclei. This term usually refers to the study of molecules made of a few atoms in the gas phase. But some consider that molecular physics is also the study of bulk properties of chemicals in terms of molecules.
- *Physical chemistry and chemical physics*: Chemistry investigated via physical methods like laser techniques, scanning tunneling microscope, etc. The formal distinction between both fields is that physical chemistry is a branch of chemistry while chemical physics is a branch of physics. In practice this distinction is quite vague.
- *Many-body theory*: The discipline studying the effects which appear in systems with large number of constituents. It is based on quantum physics—mostly second quantization formalism—and quantum electrodynamics.

Hence, the theoretical chemistry discipline is sometimes seen as a branch of those fields of research. Nevertheless, more recently, with the rise of the density functional theory and other methods like molecular mechanics, the range of application has been extended to chemical systems which are relevant to other fields of chemistry and physics like biochemistry, condensed matter physics, nanotechnology or molecular biology.

13.2 Quantum Chemistry

Quantum chemistry is a branch of theoretical chemistry, which applies quantum mechanics and quantum field theory to address issues and problems in chemistry. The description of the electronic behavior of atoms and molecules as pertaining to their reactivity is one of the applications of quantum chemistry. Quantum chemistry lies on the border between chemistry and physics, and significant contributions have been made by scientists from both fields. It has a strong and active overlap with the field of atomic physics and molecular physics, as well as physical chemistry.

Quantum chemistry mathematically describes the fundamental behavior of matter at the molecular scale. It is, in principle, possible to describe all chemical systems using this theory. In practice, only the simplest chemical systems may realistically be investigated in purely quantum mechanical terms, and approximations must be made for most practical purposes (*e.g.*, Hartree-Fock, post Hartree-Fock or Density functional theory, see computational chemistry for more details). Hence a detailed understanding of quantum mechanics is not necessary for most chemistry, as the important implications of the theory (principally the orbital approximation) can be understood and applied in simpler terms.

In quantum mechanics (several applications in computational chemistry and quantum chemistry), the Hamiltonian, or the physical state, of a particle can be expressed as the sum of two operators, one corresponding to kinetic energy and the other to potential energy. The Hamiltonian in the Schrödinger wave equation used in quantum chemistry does not contain terms for the spin of the electron.

Solutions of the Schrödinger equation for the hydrogen atom gives the form of the wave function for atomic orbitals, and the relative energy of the various orbitals. The orbital approximation can be used to understand the other atoms *e.g.* helium, lithium and carbon.

History

The *history of quantum chemistry* essentially began with the 1838 discovery of cathode rays by Michael Faraday, the 1859 statement of the black body radiation problem by Gustav Kirchhoff, the 1877 suggestion by Ludwig Boltzmann that the energy states of a physical system could be discrete, and the 1900 quantum hypothesis by Max Planck that any energy radiating atomic system can theoretically be divided into a number of discrete energy elements *å* such that each of these energy elements is proportional to the frequency *í* with which they each individually radiate energy, as defined by the following formula:

$$\in = h\nu$$

where h is a numerical value called Planck's Constant. Then, in 1905, to explain the photoelectric effect (1839), *i.e.*, that shining light on certain materials can function to eject electrons from the material, Albert Einstein postulated, based on Planck's quantum hypothesis, that light itself consists of individual quantum particles, which later came to be called photons (1926). In the years to follow, this theoretical basis slowly began to be applied to chemical structure, reactivity, and bonding.

Electronic Structure

The first step in solving a quantum chemical problem is usually solving the Schrödinger equation (or Dirac equation in relativistic quantum chemistry) with the electronic molecular Hamiltonian. This is called determining the *electronic structure* of the molecule. It can be said that the electronic structure of a molecule or crystal implies essentially its chemical properties.

Wave Model

The foundation of quantum mechanics and quantum chemistry is the *wave model*, in which the atom is a small, dense, positively charged nucleus surrounded by electrons. Unlike the earlier Bohr model of the atom, however, the wave model describes electrons as "clouds" moving in orbitals, and their positions are represented by probability distributions rather than discrete points. The strength of this model lies in its predictive power. Specifically, it predicts the pattern of chemically similar elements found in the periodic table. The wave model is so named because electrons exhibit properties (such as interference) traditionally associated with waves.

Valence Bond

Although the mathematical basis of quantum chemistry had been laid by Schrödinger in 1926, it is generally accepted that the first true calculation in quantum chemistry was that of the German physicists Walter Heitler and Fritz London on the hydrogen (H_2) molecule in 1927. Heitler and London's method was extended by the American theoretical physicist John C. Slater and the American theoretical chemist Linus Pauling to become the *Valence-Bond (VB)* [or *Heitler-London-Slater-Pauling (HLSP)*] method. In this method, attention is primarily devoted to the pairwise interactions between atoms, and this method therefore correlates closely with classical chemists' drawings of bonds.

Molecular Orbital

An alternative approach was developed in 1929 by Friedrich Hund and Robert S. Mulliken, in which electrons are described by mathematical functions delocalized over an entire molecule. The *Hund-Mulliken* approach or *molecular orbital (MO) method* is less intuitive to chemists, but has turned out capable of predicting spectroscopic properties better than the VB method. This approach is the conceptional basis of the *Hartree-Fock method* and further post Hartree-Fock methods.

Density Functional Theory

The *Thomas-Fermi model* was developed independently by Thomas and Fermi in 1927. This was the first attempt to describe many-electron systems on the basis of electronic density instead of wave functions, although it was not very successful in the treatment of entire molecules. The method did provide the basis for what is now known as *density functional theory*. Though this method is less developed than post Hartree-Fock methods, its lower computational requirements allow it to tackle larger polyatomic molecules and even macromolecules, which has made it the most used method in computational chemistry at present.

Chemical Dynamics

A further step can consist of solving the Schrödinger equation with the total molecular Hamiltonian in order to study the motion of molecules. Direct solution of the Schrödinger equation is called *quantum molecular dynamics*, within the semiclassical approximation *semiclassical molecular dynamics*, and within the classical mechanics framework *molecular dynamics (MD)*. Statistical approaches, using for example Monte Carlo methods, are also possible.

Adiabatic Chemical Dynamics

In *adiabatic dynamics*, interatomic interactions are represented by single scalar potentials called potential energy surfaces. This is the Born-Oppenheimer approximation introduced by Born and Oppenheimer in 1927. Pioneering applications of this in chemistry were performed by Rice and Ramsperger in 1927 and Kassel in 1928, and generalized into the RRKM theory in 1952 by Marcus who took the transition state theory developed by Eyring in 1935 into account. These methods enable simple estimates of unimolecular reaction rates from a few characteristics of the potential surface.

Non-adiabatic Chemical Dynamics

Non-adiabatic dynamics consists of taking the interaction between several coupled potential energy surface (corresponding to different electronic quantum states of the molecule). The coupling terms are called *vibronic couplings*. The pioneering work in this field was done by Stueckelberg, Landau, and Zener in the 1930s, in their work on what is now known as the Landau-Zener transition. Their formula allows the transition probability between two diabatic potential curves in the neighborhood of an avoided crossing to be calculated.

Quantum Chemistry and Quantum Field Theory

The application of quantum field theory (QFT) to chemical systems and theories has become increasingly common in the modern physical sciences. One of the first and most fundamentally explicit appearances of this is seen in the theory of the photomagneton. In this system, plasmas, which are ubiquitous in both physics and chemistry, are studied in order to determine the basic quantization of the underlying bosonic field. However, quantum field theory is of interest in many fields of chemistry, including: nuclear chemistry, astro chemistry, sono chemistry, and quantum hydrodynamics. Field theoretic methods have also been critical in developing the ab initio Effective Hamiltonian theory of semi-empirical pi-electron methods.

13.3 Computational Chemistry

Computational chemistry is a branch of chemistry that uses computers to assist in solving chemical problems. It uses the results of theoretical chemistry, incorporated into efficient computer programs, to calculate the structures and properties of molecules and solids. While its results normally complement the information obtained by chemical experiments, it can in some cases predict hitherto unobserved chemical phenomena. It is widely used in the design of new drugs and materials.

Examples of such properties are structure (*i.e.* the expected positions of the constituent atoms), absolute and relative (interaction) energies, electronic charge distributions, dipoles and higher multipole moments, vibrational frequencies, reactivity or other spectroscopic quantities, and cross sections for collision with other particles.

The methods employed cover both static and dynamic situations. In all cases the computer time and other resources (such as memory and disk space) increase rapidly with the size of the system being studied. That system can be a single molecule, a group of molecules, or a solid. Computational chemistry methods range from highly accurate to very approximate; highly accurate methods are typically feasible only for small systems. Ab initio methods are based entirely on theory from first principles. Other (typically less accurate) methods are called empirical or semi-empirical because they employ experimental results, often from acceptable models of atoms or related molecules, to approximate some elements of the underlying theory.

Both ab initio and semi-empirical approaches involve approximations. These range from simplified forms of the first-principles equations that are easier or faster to solve, to approximations limiting the size of the system (for example, Periodic boundary conditions), to fundamental approximations to the underlying equations that are required to achieve any solution to them at all. For example, most ab initio calculations make the Born-Oppenheimer approximation, which greatly simplifies the underlying Schroedinger Equation by freezing the nuclei in place during the calculation. In principle, ab initio methods eventually converge to the exact solution of the underlying equations as the number of approximations is reduced. In practice, however, it is impossible to eliminate all approximations, and residual error inevitably remains. The goal of computational chemistry is to minimize this residual error while keeping the calculations tractable.

History

Building on the founding discoveries and theories in the history of quantum mechanics, the first theoretical calculations in chemistry were those of Walter Heitler and Fritz London in 1927. The books that were influential in the early development of computational quantum chemistry include: Linus Pauling and E. Bright Wilson's 1935 *Introduction to Quantum Mechanics—with Applications to Chemistry*, Eyring, Walter and Kimball's 1944 *Quantum*

Chemistry, Heitler's 1945 *Elementary Wave Mechanics— with Applications to Quantum Chemistry*, and later Coulson's 1952 textbook *Valence*, each of which served as primary references for chemists in the decades to follow.

With the development of efficient computer technology in the 1940s, the solutions of elaborate wave equations for complex atomic systems began to be a realizable objective. In the early 1950s, the first semi-empirical atomic orbital calculations were carried out. Theoretical chemists became extensive users of the early digital computers. A very detailed account of such use in the United Kingdom is given by Smith and Sutcliffe. The first ab initio Hartree-Fock calculations on diatomic molecules were carried out in 1956 at MIT, using a basis set of Slater orbitals. For diatomic molecules, a systematic study using a minimum basis set and the first calculation with a larger basis set were published by Ransil and Nesbet respectively in 1960. The first polyatomic calculations using Gaussian orbitals were carried out in the late 1950s. The first configuration interaction calculations were carried out in Cambridge on the EDSAC computer in the 1950s using Gaussian orbitals by Boys and coworkers. By 1971, when a bibliography of ab initio calculations was published, the largest molecules included were naphthalene and azulene. Abstracts of many earlier developments in ab initio theory have been published by Schaefer.

In 1964, Hückel method calculations (using a simple linear combination of atomic orbitals (LCAO) method for the determination of electron energies of molecular orbitals of ð electrons in conjugated hydrocarbon systems) of molecules ranging in complexity from butadiene and benzene to ovalene, were generated on computers at Berkeley and Oxford. These empirical methods were replaced in the 1960s by semi-empirical methods such as CNDO.

In the early 1970s, efficient ab initio computer programs such as ATMOL, GAUSSIAN, IBMOL, and POLYAYTOM, began to be used to speed up ab initio calculations of molecular orbitals. Of these four programs, only GAUSSIAN, now massively expanded, is still in use, but many other programs are now in use. At the same time, the methods of molecular mechanics, such as MM2, were developed, primarily by Norman Allinger.

One of the first mentions of the term "computational chemistry" can be found in the 1970 book *Computers and Their Role in the Physical Sciences* by Sidney Fernbach and Abraham Haskell Taub, where they state "It seems, therefore, that 'computational chemistry' can finally be more and more of a reality." During the 1970s, widely different methods began to be seen as part of a new emerging discipline of *computational chemistry*. The *Journal of Computational Chemistry* was first published in 1980.

Concepts

The term *theoretical chemistry* may be defined as a mathematical description of chemistry, whereas *computational chemistry* is usually used when a mathematical method is sufficiently well developed that it can be automated for implementation on a computer. Note that the words *exact* and *perfect* do not appear here, as very few aspects of chemistry can be computed exactly. However, almost every aspect of chemistry can be described in a qualitative or approximate quantitative computational scheme.

Molecules consist of nuclei and electrons, so the methods of quantum mechanics apply. Computational chemists often attempt to solve the non-relativistic Schrödinger equation, with relativistic corrections added, although some progress has been made in solving the fully relativistic Dirac equation. In principle, it is possible to solve the Schrödinger equation in either its time-dependent or time-independent form, as appropriate for the problem in hand; in practice, this is not possible except for very small systems. Therefore, a great number of approximate methods strive to achieve the best trade-off between accuracy and computational cost. Accuracy can always be improved with greater computational cost. Significant errors can present themselves in *ab initio* models comprising of many electrons, due to the computational expense of full relativistic-inclusive methods. This complicates the study of molecules interacting with high atomic mass unit atoms, such as transitional metals and their catalytic properties. Present algorithms in computational chemistry can routinely calculate the properties of molecules that contain up to about 40 electrons with sufficient accuracy. Errors for energies can be less than a few kJ/mol. For geometries, bond lengths can be predicted within a few picometres and bond angles within 0.5 degrees. The treatment of larger molecules that contain a few dozen electrons is computationally tractable by approximate methods such as density functional theory (DFT). There is some dispute within the field whether or not the latter methods are sufficient to describe complex chemical reactions, such as those in biochemistry. Large molecules can be studied by semi-empirical approximate methods. Even larger molecules are treated by classical mechanics methods that employ what are called molecular mechanics. In QM/MM methods, small portions of large complexes are treated quantum mechanically (QM), and the remainder is treated approximately (MM).

In theoretical chemistry, chemists, physicists and mathematicians develop algorithms and computer programs to predict atomic and molecular properties and reaction paths for chemical reactions. Computational chemists, in contrast, may simply apply existing computer programs and methodologies to specific chemical questions. There are two different aspects to computational chemistry:

- ❖ Computational studies can be carried out in order to find a starting point for a laboratory synthesis, or to assist in understanding experimental data, such as the position and source of spectroscopic peaks.
- ❖ Computational studies can be used to predict the possibility of so far entirely unknown molecules or to explore reaction mechanisms that are not readily studied by experimental means.

Thus, computational chemistry can assist the experimental chemist or it can challenge the experimental chemist to find entirely new chemical objects.

Several major areas may be distinguished within computational chemistry:

- ❖ The prediction of the molecular structure of molecules by the use of the simulation of forces, or more accurate quantum chemical methods, to find stationary points on the energy surface as the position of the nuclei is varied.
- ❖ Storing and searching for data on chemical entities.
- ❖ Identifying correlations between chemical structures and properties.
- ❖ Computational approaches to help in the efficient synthesis of compounds.
- ❖ Computational approaches to design molecules that interact in specific ways with other molecules (*e.g.* drug design and catalysis).

Methods

A single molecular formula can represent a number of molecular isomers. Each isomer is a local minimum on the energy surface (called the potential energy surface) created from the total energy (*i.e.*, the electronic energy, plus the repulsion energy between the nuclei) as a function of the coordinates of all the nuclei. A stationary point is a geometry such that the derivative of the energy with respect to all displacements of the nuclei is zero. A local (energy) minimum is a stationary point where all such displacements lead to an increase in energy. The local minimum that is lowest is called the global minimum and corresponds to the most stable isomer. If there is one particular coordinate change that leads to a decrease in the total energy in both directions, the stationary point is a transition structure and the coordinate is the reaction coordinate. This process of determining stationary points is called geometry optimization.

The determination of molecular structure by geometry optimization became routine only after efficient methods for calculating the first derivatives of the energy with respect to all atomic coordinates became available. Evaluation of the related second derivatives allows the prediction of vibrational frequencies if harmonic motion is estimated. More importantly, it allows for the characterization of stationary points. The frequencies are related to the eigenvalues of the Hessian matrix, which contains second derivatives. If the eigenvalues are all positive, then the frequencies are all real and the stationary point is a local minimum. If one eigenvalue is negative (*i.e.*, an imaginary frequency), then the stationary point is a transition structure. If more than one eigenvalue is negative, then the stationary point is a more complex one, and is usually of little interest. When one of these is found, it is necessary to move the search away from it if the experimenter is looking solely for local minima and transition structures.

The total energy is determined by approximate solutions of the time-dependent Schrödinger equation, usually with no relativistic terms included, and by making use of the Born-Oppenheimer approximation, which allows for the separation of electronic and nuclear motions, thereby simplifying the Schrödinger equation. This leads to the evaluation of the total energy as a sum of the electronic energy at fixed nuclei positions and the repulsion energy of the nuclei. A notable exception are certain approaches called direct quantum chemistry, which treat electrons and nuclei on a common footing. Density functional methods and semi-empirical methods are variants on the major theme. For very large systems, the relative total energies can be compared using molecular mechanics. The ways of determining the total energy to predict molecular structures are:

Ab initio Methods

The programs used in computational chemistry are based on many different quantum-chemical methods that solve the molecular Schrödinger equation associated with the molecular Hamiltonian. Methods that do not include any empirical or semi-empirical parameters in their equations—being derived directly from theoretical principles, with no inclusion of experimental data—are called *ab initio* methods. This does not imply that the solution is an exact one; they are all approximate quantum mechanical calculations. It means that a particular approximation is rigorously defined on first principles (quantum theory) and then solved within an error margin that is qualitatively known beforehand. If numerical iterative methods have to be employed, the aim is to iterate until full machine accuracy is obtained (the best that is possible with a finite word length on the computer, and within the mathematical and/or physical approximations made).

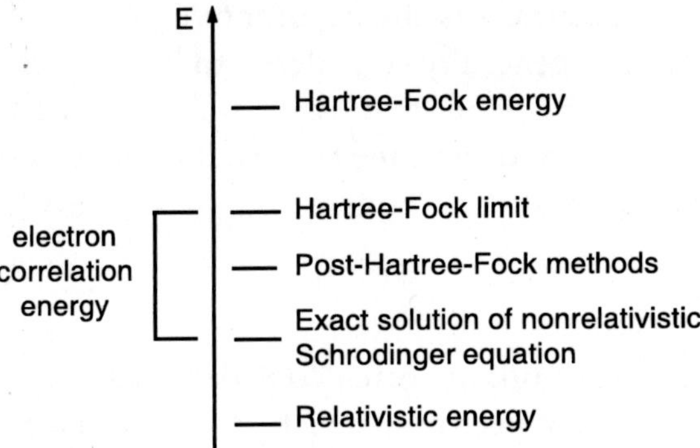

Fig. 13.1: Diagram illustrating various *ab initio* electronic structure methods in terms of energy. Spacings are not to scale.

The simplest type of *ab initio* electronic structure calculation is the Hartree-Fock (HF) scheme, an extension of molecular orbital theory, in which the correlated electron-electron repulsion is not specifically taken into account; only its average effect is included in the calculation. As the basis set size is increased, the energy and wave function tend towards a limit called the Hartree-Fock limit. Many types of calculations (known as post-Hartree-Fock methods) begin with a Hartree-Fock calculation and subsequently correct for electron-electron repulsion, referred to also as electronic correlation. As these methods are pushed to the limit, they approach the exact solution of the non-relativistic Schrödinger equation. In order to obtain exact agreement with experiment, it is necessary to include relativistic and spin orbit terms, both of which are only really important for heavy atoms. In all of these approaches, in addition to the choice of method, it is necessary to choose a basis set. This is a set of functions, usually centered on the different atoms in the molecule, which are used to expand the molecular orbitals with the LCAO ansatz. Ab initio methods need to define a level of theory (the method) and a basis set.

The Hartree-Fock wave function is a single configuration or determinant. In some cases, particularly for bond breaking processes, this is quite inadequate, and several configurations need to be used. Here, the coefficients of the configurations and the coefficients of the basis functions are optimized together.

The total molecular energy can be evaluated as a function of the molecular geometry; in other words, the potential energy surface. Such a surface can be used for reaction dynamics. The stationary points of the surface lead to predictions of different isomers and the transition structures for conversion between isomers, but these can be determined without a full knowledge of the complete surface.

A particularly important objective, called computational thermochemistry, to chemical accuracy. Chemical accuracy is the accuracy required to make realistic chemical predictions and is generally considered to be 1 kcal/mol or 4 kJ/mol. To reach that accuracy in an economic way it is necessary to use a series of post-Hartree-Fock methods and combine the results. These methods are called quantum chemistry composite methods.

Density Functional Methods

Density functional theory (DFT) methods are often considered to be *ab initio* methods for determining the molecular electronic structure, even though many of the most common functionals use parameters derived from empirical data, or from more complex calculations. This means that they could also be called semi-empirical methods. It is best to treat them as a class on their own. In DFT, the total energy is expressed in terms of the total one-electron density rather than the wave function. In this type of calculation, there is an approximate Hamiltonian and an approximate expression for the total electron density. DFT methods can be very accurate for little computational cost. The drawback is that, unlike ab initio methods, there is no systematic way to improve the methods by improving the form of the functional. Some methods combine the density functional exchange functional with the Hartree-Fock exchange term and are known as hybrid functional methods.

Semi-empirical and Empirical Methods

Semi-empirical quantum chemistry methods are based on the Hartree-Fock formalism, but make many approximations and obtain some parameters from empirical data. They are very important in computational chemistry for treating large molecules where the full Hartree-Fock method without the approximations is too expensive. The use of empirical parameters appears to allow some inclusion of correlation effects into the methods.

Semi-empirical methods follow what are often called empirical methods, where the two-electron part of the Hamiltonian is not explicitly included. For ð-electron systems, this was the Hückel method proposed by Erich Hückel, and for all valence electron systems, the Extended Hückel method proposed by Roald Hoffmann.

Molecular Mechanics

In many cases, large molecular systems can be modeled successfully while avoiding quantum mechanical calculations entirely. Molecular mechanics simulations, for example, use a single classical expression for the energy of a compound, for instance the harmonic oscillator. All constants appearing in the equations must be obtained beforehand from experimental data or *ab initio* calculations.

The database of compounds used for parameterization, *i.e.*, the resulting set of parameters and functions is called the force field, is crucial to the success of molecular mechanics calculations. A force field parameterized against a specific class of molecules, for instance proteins, would be expected to only have any relevance when describing other molecules of the same class.

These methods can be applied to proteins and other large biological molecules, and allow studies of the approach and interaction (docking) of potential drug molecules.

Methods for Solids

Computational chemical methods can be applied to solid state physics problems. The electronic structure of a crystal is in general described by a band structure, which defines the energies of electron orbitals for each point in the Brillouin zone. Ab initio and semi-empirical calculations yield orbital energies, therefore they can be applied to band structure calculations. Since it is time-consuming to calculate the energy for a molecule, it is even more time-consuming to calculate them for the entire list of points in the Brillouin zone.

Chemical Dynamics

Once the electronic and nuclear variables are separated (within the Born-Oppenheimer representation), in the time-dependent approach, the wave packet corresponding to the nuclear degrees of freedom is propagated via the time evolution operator (physics) associated to the time-dependent Schrödinger equation (for the full molecular Hamiltonian). In the complementary energy-dependent approach, the time-independent Schrödinger equation is solved using the scattering theory formalism. The potential representing the interatomic interaction is given by the potential energy surfaces. In general, the potential energy surfaces are coupled via the vibronic coupling terms.

The most popular methods for propagating the wave packet associated to the molecular geometry are:

- The split operator technique,
- The Multi-Configuration Time-Dependent Hartree method (MCTDH),
- The semiclassical method.

Molecular dynamics (MD) examines (using Newton's laws of motion) the time-dependent behavior of systems, including vibrations or Brownian motion, using a classical mechanical description. MD combined with density functional theory leads to the Car-Parrinello method.

Interpreting Molecular Wave Functions

The Atoms in Molecules model developed by Richard Bader was developed in order to effectively link the quantum mechanical picture of a molecule, as an electronic wavefunction, to chemically useful older models such as the theory of Lewis pairs and the valence bond model. Bader has demonstrated that these empirically useful models are connected with the topology of the quantum charge density. This method improves on the use of Mulliken population analysis.

Software Packages

There are many self-sufficient software packages used by computational chemists. Some include many methods covering a wide range, while others concentrating on a very specific range or even a single method. Details of most of them can be found in:

- Quantum chemistry computer programs supporting several methods.
- Density functional theory programs.
- Molecular mechanics programs.
- Semi-empirical programs.
- Solid state system programs with periodic boundary conditions.
- Valence Bond programs.

Bibliography

"Brute Force Breaks Bonds", Chemical & Engineering News, 22 Mar 2007.

"Radioactivity", *Encyclopædia Britannica* (2006). Encyclopædia Britannica Online. 18 Dec. 2006.

"Study links traffic pollution to thousands of deaths", *The Guardian*, London, UK: Guardian Media Group, 2008-04-15. Retrieved on 2008-04-15. (English).

"Safety of Manufactured Nanomaterials: About," OECD Environment Directorate, OECD.org, 18 July 2007 <http://www.oecd.org/about/0,3347,en_2649_37015404_1_1_1_1_1,00.html>.

A. Szabo, N. S. Ostlund, *Modern Quantum Chemistry*, McGraw-Hill (1982).

Abdelwahed W, Degobert G, Stainmesse S, Fessi H, (2006). "Freeze-drying of nanoparticles: Formulation, process and storage considerations". *Advanced Drug Delivery Reviews*. **58** (15): 1688–1713.

Air Pollution—What it means for your health.

Air Pollution Bandings and Indexes.

Air Pollution, Heart Disease and Stroke from the website of the American Heart Organization January 5, 2008.

Alexander, (1977), *Soil Microbiology*, 2nd Ed., Wiley Interscience.

Allinger, Norman (1977). "Conformational analysis. 130. MM2. A hydrocarbon force field utilizing V1 and V2 torsional terms". *Journal of the American Chemical Society* **99**: 8127–8134. doi:10.1021/ja00467a001.

Amdur MO, Doull J, Klaassen, CD. 1993. Cassarett and Doull's Toxicology: *The Basic Science of Poisons*. New York: McGraw-Hill, Inc.

American Lung Association, June 2, 2007.

American Scientist Jan/Feb 2007.

ANALYTICAL SCIENCES 2001, VOL.17 SUPPLEMENT [1], Basic Education in Analytical Chemistry.

Andreae, M. O. (1991). Biomass burning: Its history, use and distribution, and its impact on environmental quality in global climate. In J. S. Levine (Ed.), Global biomass burning: Atmosphere, climatic, and biospheric implications (p. 8). Cambridge, MA: MIT Press.

Announcing the 2005 Canadian Green Chemistry Medal. *RSC Publishing*. Retrieved on 2006-08-04.

AP 42, Volume I.

Applications/Products. National Nanotechnology Initiative. Retrieved on 2007-10-19.

Approaches to Safe Nanotechnology: An Information Exchange with NIOSH. United States National Institute for Occupational Safety and Health. Retrieved on 2008-04-13.

Australian National Pollutant Inventory Emissions Estimation Technique Manuals.

Bard, A. J.; Faulkner, L. R. Electrochemical Methods: Fundamentals and Applications. New York: John Wiley & Sons, 2nd Edition, 2000.

BBC Weather Service.

Berg J. M., Tymoczko J. L. Stryer L. (2002). *Molecular Cell Biology*, 5th ed., W. H. Freeman. ISBN 0-7167-4955-6.

Bernard, Harold W. Jr. *The Greenhouse Effect*, Ballinger Publishing Company, Massachusetts, 1980.

Beychok, M. R. (2005). *Fundamentals Of Stack Gas Dispersion*, 4th Edition, author-published. ISBN 0-9644588-0-2. www.air-dispersion.com.

Bilstein, Roger E. *Flight in America, 1900-1983*, The John Hopkins University Press, 1984.

Bohn, McNeal, and O'Connor, 1985, Soil Chemistry, 2nd Ed, Wiley Interscience.

Bolt and Bruggenwert, 1976, Soil Chemistry. A. Basic Elements, Elsevier.

Boukallel M, Gauthier M, Dauge M, Piat E, Abadie J. (2007). "Smart microrobots for mechanical cell characterization and cell convoying.". *IEEE Trans. Biomed. Eng.* 54 (8): 1536–40. doi:10.1109/TBME.2007.891171. PMID 17694877.

Bowman D, and Hodge G (2006). "Nanotechnology: Mapping the Wild Regulatory Frontier". *Futures* 38: 1060–1073. doi:10.1016/j.futures.2006.02.017.

Boys, S. F.; Cook G. B., Reeves C. M., Shavitt, I. (1956). "Automatic fundamental calculations of molecular structure". *Nature* **178** (2): 1207. doi:10.1038/1781207a0.

Brasseur, Guy P.; Orlando, John J.; Tyndall, Geoffrey S. (1999). Atmospheric Chemistry and Global Change. Oxford University Press. ISBN 0-19-510521-4.

Buenker, R. J.; Peyerimhoff S. D. (1969). Chemical Physics Letters 3: 37.

C&En: Cover Story - Nanotechnology.

California NanoSystems Institute.

Canada-Wide Standards for Particulate Matter (PM) and Ozone.

Canadian GHG Inventory Methodologies.

Cannon, James. *A Clear View*, Inform, Inc., 1975.

Cavalcanti A, Shirinzadeh B, Freitas RA Jr., Kretly LC. (2007). "Medical Nanorobot Architecture Based on Nanobioelectronics". *Recent Patents on Nanotechnology.* 1 (1): 1–10. doi:10.2174/187221007779814745.

Chang, Raymond (2002). "Electrochemistry", *Chemistry*, 7th Edition, Mc Graw Hill. ISBN 0-07-365601-1.

Chemistry for the Environment. *Interuniversity Consortium*. Retrieved on 2007-02-15.

Christopher H. Goss, Stacey A. Newsom, Jonathan S. Schildcrout, Lianne Sheppard and Joel D. Kaufman (2004). "Effect of Ambient Air Pollution on Pulmonary Exacerbations and Lung Function in Cystic Fibrosis". *American Journal of Respiratory and Critical Care Medicine* **169**: 816-821. doi:10.1164/rccm.200306-779OC.

Christopher J. Cramer *Essentials of Computational Chemistry*, John Wiley & Sons (2002).

Clausius, R. (1865). *The Mechanical Theory of Heat – with its Applications to the Steam Engine and to Physical Properties of Bodies*. London: John van Voorst, 1 Paternoster Row. MDCCCLXVII.

Coal Combustion - ORNL Review Vol. 26, No. 3&4, 1993.

Committee on Environmental Health (2004). "Ambient Air Pollution: Health Hazards to Children". *Pediatrics* **114** (6): 1699-1707. doi: 10.1542/peds.2004-2166.

Cosmic origins of Uranium.

Cresser, Killham, and Edwards, 1993, Soil Chemistry and its applications, Cambridge.

Cristina Buzea, Ivan Pacheco, and Kevin Robbie "Nanomaterials and Nanoparticles: Sources and Toxicity" Biointerphases 2 (1007) MR17-MR71.

Critical Reviews in Toxicology, A peer-reviewed academic research journal covering all aspects of toxicology, edited by Dr. Roger O. McClellan.

Crum, L. A. Physics Today 1994, 47, 22.

Current Air Pollution Bulletin.

D. Rogers *Computational Chemistry Using the PC, 3rd Edition*, John Wiley & Sons (2003).

D. Young *Computational Chemistry: A Practical Guide for Applying Techniques to Real World Problems*, John Wiley & Sons (2001).

D. A. Crossley, *Roles of Microflora and fauna in soil systems*, International Symposium on Pesticides in Soils, Feb. 25, 1970, University of Michigan.

Das S, Gates AJ, Abdu HA, Rose GS, Picconatto CA, Ellenbogen JC. (2007). "Designs for Ultra-Tiny, Special-Purpose Nanoelectronic Circuits.". *IEEE Transactions on Circuits and Systems I* 54 (11): 2528–2540. doi:10.1109/TCSI.2007.907864.

David Young's Introduction to Computational Chemistry.

Davis and Hayes, 1986, Geochemical Processes at Mineral Surfaces, American Chemical Soc.

Davis, Devra (2002). *When Smoke Ran Like Water: Tales of Environmental Deception and the Battle Against Pollution*. Basic Books. ISBN 0-465-01521-2.

DeBell, Garret, editor. *The Environmental Handbook*, Ballantine Books, Inc., New York, 1970.

Decommissioning costs of WWER-440 nuclear power plants. International Atomic Energy Agency (November 2002). Retrieved on 2008-06-06.

Degler, Stanley E. *Federal Pollution Control Programs*: water, air, and solid wastes, Washington, D.C. 1971.

Directive 2001/81/EC of the European Parliament and of the Council of 23 October 2001 on national emission ceilings for certain atmospheric pollutants.

Dixon and Weed, 1989, Minerals in Soil Environments, Soil Sci. Soc. America.

Dukhin, A. S. and Goetz, P. J. "Ultrasound for characterizing colloids", Elsevier, 2002.

EcoScale, a semi-quantitative tool to select an organic preparation based on economical and ecological parameters. Van Aken K, Strekowski L, Patiny L Beilstein Journal of Organic Chemistry, **2006** 2:3 (3 March 2006) Article.

Einhorn, C.; Einhorn, J. Luche, J. L. Synthesis 1989, 787.

Elder, A. (2006). Tiny Inhaled Particles Take Easy Route from Nose to Brain.

Electrochemistry. *Corrosion*. Retrieved on January 28, 2006.

Electrochemistry. *General Chemistry II by Dr. Michael Blaber*. Retrieved on January 30, 2006.

Ellis LA, Roberts DJ (1997). "Chromatographic and hyphenated methods for elemental speciation analysis in environmental media". *Journal of chromatography. A* **774** (1-2): 3–19. PMID 9253184.

Esposito, John C. and Larry Silverman. *Vanishing Air*, Grossman Publishing Co., Inc. 1982.

Essington, 2003, Soil and Water Chemistry: An Integrative Approach, CRC Press.

Estimated deaths & DALYs attributable to selected environmental risk factors, by WHO Member State, 2002.

EU-wide centralised geological waste disposal sites.

F. Jensen *Introduction to Computational Chemistry*, John Wiley & Sons (1999).

Facing up to "invisible pollution".

Fernbach, Sidney; Taub, Abraham Haskell (1970). *Computers and Their Role in the Physical Sciences*. Routledge. ISBN 0677140304.

Finlayson-Pitts, Barbara J.; Pitts, James N., Jr.; (2000) Chemistry of the Upper and Lower Atmosphere. Academic Press. ISBN 0-12-257060-X.

Flanagan, R. (1996, October). Engineering a cooler planet. Earth, 5, 34-39.

Flint, E. B.; Suslick, K. S. Science. 1991, 253, 1397.

Foster, Robert J. *Earth Science*, The Benjamin Cummings Publishing Co., Inc. 1982.

Ghalanbor Z, Marashi SA, Ranjbar B (2005). "Nanotechnology helps medicine: nanoscale swimmers and their future applications". *Med Hypotheses* **65** (1): 198–199. doi:10.1016/j.mehy.2005.01.023. PMID 15893147.

Gilbert SG. *A Small Dose of Toxicology – The Health Effects of Common Chemicals.* CRC Press, Boca Raton, February 2004, p 266.

GlobalSecurity.org, Chelyabinsk-65/Ozersk, retrieved September 2007.

Green & Sustainable Chemistry Network, Japan. *Green & Sustainable Chemistry Network*. Retrieved on 2006-08-04.

Guetens G, De Boeck G, Highley MS, Wood M, Maes RA, Eggermont AA, Hanauske A, de Bruijn EA, Tjaden UR (2002). "Hyphenated techniques in anticancer drug monitoring. II. Liquid chromatography-mass spectrometry and capillary electrophoresis-mass spectrometry". *Journal of chromatography. A* **976** (1-2): 239–47. PMID 12462615.

Guetens G, De Boeck G, Wood M, Maes RA, Eggermont AA, Highley MS, van Oosterom AT, de Bruijn EA, Tjaden UR (2002). "Hyphenated techniques in anticancer drug monitoring. I. Capillary gas chromatography-mass spectrometry". *Journal of chromatography. A* **976** (1-2): 229–38. PMID 12462614.

Harter, 1986, Adsorption Phenomena, Van Nostrand Reinhold.

Helv. Chim. Acta vol. 83 (2000), pp. 1766. [1].

Hodgeson E, Levi PE. 1987. A Textbook of Modern Toxicology. New York: Elsevier Science Publishing Co., Inc.

Holland WW, Reid DD. The urban factor in chronic bronchitis. Lancet. 1965;I:445-448.

Holt RM, Newman MJ, Pullen FS, Richards DS, Swanson AG (1997). "High-performance liquid chromatography/NMR spectrometry/mass spectrometry: further advances in hyphenated technology". *Journal of mass spectrometry : JMS* **32** (1): 64–70. doi:10.1002/(SICI)1096-9888(199701)32:1<64::AID-JMS450>3.0.CO;2-7. PMID 9008869

How is high-level nuclear waste managed in Canada?. *The Canadian Nuclear FAQ*. Retrieved on June 28, 2006.

http://reports.eea.eu.int/EMEPCORINAIR4/en European Environment Agency's 2005 Emission Inventory Guidebook]

http://www.cppa.utah.edu/publications/environment/nuclear_waste_summary.pdf

http://www.gnep.energy.gov/pdfs/GNEP_SOP.pdf

http://www.iaea.org/Publications/Magazines/Bulletin/Bull413/article1.pdf

http://www.ias.ac.in/currsci/dec252001/1534.pdf

Hu, Z. (2008). Too much technology may be killing bacteria.

IAEA Waste Management Database: Report 3 - L/ILW-LL. International Atomic Energy Agency (2000-03-28). Retrieved on 2008-06-06.

International Atomic Energy Agency, *The radiological accident in Goiânia*, 1988, retrieved September 2007.

International Atomic Energy Agency, *The radiological accident in Goiânia*, 1988, retrieved September 2007.

Ionizing & Non-Ionizing Radiation.

Issues relating to safety standards on the geological disposal of radioactive waste. International Atomic Energy Agency (2001-06-22). Retrieved on 2008-06-06.

J. Sunyer (2001). "Urban air pollution and Chronic Obstructive Pulmonary disease: a review". *European Respiratory Journal* **17**: 1024-1033.

Jackson, A. J. *Air Travel*, Macdonald Educational, 1979.

John E. Moulder. Static Electric and Magnetic Fields and Human Health.

Kahn, Jennifer (2006). "Nanotechnology". *National Geographic* **2006** (June): 98–119.

Klotz, I. (1950). *Chemical Thermodynamics*. New York: Prentice-Hall, Inc.

Krivtsov, A. I., 2006, Geoenvironmental Problems of Mineral Resources Development, in *Geology and Ecosystems*, Zekster (Ru), Marker (UK), Ridgeway (UK), Rogachevskayarochmaninoff (Ru), & Vartanyan (Ru), 2006 Springer Inc.

Kubik T, Bogunia-Kubik K, Sugisaka M. (2005). "Nanotechnology on duty in medical applications". *Curr Pharm Biotechnol.* **6** (1): 17–33. PMID 15727553.

Kwan-Hoong Ng (20th – 22nd October 2003). "[http://www.who.int/peh-emf/meetings/archive/en/keynote3ng.pdf Non-Ionizing Radiations – Sources, Biological Effects, Emissions and Exposures]". *Proceedings of the International Conference on Non-Ionizing Radiation at UNITEN* ICNIR2003 Electromagnetic Fields and Our Health.

L. Brown, Theodore; H. Eugene LeMay, Jr., Bruce E. Bursten, Julia R. Burdge (2003). "Electrochemistry", *Chemistry*, 9th Edition, US: Pearson Education. ISBN 0-13-066997-0.

L. R. MacGillivray, J. L. Reid and J. A. Ripmeester (2000). "Supramolecular Control of Reactivity in the Solid State Using Linear Molecular Templates". *J. Am. Chem. Soc.* **122** (32): 7817–7818. doi:10.1021/ja001239i.

Laidler, Keith; John H. Meiser, Bryan C. Sanctuary (May 2002). "Electrochemistry", in Keith Laidler: *Physical Chemistry*, 4th Edition, Boston, U.S.: Houghton Mifflin Company College Division. ISBN 061815292X.

LAQM Air Quality Management Areas.

Leary SP, Liu CY, Apuzzo MLJ. (2006). "Toward the Emergence of Nanoneurosurgery: Part III-Nanomedicine: Targeted Nanotherapy, Nanosurgery, and Progress Toward the Realization of Nanoneurosurgery.". *Neurosurgery* **58** (6): 1009–1026. doi:10.1227/01.NEU.0000217016.79256.16.

Leighton, T. G. The Acoustic Bubble; Academic Press: London, 1994, pp.531-555.

Leighton, T. G. The Acoustic Bubble; Academic Press: London, 1994, pp.531-555.

Levine, J. S. (Ed.). (1991). Global biomass burning: Atmosphere, climatic, and biospheric implications. Cambridge, MA: MIT Press. Particularly good chapters are the Introduction, and Chapters 1 and 55.

Levins CG, Schafmeister CE. *The synthesis of curved and linear structures from a minimal set of monomers.* Journal of Organic Chemistry, **70**, p. 9002, 2005. doi:10.1002/chin.200605222.

Lindsay, Willard L. 1979, Chemical Equilibria in Soils, Wiley Interscience.

Lobert, J. M. (1991). Experimental evaluation of biomass burning emissions: Nitrogen and carbon containing compounds. In J. S. Levine (Ed.), Global biomass burning: Atmosphere, climatic, and biospheric implications (p. 293). Cambridge, MA: MIT Press.

Lodish H, Berk A, Matsudaira P, *et al* (2004). *Molecular Cell Biology*, 5th ed., WH Freeman: New York, NY. ISBN 978-0716743668.

Lubick, N. (2008). Silver socks have cloudy lining.

Luche, J. L.; Compets. Rendus. Serie. IIB 1996, 323, 203, 307.

M. I. Ojovan, W. E. Lee. *An Introduction to Nuclear Waste Immobilisation*, Elsevier Science Publishers B. V., Amsterdam, 315pp. (2005).

McAlester, A. Lee. *The Earth*, Prentice-Hall, Inc., New Jersey, 1973.

McBride, Murray M. 1994. Environmental Chemistry of Soils, Oxford.

McMurry, John; Robert C. Fay (March 2004). "Electrochemistry", *Chemistry*, 3rd Edition, Prentice Hall. ISBN 0-13-056765-5.

Michael Faraday. *Biography*. Retrieved on January 30, 2006.

Michael Hogan, Leda Patmore, Gary Latshaw and Harry Seidman *Computer modelng of pesticide transport in soil for five instrumented watersheds*, prepared for the U.S. Environmental Protection Agency Southeast Water laboratory, Athens, Ga. by ESL Inc., Sunnyvale, California (1973).

Michael Kymisis, Konstantinos Hadjistavrou (2008). "Short-Term Effects Of Air Pollution Levels On Pulmonary Function Of Young Adults". *The Internet Journal of Pulmonary Medicine* 9(2).

Milan Bier (ed.) (1959). *Electrophoresis. Theory, Methods and Applications*, 3rd printing, Academic Press, 225. LCC 59-7676. OCLC 1175404.

Millero, Frank J. (2005). *Chemical Oceanography*, 3, CRC Press. ISBN 0849322804.

Monterey Bay Aquarium Research Institute (MBARI) (2005-06-09). ""Sinkers" provide missing piece in deep-sea puzzle". Press release. Retrieved on 2007-10-07.

N. Taniguchi, "On the Basic Concept of 'Nano-Technology'," *Proc. Intl. Conf. Prod.* London, Part II, British Society of Precision Engineering, 1974.

Nanosystems: Molecular Machinery, Manufacturing, and Computation. 2006, ISBN 0-471-57518-6.

Nanotechnology: Developing Molecular Manufacturing.

Narayan RJ, Kumta PN, Sfeir C, Lee D-H, Olton D, Choi D. (2004). "Nanostructured Ceramics in Medical Devices: Applications and Prospects.". *JOM* **56** (10): 38–43. doi:10.1007/s11837-004-0289-x.

National Policy Analysis #396: The Separations Technology and Transmutation Systems (STATS) Report: Implications for Nuclear Power Growth and Energy Sufficiency - February 2002.

National Research Council (1995). *Technical Bases for Yucca Mountain Standards.* Washington, D.C.: National Academy Press. cited in in The Status of Nuclear Waste Disposal. The American Physical Society (January 2006). Retrieved on 2008-06-06.

National Research Council: Committee on Air Quality Management in the United States, Board on Environmental Studies and Toxicology, Board on Atmospheric Sciences and Climate, Division on Earth and Life Studies (2004). *Air Quality Management in the United States.* National Academies Press. ISBN 0-309-08932-8.

Negraa, Christine; Donald S. Rossa and Antonio Lanzirottib (2005). "Oxidizing Behavior of Soil Manganese: Interactions among Abundance, Oxidation State, and pH". *Soil Science Society of America Journal* **1** (69): 97–95. Retrieved on 2006-11-11.

Nielsen, John. "The Killer Fog of '52: Thousands died as Poisonous Air Smothered London", National Public Radio, 2002-12-12.

Nierenberg, W. A. (Ed.). (1992). The encyclopedia of earth system science. (Vol. 1). New York: Academic Press.

Nuclear Information and Resource Service,Radioactive Waste Project, retrieved September 2007.

Obert, Edward F. *Internal Combustion Engines and Air Pollution*, In text Educational Publisher, New York, 1973.

Ott, Bevan J.; Boerio-Goates, Juliana (2000). *Chemical Thermodynamics – Principles and Applications*. Academic Press. ISBN 0-12-530990-2.

P. Thurn, E. Bücheler: Einführung in die radiologische Diagnostik, Stuttgart: Thieme, 8. Aufl. 1986.

Parlini, A. (2008.) New nanotech products hitting the market at a rate of 3-4 per week.

Paul von Ragué Schleyer (Editor-in-Chief). *Encyclopedia of Computational Chemistry*. Wiley, **1998**. ISBN 0-471-96588-X.

Paull, J. & Lyons, K. (2008) , Nanotechnology: The Next Challenge for Organics, Journal of Organic Systems, 3(1) 3-22.

Pestman, J. M.; Engberts, J.B.F.N.; de Jong, F. Jong. Recl. Trav. Chim. Pays-Bas. 1994, 113, 533.

Photochem. Photobiol. Sci., 2004, 3, 337 - 340, DOI: 10.1039/b316210a [2].

Pink, Daniel H. "Investing in Tomorrow's Liquid Gold", Yahoo, April 19, 2006.

Polluted Cities: The Air Children Breathe. World Health Organization.

Pople, John A.; David L. Beveridge (1970). *Approximate Molecular Orbital Theory*. New York: McGraw Hill.

Preuss, H. (1968). International Journal of Quantum Chemistry 2: 651.

Public Health and Environmental Radiation Protection Standards for Yucca Mountain, Nevada; Proposed Rule. Environmental Protection Agency (2005-08-22). Retrieved on 2008-06-06.

Public Law 107-303, November 27, 2002.

Putterman, S. J. Sci. Am. February 1995, p. 46.

R. Dronskowski *Computational Chemistry of Solid State Materials*, Wiley-VCH (2005).

Rainer Stegmann, *Treatment of Contaminated Soil: Fundamentals, Analysis, Applications*, Springer Verlag, Berlin 2001.

Removal of Silicon from High Level Waste Streams via Ferric Flocculation.

Reviews in Computational Chemistry vol 1, preface.

Revised 1996 IPCC Guidelines for National Greenhouse Gas Inventories (reference manual).

Richards, W. G.; Walker T. E. H and Hinkley R. K. (1971). *A bibliography of ab initio molecular wave functions*. Oxford: Clarendon Press.

Robyt, John F. (1990). *Biochemical Techniques Theory and Practice*. Waveland Press. ISBN 0-88133-556-8.

Rowe G, Horlick-Jones T, Walls J, Pidgeon N, (2005). "Difficulties in evaluating public engagement initiatives: reflections on an evaluation of the UK GM Nation?". *Public Understanding of Science*. 14: 333.

Royal Society and Royal Academy of Engineering (2004). "*Nanoscience and nanotechnologies: opportunities and uncertainties*". Retrieved on 2008-05-18.

S. K. Gupta, C. T. Kincaid, P. R. Mayer, C. A. Newbill and C. R. Cole, ''A multidimensional finite element code for the analysis of coupled fluid, energy and solute transport'', Battelle Pacific Northwest Laboratory PNL-2939, EPA contract 68-03-3116 (1982).

SAPIERR-2 program.

Schaefer, Henry F. III (1972). *The electronic structure of atoms and molecules*. Reading, Massachusetss: Addison-Wesley Publishing Co., 146.

Schaefer, Henry F. III (1984). *Quantum Chemistry*. Oxford: Clarendon Press.

Schiff, Barry. *Flying*, Golden Press, New York, 1971.

Schulthess, C. P. 2005. Soil Chemistry with Applied Mathematics. Trafford Publishing, Victoria, BC, Canada.

Sea-based Nuclear Waste Solutions.

Sedjo, Roger.1993. The Carbon Cycle and Global Forest Ecosystem. Water, Air, and Soil Pollution 70, 295-307. (via Oregon Wild Report on Forests, Carbon, and Global Warming).

Seinfeld, John H.; Pandis, Spyros N. (2006). Atmospheric Chemistry and Physics - From Air Pollution to Climate Change (2nd Ed.). John Wiley and Sons, Inc. ISBN 0471828572.

Shetty RC (2005). "Potential pitfalls of nanotechnology in its applications to medicine: immune incompatibility of nanodevices". *Med Hypotheses* **65**(5): 998–9. doi:10.1016/j.mehy.2005.05.022. PMID 16023299.

Shim, J., Dutta, P., Ivory, C. F., Modeling and simulation of IEF in 2-D microgeometries, Electrophoresis, 2007, 28, 527-586.

Simi Chakrabarti. "20th anniversary of world's worst industrial disaster", Australian Broadcasting Corporation.

Skoog, D. A.; West, D. M.; Holler, F. J. Fundamentals of Analytical Chemistry New York: Saunders College Publishing, 5th Edition, 1988.

Small Sizes that Matter: Opportunities and Risks of Nanotechnologies, Joint report of the Allianz Center for Technology and the OECD International Futures Programme, ed. Dr. Christoph Lauterwasser, OECD.org 18 July 2007 <http://www.oecd.org/dataoecd/37/19/37770473.pdf> (28).

Small Sizes that Matter: Opportunities and Risks of Nanotechnologies, Joint report of the Allianz Center for Technology and the OECD International Futures Programme, ed. Dr. Christoph Lauterwasser, OECD.org 18 July 2007 <http://www.oecd.org/dataoecd/37/19/37770473.pdf> (30-32).

Smith, Brian, E. (2004). *Basic Chemical Thermodynamics*. Oxford University Press. ISBN 1-86094-446-9.

Smith, S. J.; Sutcliffe B. T., (1997). "The development of Computational Chemistry in the United Kingdom". *Reviews in Computational Chemistry* **70**: 271–316.

Sonon, L. S. , M. A. Chappell and V. P. Evangelou (2000) The History of Soil Chemistry. Url accessed on 2006-04-11.

Space Disposal of Nuclear Wastes Eric E. Rice Battelle Memorial Institute.

Sparks, D. L. 2003, Environmental Soil Chemistry, Academic Press.

Sparks, D. L. 1989, Kinetics of Soil Chemical Processes, Academic Press.

Sparks, D. L. 1999, Soil Physical Chemistry, CRC Press.

Sposito, G. 1984, The Surface Chemistry of Soils, Oxford Press.

Sposito, G., 1989,The Chemistry of Soils, Oxford University Press.

Starting from [1] Pollution - Definition from the Merriam-Webster Online Dictionary.

Streitwieser, A.; Brauman J. I. and Coulson C. A. (1965). *Supplementary Tables of Molecular Orbital Calculations*. Oxford: Pergamon Press.

Survey & Identification of NORM Contaminated Equipment.

Suslick, K. S.; Casadonte, D. J. J. Am. Chem. Soc. 1987, 109, 3459.

Suslick, K. S.; Doktycz, S. J. Adv. Sonochem. 1990, 1, 197-230.

Suslick, K. S.; Hammerton, D. A.; Cline, R. E., Jr. J. Am. Chem. Soc. 1986, 108, 5641.

T. Clark *A Handbook of Computational Chemistry*, Wiley, New York (1985).

Taking the Oxford air adds up to a 60-a-day habit (a newspaper article in The Guardian.

Talanta Volume 51, Issue 5, p921-933 [2], Review of analyticalnext term measurements facilitated by drop formation technology.

Talanta, Volume 36, Issues 1-2, January-February 1989, Pages 1-9 [4] History of analytical chemistry in the U.S.A.

Tan, Kim H., 1993, Principles of Soil Chemistry, 2nd Ed., Marcel Dekker.

Terms of Reference, Working Group on the Revision of National Emissions Ceilings and Policy InstrumentsPDF (24.4 KiB).

Testimony of David Rejeski for U.S. Senate Committee on Commerce, Science and Transportation Project on Emerging Nanotechnologies. Retrieved on 2008-3-7.

The 12 Principles of Green Chemistry. *United States Environmental Protection Agency*. Retrieved on 2006-07-31.

The Department for Environment, Food & Rural Affairs (DEFRA): Air Pollution.

The Faraday law of electrochemistry. *Faraday laws of electrochemistry*. Retrieved on January 30, 2006.

The Nobel Prize in Chemistry 2005. *The Nobel Foundation*. Retrieved on 2006-08-04.

The Nobel Prize in Physics 2007. Nobelprize.org. Retrieved on 2007-10-19.

The Presidential Green Chemistry Awards. *United States Environmental Protection Agency*. Retrieved on 2006-07-31.

TrAC Trends in Analytical Chemistry Volume 21, Issues 9-10, Pages 547-557 [3], History of gas chromatography.

Transmutation being banned in the US since 1977.

Troubleshooting DNA agarose gel electrophoresis. Focus 19:3 p.66 (1997).

Turk, Amos and Jonathan, *Environmental Science*, W. B. Saunders Company, Philadelphia, PA. 1974.

Turner, D. B. (1994). *Workbook of atmospheric dispersion estimates: an introduction to dispersion modeling*, 2nd Edition, CRC Press. ISBN 1-56670-023-X. www.crcpress.com.

U.S. Department of Energy Environmental Management - "Department of Energy Five Year Plan FY 2007-FY 2011 Volume II." Retrieved on 8 April 2007.

U.S. Geological Survey, Radioactive Elements in Coal and Fly Ash: Abundance, Forms, and Environmental Significance, *Fact Sheet* FS-163-1997, October 1997, retrieved September 2007.

UK Air Quality Archive.

UK National Air Quality Objectives.

United Kingdom's emission factor database.

Uranium Information Centre, Synroc, *Nuclear Issues Briefing Paper* 21, retrieved September 2007.

Wall Street Journal article, July 20, 2007.

Wall Street Journal article, May 23, 2006.

Warneck, Peter (2000). Chemistry of the Natural Atmosphere (2nd Ed.). Academic Press. ISBN 0-12-735632-0.

Wayne, Richard P. (2000). Chemistry of Atmospheres (3rd Ed.). Oxford University Press. ISBN 0-19-850375-X.

Weiss, R. (2008). Effects of Nanotubes May Lead to Cancer, Study Says.

West, Larry. "World Water Day: A Billion People Worldwide Lack Safe Drinking Water", About, March 26, 2006.

Wild, 1988, Russell's Soil Conditions and Plant Growth, 11th Ed., Longman.

Wilkins CL (1983). "Hyphenated techniques for analysis of complex organic mixtures". *Science* **222** (4621): 291–6. PMID 6353577.

William Hill, John; Ralph H. Petrucci, Terry McCreary, Scott S. Perry (March 2004). "Electrochemistry", *General Chemistry: An Integrated Approach*, 7th Edition, Pearson Education, 1200. ISBN 0-13-140283-8.

Williamson, Samuel J. *Fundamentals of Air Pollution*, Addison-Wesley Publishing Co., New York, 1973.

Wireless nanocrystals efficiently radiate visible light.

Wolt, Jeffrey D. 1994, Soil Solution Chemistry: Applications to Environmental Science and Agriculture, John Wiley and Sons, Inc.

Wood, R. W.; Loomis, A. L. The Physical and Biological Effects of High Frequency Sound Waves of Great Intensity. Philos. Mag. 1927, 4, 414.

World Bank Statistics.

Zoidis, John D. (1999). "The Impact of Air Pollution on COPD". *RT: For Decision Makers in Respiratory Care*.

Zsigmondy, R. "Colloids and the Ultramicroscope", J. Wiley and Sons, NY, (1914).

Index